E-Cigarettes

Yin-Hui Leong
Ahmad Shalihin Mohd Samin
Editors

E-Cigarettes

Risks, Research and Challenges

Springer

Editors
Yin-Hui Leong
Malaysia National Poison Centre
Universiti Sains Malaysia
Penang, Malaysia

Ahmad Shalihin Mohd Samin
Malaysia National Poison Centre
Universiti Sains Malaysia
Penang, Malaysia

ISBN 978-981-96-5086-6 ISBN 978-981-96-5087-3 (eBook)
https://doi.org/10.1007/978-981-96-5087-3

This Springer imprint is published by the registered company Springer Nature Singapore Pte Ltd.
The registered company address is: 152 Beach Road, #21-01/04 Gateway East, Singapore 189721, Singapore

Contents

About the Editors

Yin-Hui Leong is a Senior Lecturer at the Malaysia National Poison Centre, Universiti Sains Malaysia, Penang, Malaysia. Her areas of specialization include public health, food analysis, food toxicology, and mycology. She has made substantial contributions to the scientific community in the study of persistent organic pollutants (POPs) such as dioxins and dioxin-like PCBs, mycotoxins, and other toxins. She has published numerous papers in peer-reviewed national and international journals, as well as several chapters in various books. As an expert and representative from Malaysia in the TobLabNet (Tobacco Laboratory Network) of the World Health Organization, she is currently conducting research on tobacco and electronic cigarettes, funded by both the World Health Organization and the Malaysian Ministry of Health.

Ahmad Shalihin Mohd Samin completed his BSc and MSc in Toxicology at Universiti Sains Malaysia. Recently, he also completed his PhD in Toxicology. He has been recognized as a professional technologist in chemical technology by the Malaysian Board of Technology (MBOT) since 2019. He is a principal researcher at the Malaysia National Poison Centre, Universiti Sains Malaysia, Penang, Malaysia. Currently, he is involved in a toxicological study on environmental health, focusing on the chemical carcinogen dioxins and their impact on Malaysian firefighters. In addition to his work on dioxins, he serves as the Principal Investigator for research on the availability of vape products in Malaysian retail. This project is funded by the World Health Organization, Western Pacific Region (WHO WPRO).

History and Evolution of E-cigarettes 1

Mohd Fadhli Razali

Abstract

In the early 2000s, a Chinese pharmacist, Hon Lik, created the modern e-cigarette (e-cig) as an alternative to conventional cigarettes after his father died from lung cancer. An e-cig is a battery-powered device that heats liquid solutions to produce an aerosol for inhalation, allowing the user the feeling of smoking without burning tobacco. The first patented device in 2004 was created to aid smoking cessation and to evade smoke-free laws. By 2013, key tobacco companies were touting e-cigs as a harm-reduction product. E-cig devices have evolved through four generations: from cigarette-like "cigalike," to larger and more customizable vape pens/box mods to the recent sleek pod systems like JUUL with high-nicotine salt formulations. Shenzhen, China, has grown to become the international center of all vaping things. In the competitive environment, consolidation is visible as British American Tobacco, Altria, Imperial Brands, and most recently RELX have all been acquiring e-cig brands (like Vuse). Vape culture has taken hold of a new subculture of users, transporting what began as cessation aids into an indulgent pastime or lifestyle akin to smoking for pleasure.

Keywords

E-cigarettes · Vaping · Tobacco harm-reduction · Nicotine · Product innovation

M. F. Razali (✉)
Malaysia National Poison Centre, Universiti Sains Malaysia, George Town, Malaysia
e-mail: fadhli_razali@usm.my

Y.-H. Leong, A. S. Mohd Samin (eds.), *E-Cigarettes*,
https://doi.org/10.1007/978-981-96-5087-3_1

1.1 Introduction

E-cigarettes (e-cigs), vapes, or ENDS (electronic nicotine delivery systems) were introduced as a substitute for conventional cigarettes to aid smoking cessation. E-cigs have been designed to simulate similar sensations without any perceived adverse health effects, attributed mainly to combusting tobacco that produces thousands of toxic substances and carcinogens. The evolution and rise in e-cigs reflect changing global shifts in how society views cigarettes, public health, and technology.

While the first generation of smokeless, non-tobacco cigarettes date to the mid-last century, it was not until the early 2000s that sufficient technology and market readiness aligned e-cigs to be developed as they are known today. Hon Lik, a Chinese pharmacist, is credited with the invention of the modern e-cig because of his personal smoking experience and subsequent lung cancer death. E-cig has paved the way toward defining a prototype, the basis of everything that was to follow in what is now, and without question, is among the largest and most diverse markets on earth.

E-cigs have evolved swiftly in product types and technology. The first-generation e-cigs, including cigalike, were designed to imitate the size and appearance of conventional cigarettes and dispensed nicotine in an aerosol form when inhaled. It has since evolved to the present-day pod-based system offering improved delivery convenience to satisfy consumer preferences and regulatory amendments. Through this rapid evolution of technology and users' preferences over time, each new generation of e-cigs showcased enhanced battery life, vapor production, and users' overall experience.

E-cig use has led to the development of a new vape culture as well. Vaping has produced a singular, sprawling category of nicotine use of its own, replacing one phalanx for another in the public health discussion on smoking cessation and restoring focus on e-cigs as an individual hobby and lifestyle object. Vape shops, online forums, and social media platforms are used for sharing knowledge, experiences, and innovations, contributing to the rapid spread and acceptance across diverse demographics.

A debate over whether e-cigs are safe and how they should be regulated remains open-ended. Proponents argue that e-cigs provide a reduction in the harms of conventional tobacco products and can be helpful to quit smoking. Critics, however, are concerned that e-cigs could attract non-smokers, particularly youths, hence a pathway for nicotine addiction among the new generation.

Global responses to e-cigs from the regulatory standpoint are diverse and they vary depending on public health goals, cultural attitudes on smoking (as well as tobacco lobbying), and often, in extremely divergent ways even within a country. In some countries (UK, Canada, New Zealand), e-cigs are viewed as a potential tool for smoking harm reduction; in other countries, they are heavily regulated due to concerns over health risks and youth uptake.

While we explore the past and development of e-cigs, it is crucial to contextualize what these products mean within a broader historical narrative on tobacco control, technological innovation, and ever-changing culture. E-cig isn't just a tale about a new product; its mere existence and potential repercussions prompt the need to reevaluate existing notions and concepts relating to smoking, health, technology, and the capacity of technology to influence and transform our choices and behaviors.

1.1.1 Invention of the First E-cig and Its Patent History

A popular example of an early e-cig was its first creation in 1963, patented by an American engineer, Herbert A. Gilbert. The modern e-cig was not born until Hon Lik, a Chinese pharmacist and inventor, himself a heavy smoker, desired a safer and cleaner alternative from conventional tobacco cigarettes. The death of his father, a heavy smoker of lung cancer, motivated Hon Lik to create an e-cig.

E-cig was created to simulate the act of smoking a "real" cigarette without burnt tobacco leaves. Instead, it heats a liquid solution containing nicotine using a battery-powered heating element. In this process, an aerosol—often called vapor—is formed and inhaled. The US patent application describes an e-cig device as "an electronic atomization cigarette that functions as substitutes for quitting smoking and cigarette substitutes" (patent No. 8,490,628 B2) (Grana et al. 2014).

By 2013, major multinational tobacco companies had infiltrated the e-cig market. E-cig was marketed as a healthier alternative to tobacco smoking, useful to quit smoking, less cigarette consumption, and to circumvent smoke-free laws by enabling users to "smoke anywhere" (Grana et al. 2014).

In 2015, Juul e-cig was introduced in the US market. Unlike many other contemporary e-cig companies, the founders behind Juul® e-cig approached their product launch like a conventional high-tech start-up company, unlike a tobacco company (Sarles et al. 2021). In China, Shenzhen is the center of e-cig production, with a complete business ecosystem built around it. The development process of the e-cig industry in Shenzhen is accompanied by the transformation of the development pattern from imitation to innovation (Xu et al. 2022).

Despite the rapid market penetration of e-cigs, there are many unanswered questions about their safety, efficacy for harm reduction and cessation, and total impact on public health. E-cig products are ever-changing and much of the findings from studies on older products may not be relevant to the assessment of newer varieties.

1.2 Evolution of E-cigs

1.2.1 First-Generation E-cigs (Cigalike)

First-generation e-cigs, also known as "cigalike," were designed to resemble conventional cigarettes in appearance and function (Fig. 1.1). They were disposable and had a limited battery life. The basic components of a first-generation e-cig included a battery, a heating element (also known as an atomizer), and a cartridge that held the e-liquid. The e-liquid typically contains nicotine, propylene glycol or vegetable glycerin, and flavorings (Rüther et al. 2018). These early models were

Fig. 1.1 The evolution of e-cigs according to product generation and characteristics. (Department of Health et al. 2020). Image showing four generations of vaping devices. From left to right: First generation is a disposable cigalike resembling a traditional cigarette. Second generation is a medium-size rechargeable and refillable vape pen. Third generation is a large-size tank or mod device. Fourth generation includes nicotine salt pod mods, featuring a sleek black stick and a blue oval-shaped device. Each generation is labeled above the devices

often criticized for their poor performance, as they could not deliver nicotine as effectively as conventional cigarettes (Heide et al. 2021). The main objective of first-generation e-cigs was to provide a smoking alternative that mimicked the experience of smoking a conventional cigarette but without the combustion of tobacco leaves.

Users of cigalike need only to inhale from the cartridge end. The action activates the battery, which in turn powers the atomizer to heat the e-liquid and create vapor. This is known as automatic activation. Some cigalike, however, may have a manual activation method where a button is pressed to activate the heating process. Once the e-liquid in the cartridge is used up, it can be replaced with a new pre-filled cartridge or in some cases, refilled with e-liquid. The battery component of cigalike is rechargeable, commonly via a USB charger.

Cigalike were the first form of e-cigs introduced in the market and gained popularity for their similarity to conventional cigarettes and ease of use, making it a popular choice for smokers looking to switch to a potentially less harmful smoking alternative. However, they were generally considered to offer a less satisfying experience owing to their smaller size, lower battery capacity, and less powerful vapor production compared to newer generations of e-cigs, such as vape pens and mods. Despite their limitations, cigalike continued to be used because of their convenience and familiarity, particularly among new vapers and those who prefer a device that closely mimics the experience of smoking a conventional cigarette.

The customization options of these first-generation e-cigs are limited compared to newer models. They were not designed for a customized nicotine delivery experience, unlike the second-generation devices, which were designed to accommodate larger atomizers and batteries for a more personalized experience (King et al. 2018). The first-generation e-cigs were typically pre-filled cartridges with no option to adjust the heating temperature of the atomizer or to use different types of cartridges (Shastry and Langdorf 2016).

1.2.2 Second-Generation E-cigs (Vape Pens)

Second-generation e-cigs introduced larger devices, often referred to as vape pens. They are so-called because of their size and shape to resemble a pen or a small tube. The device had larger rechargeable batteries including e-liquid refill. Vape pens were designed with an improved nicotine delivery system compared to the first generation (Hickman et al. 2022).

Vape pens consist of several key components:

- Battery: The battery is the biggest component of the vape pen and provides power to the heating element. It is normally rechargeable via a USB and offers longer battery life compared to cigalike.
- Tank or Clearomizer: This component holds the e-liquid and is usually transparent for the user to see the amount of e-liquid remaining. The tank includes a mouthpiece from which the user inhales the vapor.
- Atomizer: Within the tank, is an atomizer containing a coil and wicking material. The coil is heated up when activated, vaporizing the e-liquid absorbed by the wick.
- Control Buttons: Many vape pens have one or more buttons for switching the device on or off, and for adjustment of the power output for a customized vaping experience.

To use a vape pen, a user fills the tank with the preferred e-liquid, ensuring not to overfill it. After the tank is filled, a few minutes are required to allow the e-liquid to saturate the wick completely. The battery is activated with a press of the button to heat the coil, vaporizing the e-liquid. The vapor is inhaled through the mouthpiece. Some vape pens offer variable voltage or wattage settings, allowing users to customize the intensity of the vapor and the flavor experience. Vape pen battery is rechargeable when needed. The unit's atomizer coils are replaceable, making it possible to maintain and extend the life of a vape pen.

The enhanced performance of vape pens eventually surpassed cigalike, offering a better vaping experience with more vapor and flavors. They were appreciated for the balance between ease of use and customization options. Many cigalike consumers switched to vape pens seeking more control over their vaping experience with the added advantage of a portable and uncomplicated device. Vape pens came in various designs, colors, and features that catered to a broad range of preferences and wishes of the vapers. The upgraded device was considered superior to cigalike in terms of performance and customization yet less complex and possibly less intimidating than more advanced mods.

1.2.3 Third-Generation E-cigs (Box Mods and Mechanical Mods)

Third-generation e-cig introduced even larger devices with customizable settings that included "pods" and "mods" or "box mods," and larger, refillable tanks for e-liquid (Hickman et al. 2022). Users could adjust the voltage, wattage, and temperature to control the amount of vapor produced.

1.2.3.1 Box Mod

Box mod is among the third-generation e-cigarette devices. It is named for its box-like shape and acknowledged for its modifiability, hence the term "mod" also known as "tank" for its bigger capacity.

Its primary components include a casing, which houses the electronic components and provides the device's structure, a battery or batteries for power, a 510 connector to attach an atomizer, and control buttons for operation and adjustment settings. The atomizer, which can be a tank or a rebuildable type, is where the e-liquid is heated to produce vapor. Control buttons allow the user to turn the device on and off as well as adjust the power output in terms of wattage or voltage. Some box mods also feature a display screen for monitoring and setting adjustment. The design and components of box mods can vary significantly, with some models featuring advanced capacities such as temperature control, variable wattage, and even software upgrades. The design of the atomizer can also vary with different coils and wicks impacting the performance and characteristics of the vapor produced.

Box mods, advanced personal vaporizers, offer a customizable vaping experience through adjustable power outputs and interchangeable atomizers. Before a box mod is used, one must ensure the device is adequately powered, either with new or recharged batteries. The user then affixes an atomizer, which can range from tanks to rebuildable atomizers, to the 510 connector located on the mod's top. If required, the atomizer is filled with e-liquid. The box mod is activated typically by a series of button presses, and the power output (wattage or voltage) is adjusted to the user's preference. Higher settings yield increased vapor and flavor production, but exceeding the atomizer's recommended settings can damage the device or result in a burnt taste. The vaping process is initiated by pressing the fire button and inhaling from the mouthpiece.

Box mods and other e-cigs rose in popularity as they are often perceived as less harmful than conventional tobacco cigarettes. They are used by diverse individuals, including those attempting to quit smoking tobacco, and adolescents who have not previously smoked combustible cigarettes.

1.2.3.2 Mechanical Mods

Mechanical mods, commonly referred to as mech mods, are another variation of e-cig known for their simplicity and raw power. Third-generation e-cigs are favored by experienced vapers for their durability and direct battery-to-coil connection. Mechanical mods are typically cylindrical in shape, although some are box-shaped. They have the following components:

- Battery Housing: Main body of the mod that houses the battery. Generally made of metal, such as stainless steel, brass, or copper, for durability and good electrical conductivity.
- Battery: Mechanical mods use rechargeable batteries: the 18,650, 20,700, or 21,700 size. The battery provides the power to heat the coil and vaporize the e-liquid.
- Atomizer: Consists of a coil and a wick that heats the e-liquid to produce vapor. The coil is heated by the battery, and the wick, usually made of cotton, absorbs the e-liquid and transfers it to the coil.
- Fire Button: This switch completes the electrical circuit when pressed, sending power from the battery to the atomizer. In many mech mods, the fire button is located at the bottom of the device.

Mechanical mods do not have any wiring or electronics. They operate on a direct battery-to-coil connection, that is, the power output depends on the battery's voltage and the coil's resistance.

To activate a mechanical mod, a coil must be installed in the atomizer to saturate the wick in the e-liquid. A charged battery is then installed into the battery housing, and the atomizer is attached to the mod. When the fire button is activated, it completes the circuit and sends power from the battery to the atomizer, heating the coil and vaporizing the e-liquid. The user then inhales the vapor through the mouthpiece of the atomizer. Mechanical mods require a good understanding of Ohm's law and battery safety. A user needs to ensure the coil's resistance is appropriate for the battery's voltage to prevent overheating or battery failure.

Mechanical mods are popular among skilled vapers for their durability, raw power, and customized vaping experience. The device, however, is not recommended for beginners without adequate knowledge to maintain the device. If not used correctly, mechanical mods can be dangerous due to the lack of safety features. Hence, they are typically used by hobbyists and vaping enthusiasts who understand the principles of battery safety and are comfortable building their own coils.

1.2.4 Fourth-Generation E-cigs

Fourth-generation e-cigs, also known as pod-based e-cigs, represent the latest evolution in e-cig design. Pod systems, also known as pod, vapes, or pod mods, are varieties of e-cigs that have gained popularity in recent years. These devices are characterized by their sleek, compact, and often USB-shaped design, hence portable and discreet. One of the most well-known brands of fourth-generation e-cigs is JUUL which is compact and easy to use. One of the key differences is that fourth-generation e-cigs frequently use nicotine salt which provides a higher nicotine concentration and smoother experience compared to the free-base nicotine used in earlier generations (Hickman et al. 2022).

The pod systems consist of two main components:

- Battery: The battery is the body of the pod system that provides power to the device. It is usually rechargeable, often via a USB cable. The battery sizes may vary, but they are generally smaller and more lightweight than those found in vape pens or mods.
- Pod: The pod is a small, usually removable cartridge that fits into the body of the device. It contains e-liquid and a built-in coil that heats the e-liquid to produce vapor. Pods can be either pre-filled with e-liquid by the manufacturer or refillable, allowing users to fill them with e-liquid of their choice.

Some pod systems also have airflow controls and power settings, although many are designed to be as simple as possible with automatic drawing activation that activates when the user takes a puff.

To use a pod system, the pod is inserted into the device. If it is a refillable pod, the user will need to first fill it with e-liquid. Once the pod is inserted, the device is ready for use. The user inhales from the mouthpiece of the pod, activating the battery to heat and produce e-liquid vapor. When the e-liquid in the pod is used up, it can be replaced with a new pre-filled pod or replenished with new e-liquid if the pod is a refillable type. The battery in the pod system is rechargeable using a USB charger.

Pod systems have become increasingly popular as they are uncomplicated, portable, and discreet. They are smaller and more lightweight compared to most other types of e-cigs, hence easy to carry and use on the go. The use of pre-filled pods is convenient as it eliminates the need to manually refill the e-liquid into the tank. Pod systems are mostly favored by cigarette smokers transitioning to vaping since they offer a similar draw and nicotine like conventional cigarettes, especially when used with nicotine salt e-liquids. They are also popular among experienced vapers for their convenience and portability.

However, the closed system of pre-filled pods can limit the choice of flavors and nicotine strengths, while the smaller size pod systems can mean shorter battery life and less vapor production compared to larger devices. Despite these limitations, pod systems continue to be a popular choice for both new and experienced vapers.

1.3 Dominant Players in the E-cig Industry

The e-cig industry has seen significant growth and consolidation in recent years, with several major players dominating the market. Mergers and acquisitions have played a crucial role in shaping the industry landscape. Some of the early e-cig brands are Ruyan, NJOY, and Blu. Ruyan was the first company to commercialize e-cigs, while NJOY, an American company, was one of the significant players to introduce e-cigs to the US market in 2006 (Laugesen 2015). Blu, another early e-cig brand, founded in 2009 was later acquired by Lorillard Tobacco Company in 2012 (Smith et al. 2015).

British American Tobacco (BAT), like other multinational tobacco companies, entered the e-cig market in response to the declining sales of conventional cigarettes and the growing demand for alternative nicotine delivery systems. The company offers products such as Vype, which is part of their "next-generation products" portfolio. In Ontario, Canada, BAT promoted Vype through retail merchandising, placing promotions at petrol pumps and convenience stores like Esso and Circle (Dewhirst 2019). BAT acquired Reynolds American, the maker of the Vuse brand of e-cigs, in a $49 billion deal in 2017, creating the world's largest listed tobacco company (McDaniel and Malone 2012).

Altria Group, one of the largest tobacco companies in the world, has made significant investments in the e-cig market, most notably through its acquisition of a 35% stake in Juul Labs in December 2018 for $12.8 billion (Gaidarov and Asher 2022). Juul, at the time of the investment, was a leading e-cig brand in the United States, holding a substantial portion of the market share. Altria's involvement in the e-cig industry represents a strategic move to diversify its product portfolio in response to the declining cigarette smoking rates and the increasing popularity of vaping products. By investing in Juul, Altria not only gained a significant share in the e-cig market but also positioned itself to influence the future direction of the vaping industry. Altria's stake in Juul was also seen as a way for the company to hedge against the declining sales of combustible cigarettes and to participate in the growing market for reduced-risk products. However, investment has come with challenges, including lawsuits and regulatory pressures, as public health advocates and regulators grapple with how to manage the rise of e-cigs and their impact on public health (Gaidarov and Asher 2022).

Imperial Brands (formerly Imperial Tobacco) acquired Blu e-cigs from Lorillard in 2014 as part of a $7.1 billion deal, expanding its presence in the e-cig market (Imperial Tobacco's Cooper Adds Blu to Take E-Cigarette Leap – Bloomberg 2014).

RELX Technology, a leading e-cig company in Asia was founded in 2018 in China. Established by Kate Wang, a former Uber China executive, RELX expanded rapidly to become a major player in the e-cig market, particularly in Asia, with a presence in over 50 countries and regions (Corporation Service Company 2022; Tan 2021). In Malaysia, RELX is among the leading e-cigs brands in the market from a baseline report (unpublished data).

These examples illustrate in Table 1.1 the significant role of mergers and acquisitions in shaping the e-cig industry and the dominance of major players in the market. The industry's landscape is continually evolving with new players entering the market and existing players expanding their product offerings and market presence.

Table 1.1 Summary of dominant players in the e-cigarette industry

Manufacturer	Origin	Brand	References
Ruyan (now owned by Imperial Tobacco Group)	Founded in 2003 by Hon Lik, a Chinese pharmacist	Ruyan e-cig, considered the first commercially successful modern e-cig	Laugesen (2015)
British American Tobacco (BAT)	Entered the e-cig market in 2013	Vype (launched in 2013), Vuse (acquired through Reynolds American in 2017)	Dewhirst (2019)
Altria Group	Made a significant investment in the e-cig market in 2018	Acquired a 35% stake in Juul Labs for $12.8 billion	Gaidarov and Asher (2022)
Imperial Brands (formerly Imperial Tobacco)	Entered the e-cig market through acquisitions	Acquired Blu eCigs from Lorillard in 2014	Imperial Tobacco's Cooper Adds Blu to Take E-Cigarette Leap—Bloomberg (2014)
RELX Technology	Founded in 2018 in China by Kate Wang	RELX is a leading e-cig brand in Asia with presence in over 50 countries	Tan (2021)

1.4 The Rise of Vape Culture

The surge in vape culture and its impact on the e-cig industry is a multifaceted issue encompassing changes in consumer behavior, technological advancements, and regulatory challenges. The initial development of vaping as a smoking cessation tool has evolved into a cultural phenomenon with its own community, practices, and controversies. This transformation is supported by the emergence of a vaping identity among young adults, associated with lower trust in health experts and lower perceptions of e-cig harm, as well as lower intentions to abstain from vaping (Lee et al. 2023). The industry has also seen innovation, particularly in product flavors, which is key to sustainable product innovation (Izzani and Hermawan 2021).

Vape shops and lounges have become social spaces akin to coffee shops, fostering a community atmosphere (Lee et al. 2023). Vape conventions and trade shows offer networking opportunities and a sense of identity among vapers. The artistic and competitive aspects of vaping, such as coil wrapping and vaping competitions, have further embedded vaping into popular culture (Lee et al. 2023).

The growth of the vaping industry is influenced by targeted demographics and the community of enthusiasts, despite challenges in regulation and public perception (Lee et al. 2023). The industry's response to regulations, such as the promotion of flavored e-cig products on social media indicates a strategic approach to maintain consumer interest despite regulatory hurdles (Kostygina et al. 2022).

In summary, the rise of vape culture has significantly shaped the e-cig industry, transforming vaping from a cessation tool to a cultural phenomenon with its own dedicated community and practices. This culture has contributed to the industry's growth by fostering a community of enthusiasts, expanding the market through targeted demographics, and navigating the challenges of regulation and public perception (Lee et al. 2023; Izzani and Hermawan 2021).

1.5 Conclusion

The evolution of e-cigs since its inception in the 1960s to the present day reflects significant technological and cultural shifts within the smoking and vaping landscape. The journey began with Herbert A. Gilbert's concept, which was later actualized by Hon Lik in response to a personal loss and a desire for a safer smoking alternative. This innovation has since proliferated into a diverse market of products, ranging from first-generation cigalike to fourth-generation pod-based systems, each offering varying levels of customization, efficiency, and user experience. The industry's growth is marked by the entry of major tobacco companies and the emergence of dominant players through strategic mergers and acquisitions, highlighting the commercial potential of e-cigs. Concurrently, the rise of vape culture has transformed vaping from a cessation tool into a global phenomenon with its own community, practices, and controversies. Despite advancements, the e-cig industry faces ongoing challenges regarding safety, efficacy, and regulation. As the market continues to evolve, it remains to be seen how these issues will be addressed to balance public health concerns with the demand for alternative nicotine delivery systems.

References

Corporation Service Company (2022) Case study RELX

Department of Health, U., Services, H., for Disease Control, C., Center for Chronic Disease Prevention, N., Promotion, H., & on Smoking, O (2020) Smoking cessation: a report of the surgeon general

Dewhirst T (2019) British American tobacco (BAT) and retail merchandising: vype e-cigarette promotion in Ontario, Canada. Tob Control 28(2):E164–E167. https://doi.org/10.1136/tobaccocontrol-2019-054957

Gaidarov O, Asher R (2022) External dynamics contextualizing the FDA's role in E-cigarette regulation. Am J Bioeth 22(10):32–34. https://doi.org/10.1080/15265161.2022.2110969

Grana R, Benowitz N, Glantz SA (2014) E-cigarettes: a scientific review. Circulation 129(19):1972–1986. https://doi.org/10.1161/CIRCULATIONAHA.114.007667

Heide JM, Adam T, Jacobs EM, Wolter JM, Ehlert S, Walte Dr-Ing A, Zimmermann R (2021) Puff-resolved analysis and selected quantification of chemicals in the gas phase of E-Cigarettes, Heat-not-Burn devices and conventional cigarettes using single photon ionization time-of-flight mass spectrometry (SPI-TOFMS): a comparative study. https://doi.org/10.1093/ntr/ntab091/6276476

Hickman E, Payton A, Duffney P, Wells H, Ceppe AS, Brocke S, Bailey A, Rebuli ME, Robinette C, Ring B, Rager JE, Alexis NE, Jaspers I (2022) Biomarkers of airway immune homeostasis differ significantly with generation of E-cigarettes. Am J Respir Crit Care Med 206(10):1248–1258. https://doi.org/10.1164/rccm.202202-0373OC

Imperial Tobacco's Cooper Adds Blu to Take E-Cigarette Leap – Bloomberg (2014). https://www.bloomberg.com/news/articles/2014-07-15/imperial-tobacco-ceo-cooper-adds-blu-to-take-e-cigarette-leap?embedded-checkout=true

Izzani KM, Hermawan D (2021) The value of innovation in tobacco processing products: an evidence from bandung E-cigarette industry. Int J Creative Bus Manag 1(1):1. https://doi.org/10.31098/ijcbm.v1i1.4262

King AC, Smith LJ, McNamara PJ, Cao D (2018) Second generation electronic nicotine delivery system vape pen exposure generalizes as a smoking cue. Nicotine Tobacco Res 20(2):246–252. https://doi.org/10.1093/ntr/ntw327

Kostygina G, Tran H, Schillo B, Silver NA, Emery SL (2022) Industry response to strengthened regulations: amount and themes of flavoured electronic cigarette promotion by product vendors and manufacturers on Instagram. https://doi.org/10.1136/tobaccocontrol-2022-057490

Laugesen M (2015) Nicotine and toxicant yield ratings of electronic cigarette brands in New Zealand. N Z Med J 128(1411):77

Lee DN, Liu J, Oduguwa K, Stevens EM (2023) Association of vaping identity and e-cigarette perceptions among young adults who vape. Subst Use Misuse 58(11):1418–1424. https://doi.org/10.1080/10826084.2023.2223286

McDaniel PA, Malone RE (2012) British American Tobacco's partnership with Earthwatch Europe and its implications for public health. Glob Public Health 7(1):14–28. https://doi.org/10.1080/17441692.2010.549832

Rüther T, Hagedorn D, Schiela K, Schettgen T, Osiander-Fuchs H, Schober W (2018) Nicotine delivery efficiency of first- and second-generation e-cigarettes and its impact on relief of craving during the acute phase of use. Int J Hyg Environ Health 221(2):191–198. https://doi.org/10.1016/j.ijheh.2017.10.012

Sarles E, Hensel E, Robinson RJ (2021) Development timeline of the Juul® E-cigarette reveals a reactive approach of Government Agencies to Emerging Technologies. https://doi.org/10.2196/preprints.27329

Shastry S, Langdorf MI (2016) Electronic vapor cigarette battery explosion causing shotgun-like superficial wounds and contusion. West J Emerg Med 17(2):177–180. https://doi.org/10.5811/westjem.2016.1.29410

Smith DM, Bansal-Travers M, O'Connor RJ, Goniewicz ML, Hyland A (2015) Associations between perceptions of e-cigarette advertising and interest in product trial amongst US adult smokers and non-smokers: results from an internet-based pilot survey electronic cigarettes: towards evidence-based regulation Dr Maciej L. Goniewicz. Tob Induc Dis 13(1):14. https://doi.org/10.1186/s12971-015-0039-6

Tan JT (2021) Who is China's vape and e-cigarette billionaire Kate Wang? The Relx Technology CEO is richer than JD.com's Richard Liu and Baidu's Robin Li – and might even challenge Yang Huiyan for the 'richest woman in China' title | South China Morning Post. https://www.scmp.com/magazines/style/news-trends/article/3119727/who-millennial-billionaire-wang-ying-rlx-technology-ceo

Xu Y, Song X, Li X, Wang Z, Zhang Y (2022) Research on the ecological deconstruction of E-cigarette industrial clusters in Shenzhen, China, and a niche analysis of related enterprises. Sustainability (Switzerland) 14(9). https://doi.org/10.3390/su14095606

Anatomy of E-cigarettes

2

Mohd Fadhli Razali

Abstract

Electronic cigarettes (e-cigs) have quickly become popular as an alternative to conventional tobacco products. This chapter explores the structure and main components of e-cig devices. The atomizer is a key part that heats the e-liquid to create aerosols inhaled by users. It includes a resistance wire coil, a wick for e-liquid transport, and a housing. The coil's resistance is crucial because it affects temperature and aerosol mass, which potentially influence harmful substances in the vapor. E-cig batteries vary in type and capacity, which affect the vaping experience; however, safety concerns arise from incidents of explosions and burns, emphasizing the need for improved regulation and design. The appealing designs of e-cigs, which are often attractive to children, raise public health issues about the normalization of vaping and potential nicotine addiction among minors. Proper disposal and recycling of e-cig devices are vital to reduce their environmental impact. Complex materials, including hazardous lithium-ion batteries, complicate recycling efforts. Emerging initiatives seek to tackle these challenges, but comprehensive solutions and regulations are necessary. This chapter highlights the importance of understanding the e-cig anatomy to inform health research, guide regulations, and develop strategies to reduce their appeal to minors and environmental impact.

Keywords

E-cigarette anatomy · Atomizer coil resistance · Battery safety · Design appeal · Recycling initiatives

---

M. F. Razali (✉)
Malaysia National Poison Centre, Universiti Sains Malaysia, George Town, Malaysia
e-mail: fadhli_razali@usm.my

Y.-H. Leong, A. S. Mohd Samin (eds.), *E-Cigarettes*,
https://doi.org/10.1007/978-981-96-5087-3_2

2.1 Introduction

Electronic cigarettes (e-cigs) have emerged as a contemporary phenomenon in the landscape of nicotine delivery systems. Their rise in popularity has been meteoric, with a diverse range of products flooding the market, and each promises a unique experience to the user. E-cigs are intricate devices that are composed of several key elements that work in unison to deliver aerosolized nicotine to the user. At the heart of these devices lies the atomizer, which is a critical component responsible for heating the e-liquid and transforming it into vapor. The atomizer is a complex assembly, which consists of a coil made of resistance wire that heats up when an electric current is applied, a wick that transports the e-liquid to the coil, and a housing that facilitates vaporization.

The coil is a pivotal element in vaping, and its resistance plays a significant role in the temperature achieved and the subsequent aerosol mass formation. The materials used in coil construction, the device's power output, and the composition of the e-liquid influence the presence of potentially harmful substances in the vapor. These devices are powered by batteries that are available in various capacities and types, which then influence the vaping experience differently. The safety of these batteries is a paramount concern because incidents of explosions and burns have been reported, highlighting the need for stringent regulation and improved design.

The design of e-cigs has evolved significantly, and manufacturers employ innovative strategies to appeal to a broad audience, including minors. Marketing tactics employed by e-cig companies often involve child-friendly themes and discreet designs and have raised alarms within the public health community.

A detailed examination of the anatomy of e-cigs aims to shed light on multifaceted issues surrounding these devices and contribute to the ongoing discourse on their role in society.

2.2 Components of E-cig Devices

E-cig devices consist of several key components that work together to vaporize a liquid solution, which typically contains nicotine, flavorings, and other chemicals (Fig. 2.1). Understanding the various parts of an e-cig is crucial to provide insights into how these devices function and identify potential health implications associated with their use.

2.2.1 Atomizer

An atomizer is a critical component of an e-cig or a vaping device. It is responsible for heating the e-liquid and turning it into vapor, which will be inhaled by the user.

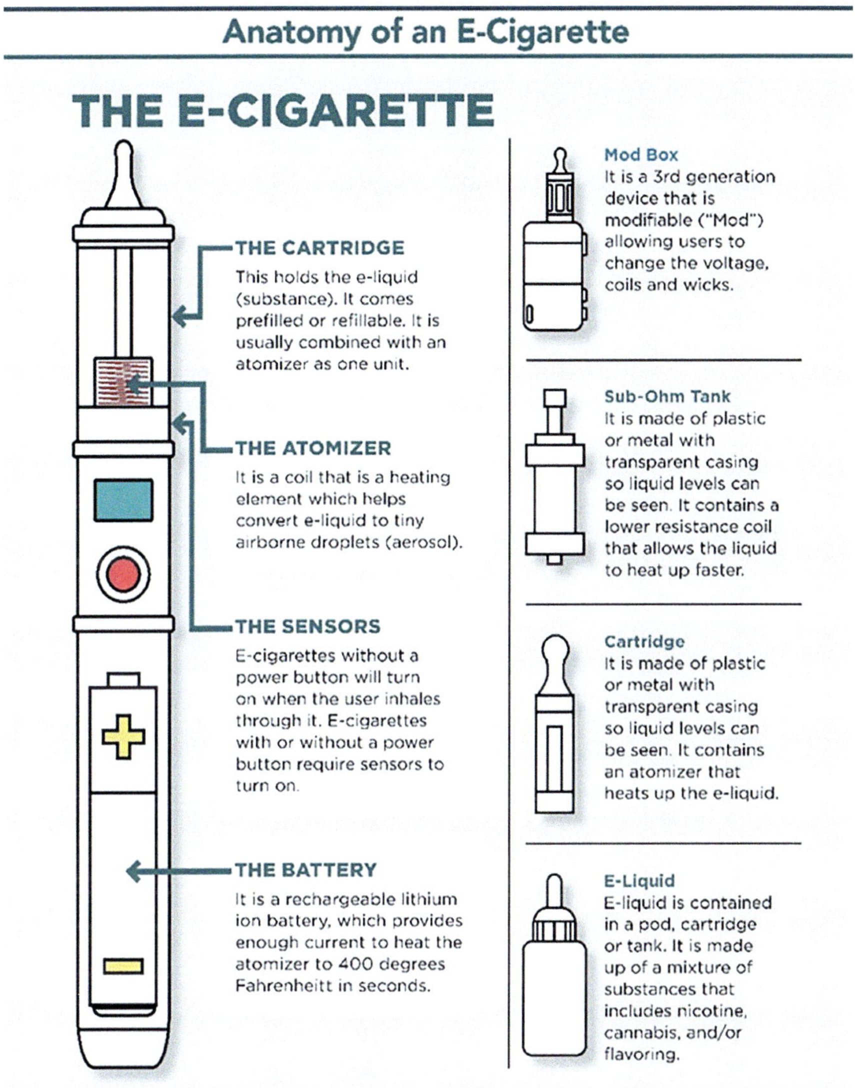

Fig. 2.1 Anatomy of an e-cigarette. (US Centers for Disease Control 2019). Diagram titled "Anatomy of an E-Cigarette" showing labeled components of an e-cigarette. The main parts include the cartridge, atomizer, sensors, and battery. The cartridge holds e-liquid and is often combined with the atomizer, which converts e-liquid to aerosol. Sensors activate the device when inhaled. The battery is rechargeable, heating the atomizer quickly. Additional components are listed on the right: Mod Box, Sub-Ohm Tank, Cartridge, and E-Liquid, each with descriptions of their functions and materials

2.2.1.1 Design and Components of Atomizer

An atomizer consists of several key parts, namely, coil, wick, housing, and connection. The coil is a small piece of resistance wire that heats up when electricity is applied from the battery. The wick is usually made of cotton, silica, or other absorbent material that is threaded through or around the coil. The wick absorbs the e-liquid and brings it into contact with the coil. The coil and wick are located inside a small metal housing. The housing has an inlet to allow e-liquid to reach the wick and an outlet where the vapor is released. When the user activates the e-cig or vaping device, power is applied to the atomizer. The coil in the atomizer heats up the e-liquid that has been absorbed by the wick. This phenomenon causes the vaporization of the e-liquid, creating an aerosol that the user inhales sers/Naili Husna/Downloads/S0306460322003458.ris of vaping, reducing nicotine concentration and restricting vaping times, coupled with behavioral counseling may be effective for cessation of electronic nicotine delivery system use.

2.2.1.2 Types of Atomizers

Several types of atomizers are used in vaping, and each type caters to different user preferences and experiences. Disposable atomizers are designed for single use and are typically found in cigalike and some pod systems. Once the coil burns out, the entire atomizer is discarded (US Centers for Disease Control 2019). Replaceable coil head atomizers feature a removable coil head that can be replaced when it burns out, allowing the rest of the atomizer to be reused. This type is commonly used in vape pens and some pod systems. Rebuildable atomizers (RBAs) allow users to build their own coils and wicks and offer a high level of customization. RBAs are often used in box and mechanical mods, making them popular among experienced vapers who enjoy tailoring their vaping experience to their specific preferences.

2.2.2 Coil

The coil in e-cigs is a critical component that heats the e-liquid and creates an aerosol inhaled by users. The coil is typically made of metals, such as stainless steel, kanthal, or nichrome (Tran et al. 2023). Coil resistance, which is measured in ohms (Ω), plays a significant role in determining the temperature of the coil and, consequently, aerosol mass formation. A low coil resistance generally leads to a higher coil temperature, and vice versa (Tran et al. 2023). The coil's temperature can also be influenced by the coil's age and the type of metal used to make the coil (Tran et al. 2023).

The coil's temperature is crucial because it affects the formation of carbonyls, a group of compounds that includes potentially harmful substances, such as formaldehyde, acetaldehyde, and acrolein. These carbonyls can be produced through thermal degradation chemistry or chemistry initiated by reactive oxygen species (ROS) (Tran et al. 2023). The temperature can also influence the thermal degradation of certain substances in the e-liquid, such as vitamin E acetate, which has been linked to e-cig or vaping product use-associated lung injury (Lynch et al. 2021; Canchola et al. 2023).

The type of e-cig device can also affect the emissions from the coil. For instance, a top-coil device generates the highest amounts of formaldehyde and carbon monoxide (CO), while a “pod” type device (e.g., JUUL) emits the highest amounts of nicotine but generates the lowest levels of carbonyl and CO compared with other tested e-cigs (Son et al. 2020). The coil’s performance can also be influenced by the e-liquid composition, device power output, and vaping topography (Son et al. 2018). Flavored e-liquids generate more carbonyls and CO than unflavored ones (Son et al. 2020).

In summary, the coil in e-cigs plays a crucial role in vaping and influences the temperature, the formation of potentially harmful substances, and the overall emissions from the device. The coil’s performance can be affected by various factors, including its resistance, age, type of metal used, e-liquid composition, and vaping parameters.

2.2.3 Battery

The battery is a crucial component because it provides the energy needed to heat the e-liquid and create vapor that will be inhaled by users. The type of battery used can influence the vaping experience, including the amount of vapor produced and the intensity of the nicotine hit. For example, powerful vape pens or mod devices can deliver a strong hit, which may be too intense for a smoker with low nicotine dependence (Mendelsohn and Beaumont 2022).

E-cigs utilize various types of batteries depending on the device’s design and functionality. Cigalike, which are first-generation e-cigs that resemble conventional tobacco cigarettes, typically employ small, often non-rechargeable batteries (Kar et al. 2019). Advanced devices, such as vape pens and mods, use large, rechargeable batteries; lithium-ion (Li-ion) batteries are popular because of their high energy density and power output. The battery life and power can vary significantly among these models (Yao and Pecht 2019). Pod vapes, which are compact rechargeable devices that often use higher nicotine salt concentrations, utilize small batteries similar to those found in cigalike. These different battery types cater to the diverse needs of e-cig users and balance factors, such as size, power output, and convenience across various designs (Mendelsohn and Beaumont 2022).

E-cig batteries can vary in capacity, and high-capacity batteries are often touted as a selling point in e-cig marketing (Chen et al. 2020). The capacity of the battery can influence the amount of vapor produced by the e-cig, with higher capacity batteries typically producing more vapor (Chen et al. 2020). However, safety concerns associated with e-cig batteries exist. Battery failure can result in explosions and burn injuries (Rossheim et al. 2019). From 2015 to 2017, about 2035 e-cig explosions and burn injuries were presented to US hospital emergency departments (Rossheim et al. 2019). Another study estimated that 2693 such injuries occurred between 2015 and 2018 (Rossheim et al. 2020). These explosions can occur in various contexts, including while a person is changing or replacing the device’s battery (Rossheim et al. 2020).

These safety concerns highlight the need for improved regulation and monitoring of e-cig devices, including their batteries (Rossheim et al. 2019). The industry should be required to address the technical problems that can lead to these explosions (Rossheim et al. 2020).

2.2.4 Tank/Cartridge

In the context of e-cigs, a tank or cartridge is a component of the device that holds the e-liquid, which is vaporized and inhaled by the user. These terms are often used interchangeably, but they refer to different types of e-cig devices.

2.2.4.1 Design and Function of Tank/Cartridge

E-cig tanks or cartridges are typically composed of a chamber to hold the e-liquid, a heating element (or coil) to vaporize the liquid, and a mouthpiece from which the user inhales the vapor (Wang 2023).

Tanks are usually associated with modifiable devices (MODs), which are refillable, reusable systems preferred by former smokers (Aherrera et al. 2023). These devices are typically large and allow users to modify their devices by replenishing the e-liquid, varying the power of the device, and changing the coil (Aherrera et al. 2023).

Cartridges are often associated with "pod" devices, which typically rely on replaceable pre-filled pods or cartridges (Snoderly et al. 2021). The e-liquid within these pods or cartridges contains nicotine salts that are less irritating and deliver nicotine at higher levels than previous e-cig generations and even conventional cigarettes (Snoderly et al. 2021).

2.2.4.2 Safety Concerns

Several safety concerns are associated with e-cig tanks and cartridges. One concern is the inhalation of metallic compounds in e-cig aerosols, which can present potential harm to users (Soulet and Sussman 2022). The concentrations of certain metals were statistically significantly higher in MOD devices, except for cobalt and nickel, which were higher in pod and disposable pod devices (Aherrera et al. 2023).

Given these potential risks, users should be aware of the type of e-cig device and the potential health implications. Studying these devices is crucial to understanding their long-term effects on health.

2.3 Ohm's Law and Its Application in Vaping

Ohm's Law, named after the German physicist Georg Simon Ohm, is a fundamental principle in the field of electrical engineering and physics. It states that the current passing through a conductor between two points is directly proportional to the voltage across the two points and inversely proportional to the resistance between them.

Ohm's Law is crucial for understanding the operation of vaping devices and ensuring their safe use. The key components of a vaping device that relate to Ohm's

Law are the battery (which provides the voltage), the coil (which has a certain resistance), and the current that flows when the device is used.

When the device is activated, the battery supplies voltage to the coil, heating it up and vaporizing the e-liquid (Zhou et al. 2018). The resistance of the coil in a vaping device is a crucial factor that determines how much heat it generates. Lower resistance coils (sub-ohm coils) will generate more heat and produce more vapor, but they will also draw more current from the battery. Higher resistance coils will generate less heat and produce less vapor, but they will draw less current (Zhou et al. 2018).

Understanding Ohm's Law is essential for the safe use of vaping devices. Using a coil with too low resistance can draw too much current from the battery, potentially causing it to overheat or even explode. Therefore, manufacturers should ensure that their device's coil resistance is appropriate for the battery's voltage (Zhou et al. 2018). Moreover, the coil's resistance can change as it heats up, which can affect the current and the device's performance. This property is known as the temperature coefficient of resistance (Zhou et al. 2018).

The resistance of a coil in a vaping device can be calculated using Ohm's Law if the voltage of the battery and the current are known. Alternatively, it can be measured directly by using a multimeter. Some vaping devices also display the coil's resistance (Zhou et al. 2018).

2.4 Allure of Unique E-cig Designs

The increasing appeal of e-cigs among minors is driven by their unique and often child-friendly designs and raises significant public health concerns. These devices, which come in shapes and sizes that mimic everyday objects, toys, or popular culture themes, not only attract young people but also contribute to the normalization of vaping. Normalization can lead to increased experimentation and use among minors, potentially ushering in a new generation of nicotine-dependent individuals.

E-cigs have evolved from their initial designs to include a wide array of shapes and styles, some of which are explicitly aimed at younger demographics. For instance, devices that resemble USB sticks, such as the JUUL, have become popular due to their discreet nature, allowing users to vape without drawing attention (Stockings et al. 2024). Other designs include vape pens adorned with cartoon characters, e-hookahs with bright LEDs, and mods that can be customized with various skins, including those with gaming themes (Stockings et al. 2024). These designs not only make the products more appealing to young people but also blur the lines between a nicotine delivery device and everyday items or hobbies.

The design and marketing strategies employed by e-cig manufacturers have a profound impact on their appeal to minors. Devices that mimic the appearance of toys and gadgets or are associated with popular culture effectively capture the interest of young people. This interest is not benign; it plays a significant role in the normalization of vaping within youth cultures. As vaping becomes normalized, the threshold for experimentation decreases, leading to increased use among minors (Stockings et al. 2024; Lepkowska 2022; Amato et al. 2021). This trend is particularly concerning given the addictive nature of nicotine and the potential of e-cigs to serve as a gateway to conventional tobacco products (Buck 2020).

The public health implications of this trend are significant. Nicotine has adverse effects on adolescent brain development as well as attention, learning, and impulse control (Skansberg 2017). Moreover, the normalization of vaping risks undoing decades of progress in tobacco control and potentially re-normalizes smoking behavior (Buck 2020). The rise in disposable e-cig use among young adults, as evidenced by a 14-fold increase in Great Britain from 2021 to 2022, underscores the urgency of addressing this issue (Tattan–Birch et al. 2023). In response to these concerns, governments and health organizations worldwide have begun to implement measures aimed at reducing the appeal and accessibility of e-cigs to minors. These measures include banning flavors and colors that appeal to young people, restricting marketing practices that target minors, and enforcing age verification for purchases (Stockings et al. 2024). However, the effectiveness of these measures remains to be fully elucidated, and continued vigilance and innovative strategies are pressing needs to protect young people from the allure of vaping (Stockings et al. 2024).

The unique designs of e-cigs play a critical role in their appeal to minors, contribute to the normalization of vaping, and potentially lead to increased experimentation and use. This trend poses significant public health challenges that require a multifaceted response that includes regulation, education, and prevention efforts. Researchers and public health professionals should continue monitoring these trends, understanding their impact on youth behavior, and developing strategies to mitigate risks associated with vaping among minors.

2.5 Disposal of E-cig Devices

E-cig recycling initiatives are an emerging area of interest due to the increasing prevalence of e-cig use and environmental concerns associated with their disposal. These initiatives encompass various aspects, including materials involved, specific programs designed for recycling, challenges faced during recycling, and the overall impact of these efforts on the environment and public health.

E-cigs consist of several components, including batteries, plastic casings, electronic circuits, and nicotine-containing cartridges or pods. The diversity of materials makes recycling e-cigs a complex process. The batteries, often lithium-ion, are hazardous and require specialized handling. Plastic components and electronic circuits can be recycled, but they need to be separated from hazardous materials first. Nicotine cartridges containing liquid nicotine pose a significant challenge due to the toxicity of nicotine, which requires careful handling and disposal (Barhdadi et al. 2021; Gómez Escobar et al. 2021).

Recycling programs for e-cigs are not as widespread as those for traditional waste materials. However, some manufacturers and environmental organizations have started to implement take-back schemes and recycling programs, specifically for e-cigs. These programs typically involve consumers returning used e-cig components to designated collection points or mailing them back to the manufacturer for recycling. For example, some companies offer incentives, such as discounts on future purchases, to encourage consumers to participate in their recycling programs. However, the effectiveness and reach of these programs are still limited, and more comprehensive, widely available e-cig recycling solutions are needed (Pazzaglia and Castellani 2023).

The recycling of e-cigs faces several challenges. The first barrier is the lack of awareness and education among consumers about the need to recycle e-cigs and the availability of recycling programs. Many users are unaware that e-cigs contain hazardous materials that should not be disposed of in regular trash. Second, the small size and complexity of e-cig components make them difficult to process in standard recycling facilities. Specialized equipment and processes are required to safely dismantle and recycle e-cigs, which can be costly and resource-intensive. Additionally, the absence of regulatory frameworks and extended producer responsibility (EPR) schemes for e-cig waste further complicates the establishment of effective recycling programs (Pazzaglia and Castellani 2023).

The impact of e-cig recycling initiatives, while still emerging, has the potential to be significant. Proper recycling of e-cigs can prevent hazardous materials from entering landfills and the environment, thereby reducing soil and water pollution. By recovering valuable materials, such as metals from batteries and plastics, recycling can reduce the demand for virgin materials, contributing to a circular economy. Moreover, effective recycling programs can raise awareness about the environmental impact of e-cigs, potentially influencing consumer behavior toward sustainable practices. However, the overall impact of these initiatives is contingent on their scale, effectiveness, and level of participation of consumers, manufacturers, and governments (Pazzaglia and Castellani 2023; Gómez Escobar et al. 2021).

Recycling initiatives for e-cigs are in their nascent stages but are crucial to mitigate the environmental impact of these products. The challenges associated with recycling e-cigs are significant and involve technical, regulatory, and behavioral aspects. However, the recycling of e-cigs becomes effective with increased awareness, the development of specialized recycling technologies, and the implementation of regulatory frameworks. The success of these initiatives will depend on collaboration among stakeholders, including manufacturers, consumers, environmental organizations, and governments, to create a sustainable approach to managing e-cig waste.

2.5.1 E-cig Battery Disposal

E-cigs contain lithium-ion batteries, which pose significant environmental hazards if not properly disposed. When these batteries end up in landfills, the risk of soil and water contamination increases because of the potential leaking of heavy metals and other toxic substances. Damaged lithium-ion batteries can cause fires in landfills. Therefore, proper recycling and disposal of e-cig batteries is critical to mitigate their environmental impact.

The core components of lithium-ion batteries in e-cigs include a cathode (positive electrode), anode (negative electrode), electrolytes, separators, steel casing, and various polymers and binding agents (Quiroga et al. 2019). The cathode is commonly made of lithium metal oxide, such as lithium cobalt oxide or lithium iron phosphate (Dyartanti et al. 2022). The anode is generally based on graphite. These electrodes contain heavy metals, such as cobalt, nickel, copper, aluminum, and iron. The electrolyte is lithium salt dissolved in an organic solvent. Common lithium salts include lithium hexafluorophosphate and lithium tetrafluoroborate. Organic solvents include ethylene carbonate, dimethyl carbonate, and diethyl carbonate (Olmedo et al. 2018).

When lithium-ion batteries end up in landfills through improper disposal, they can leak hazardous chemicals over time as they degrade and corrode (Ranpara et al. 2021). These chemicals can contaminate soil and groundwater supplies. Heavy metals, such as cobalt, nickel, copper, cadmium, and lead from the cathode and anode materials, are toxic even at low concentrations and do not degrade over time (Goros et al. 2023). Organic electrolytes are flammable and toxic (Mishra and Apte 2023).

Proper recycling and disposal of lithium-ion batteries are critical to mitigate the environmental impact of e-cig wastes (Yu et al. 2018). Batteries should be removed from e-cigs before disposal and recycled separately. Undamaged batteries in whole form should be disposed to avoid short circuits. Recycling allows recovery of battery materials, such as lithium, cobalt, nickel, copper, and aluminum, through hydrometallurgical and pyrometallurgical processes (Boissiere et al. 2020). Recycled metals can be used to manufacture new batteries. Several battery manufacturers and e-cig companies have battery take-back and recycling initiatives (Ngambo et al. 2023). However, robust recycling infrastructure and regulations are needed to manage the growing waste stream as e-cig use rises globally. Implementing extended producer responsibility can incentivize companies to facilitate battery collection and fund recycling initiatives.

E-cig lithium-ion batteries contain toxic and hazardous components that can contaminate soil and water in landfills and also pose fire risks. Proper recycling allows the recovery of valuable battery materials and reduces environmental impact. Interventions are urgently needed to build recycling infrastructure and minimize improper disposal of e-cig wastes.

2.6 Conclusion

The anatomy and design of e-cigs have evolved rapidly since their introduction, giving rise to a diverse array of products that cater to various user preferences and experiences. While the core components remain similar, the intricate interplay between the battery, atomizer, coil, and tank/cartridge systems allows for customization and personalization, appealing to novice and experienced users alike.

However, this diversity and complexity underscores the need for a comprehensive understanding of the potential risks and safety concerns associated with these devices. The materials used in their construction, the heating mechanisms, and the chemical interactions between the e-liquid and the device's components can potentially generate harmful byproducts, raising questions about their long-term impact on human health and the environment.

As the popularity of e-cigs continues to grow, particularly among younger demographics, the allure of unique and discreet designs that may contribute to the normalization of vaping should be addressed. Regulatory bodies and public health organizations must remain vigilant and implement measures to curb the appeal of these products to minors while promoting education and awareness about their potential risks.

Furthermore, the responsible disposal and recycling of e-cig devices and their components, such as lithium-ion batteries, must be prioritized to mitigate their environmental impact. Collaborative efforts between manufacturers, consumers, and regulatory agencies are essential to establish effective recycling programs and infrastructure and ensure a sustainable approach to managing the growing waste stream associated with these products.

The intricacies of e-cig anatomy, their emissions, and their potential health effects should be continuously investigated. Through rigorous scientific inquiry and a multidisciplinary approach, we can develop informed policies and strategies to address the challenges posed by this rapidly evolving technology to safeguard public health while promoting responsible innovation.

References

Aherrera A, Lin JJ, Chen R, Tehrani M, Schultze A, Borole A, Tanda S, Goessler W, Rule AM (2023) Metal concentrations in E-cigarette aerosol samples: a comparison by device type and flavor. Environ Health Perspect 131(12):127004. https://doi.org/10.1289/EHP11921

Amato MS, Bottcher MM, Cha S, Jacobs MA, Pearson JL, Graham AL (2021) "It's really addictive and I'm trapped:" a qualitative analysis of the reasons for quitting vaping among treatment-seeking young people. Addict Behav 112. https://doi.org/10.1016/j.addbeh.2020.106599

Barhdadi S, Rogiers V, Deconinck E, Vanhaecke T (2021) Toxicity assessment of flavour chemicals used in e-cigarettes: current state and future challenges. Arch Toxicol 95(8):2879–2881. https://doi.org/10.1007/s00204-021-03080-6

Beginners Guide to Vape Atomizers (2021). https://www.artisanvaporcompany.com/vaping/beginners-guide-to-vape-atomizers/

Boissiere F, Bekara F, Luca-Pozner V, Godillot C, Gandolfi S, Gibrila J, Chaput B, Herlin C (2020) Thermal and chemical burns caused by e-cigarette battery explosions. Ann Chir Plast Esthet 65(1):24–30. https://doi.org/10.1016/j.anplas.2019.12.001

Buck D (2020) Increasing adult awareness of Hawai'i Youth Vaping. https://api.semanticscholar.org/CorpusID:229556854

Canchola A, Langmo S, Meletz R, Lum M, Lin YH (2023) External factors modulating vaping-induced thermal degradation of vitamin E acetate. Chem Res Toxicol 36(1):83–93. https://doi.org/10.1021/acs.chemrestox.2c00298

Chen Y, Sun S, Zhao X, Zhou H, Wang F (2020) Objective facts or misleading hype? Associations between features of e-cigarette marketing and sales on a Chinese e-commerce platform. Int J Environ Res Public Health 17(18):1–12. https://doi.org/10.3390/ijerph17186711

Dyartanti ER, Jumari A, Paramitha T, Putri AYM, Sunaryati N (2022) Synthesis and characterization of cellulose acetate membrane from cigarette butts as gel polymer electrolytes for lithium-ion battery. Solid State Phenom 334:182–191. https://doi.org/10.4028/p-6v0288

Gómez Escobar V, Moreno González C, Arévalo Caballero MJ, Gata Jaramillo AM (2021) Initial conditioning of used cigarette filters for their recycling as acoustical absorber materials. Materials 14(15). https://doi.org/10.3390/ma14154161

Goros RA, Xu X, Li G, Zuo YY (2023) Adverse biophysical impact of e-cigarette flavors on pulmonary surfactant. Environ Sci Technol 57(42):15882–15891. https://doi.org/10.1021/acs.est.3c05896

Kar M, Emre IE, Muluk NB, Cingi C (2019) Effect of electronic cigarettes on the inner mucosa of the craniofacial region. J Craniofac Surg 30(3):E235–E238. https://doi.org/10.1097/SCS.0000000000005190

Lepkowska D (2022) Vaping: use of electronic cigarettes is on the rise among young people. Br J Child Health 3(5):220–221. https://doi.org/10.12968/chhe.2022.3.5.220

Lynch J, Lorenz L, Brueggemeyer JL, Lanzarotta A, Falconer TM, Wilson RA (2021) Simultaneous temperature measurements and aerosol collection during vaping for the analysis of Δ9-tetrahydrocannabinol and vitamin E acetate mixtures in ceramic coil style cartridges. Front Chem 9. https://doi.org/10.3389/fchem.2021.734793

Mendelsohn CP, Beaumont C (2022) What doctors should consider before prescribing e-liquids for e-cigarettes. Med J Aust 217(10):547–547. https://doi.org/10.5694/mja2.51764

Mishra P, Apte S (2023) Geo-environmental and human health impacts of spent lithium-ion battery waste and its recycling: a critical review. Int J Environ Waste Manag 32(3):286–300. https://doi.org/10.1504/IJEWM.2023.133591

Ngambo G, Hanna EG, Gannon J, Marcus H, Lomazzi M, Azari R (2023) A scoping review on e-cigarette environmental impacts. Tob Prev Cessat 9:30. https://doi.org/10.18332/TPC/172079

Olmedo P, Goessler W, Tanda S, Grau-Perez M, Jarmul S, Aherrera A, Chen R, Hilpert M, Cohen JE, Navas-Acien A, Rule AM (2018) Metal concentrations in e-cigarette liquid and aerosol samples: the contribution of metallic coils. Environ Health Perspect 126(2):027010. https://doi.org/10.1289/EHP2175

Pazzaglia A, Castellani B (2023) Review of the policy, social, operational, and technological factors affecting cigarette butt recycling potential in extended producer responsibility programs. Recycling 8(6):95. https://doi.org/10.3390/recycling8060095

Quiroga L, Asif M, Lagziel T, Bhat D, Caffrey J (2019) E-cigarette battery explosions: review of the acute management of the burns and the impact on our population. Cureus 11:e5355. https://doi.org/10.7759/cureus.5355

Ranpara A, Stefaniak AB, Fernandez E, LeBouf RF (2021) Effect of puffing behavior on particle size distributions and respiratory depositions from pod-style electronic cigarette, or vaping, products. Front Public Health 9. https://doi.org/10.3389/fpubh.2021.750402

Rossheim ME, Livingston MD, Soule EK, Zeraye HA, Thombs DL (2019) Electronic cigarette explosion and burn injuries, US emergency departments 2015-2017. Tob Control 28(4):472–474. https://doi.org/10.1136/tobaccocontrol-2018-054518

Rossheim ME, Soule EK, Thombs DL, Barnett TE, Livingston MD, Gimm G, Emechebe OC (2020) Electronic cigarette explosion and burn injuries. Tob Regul Sci 6(3):179–186. https://doi.org/10.18001/TRS.6.3.2

Skansberg A (2017) How addictive are they: effects of E-cigarette vapor on intracranial self-stimulation compared to nicotine alone. World academy of science, engineering and technology. Int J Pharm Pharm Sci 4

Snoderly HT, Nurkiewicz TR, Bowdridge EC, Bennewitz MF (2021) E-cigarette use: device market, study design, and emerging evidence of biological consequences. Int J Mol Sci 22(22). https://doi.org/10.3390/ijms222212452

Son Y, Wackowski O, Weisel C, Schwander S, Mainelis G, Delnevo C, Meng Q (2018) Evaluation of E-vapor nicotine and Nicotyrine concentrations under various E-liquid compositions, device settings, and vaping topographies. Chem Res Toxicol 31(9):861–868. https://doi.org/10.1021/acs.chemrestox.8b00063
Son Y, Bhattarai C, Samburova V, Khlystov A (2020) Carbonyls and carbon monoxide emissions from electronic cigarettes affected by device type and use patterns. Int J Environ Res Public Health 17(8). https://doi.org/10.3390/ijerph17082767
Soulet S, Sussman RA (2022) A critical review of recent literature on metal contents in E-cigarette aerosol. Toxics 10(9). https://doi.org/10.3390/toxics10090510
Stockings EA, Gardner LA, Newton NC (2024) Vaping among young people – our best defence is self-defence. Drug Alcohol Rev 43(2):355–358. https://doi.org/10.1111/dar.13746
Tattan-Birch H, Jackson SE, Kock L, Dockrell M, Brown J (2023) Rapid growth in disposable e-cigarette vaping among young adults in Great Britain from 2021 to 2022: a repeat cross-sectional survey. Addiction 118(2):382–386. https://doi.org/10.1111/add.16044
Tran LN, Chiu EY, Hunsaker HC, Wu KC, Poulin BA, Madl AK, Pinkerton KE, Nguyen TB (2023) Carbonyls and aerosol mass generation from vaping nicotine salt solutions using fourth- and third-generation E-cigarette devices: effects of coil resistance, coil age, and coil metal material. Chem Res Toxicol 36:1599. https://doi.org/10.1021/acs.chemrestox.3c00172
US Centers for Disease Control (2019) *E*-Cigarette, or vaping, products visual dictionary
Wang X (2023) E-cigarette toxicology and public health – exploring the safety of E-cigarette compared to traditional cigarette. Highl Sci Eng Technol 65:74–79. https://doi.org/10.54097/hset.v65i.11258
What Exactly Is Vaping Anyway? How Does It Work? Spinfuel (2023). https://spinfuel.com/what-exactly-is-vaping-anyway-how-does-it-work/
Why Do Vapers Use Rebuildable Atomisers? (2021). https://www.innokin.com/blog/why-do-vapers-use-rebuildable-atomisers
Yao XY, Pecht MG (2019) Tab design and failures in cylindrical li-ion batteries. IEEE Access 7:24082–24095. https://doi.org/10.1109/ACCESS.2019.2899793
Yu C, Hou H, Liu X, Han L, Yao Y, Dai Z, Li D (2018) The recovery of the waste cigarette butts for N-doped carbon anode in lithium ion battery. Front Mater 5. https://doi.org/10.3389/fmats.2018.00063
Zhou J, Zhang Q, Li T, Wang H (2018) Measurement method for wire resistance temperature coefficient based on heat transfer model. IOP Conf Ser Mater Sci Eng 452(4). https://doi.org/10.1088/1757-899X/452/4/042107

Vaping Trends: Empowering Healthcare Professionals to Address Emerging Risks

3

Keyun Look, Xinwei Koong, Ken Yien Leong, Long Chiau Ming, and Serena Leow

Abstract

Smoking cigarettes has been identified as detrimental to individuals' health, especially among individuals who are high in co-morbidity. Recent trends have indicated that individuals who smoke have switched to vaping, which is perceived as a better option and less harmful in content. Vape products or e-cigarettes (e-cigs) often contain high levels of nicotine and other harmful chemicals that are still being examined by researchers and healthcare professionals. Vaping and purchasing vape products, as well as accessories have been prevalent among young adults and minors. This chapter provides a background on vaping trends, vape products and e-cigs, as well as addresses the lack of knowledge and awareness among healthcare professionals concerning the long-term effects of vaping and the use of e-cigs. This chapter also highlights the different regulations in various countries and the need for better laws to regulate vaping, especially among young adults. By keeping up with the latest research, healthcare professionals can better educate and counsel their patients on the risks of vaping.

Keywords

Public health · Youth health risks · Addiction prevention · Health education Regulatory compliance

K. Look · X. Koong · L. C. Ming
Faculty of Medical and Life Sciences, Sunway University, Sunway City, Malaysia

K. Y. Leong
Sunway Business School, Sunway University, Sunway City, Malaysia

S. Leow (✉)
Faculty of Arts and Social Sciences, Sunway University, Sunway City, Malaysia
e-mail: serenaleow@sunway.edu.my

Y.-H. Leong, A. S. Mohd Samin (eds.), *E-Cigarettes*,
https://doi.org/10.1007/978-981-96-5087-3_3

3.1 Introduction

Smoking cigarettes has been associated with high risks, particularly for individuals with pre-existing health conditions. Recent trends indicate a significant shift from smoking cigarettes to vaping, raising significant concerns among healthcare professionals about the health implications and addiction potential of e-cigs. These products often contain, if not higher levels, of nicotine and other harmful chemicals. Traditionally, tobacco use has been proven detrimental to health, contributing to a significant rise in mortality (Zhong et al. 2023). Thus, the use of nicotine and tobacco alike are major public health concerns, particularly among minors or young adults.

The prevalence of vaping or e-cigs among the younger generation has increased due to the rise of online advertising and technological advancements in product promotion. The trend of vaping has seeped into the youth culture, influencing minors to use these products and posing health risks at a young age. Many perceive e-cigs as less harmful than conventional tobacco smoking. Some individuals believe that vaping helps in quitting tobacco smoking, leading to the use of vape products (Abu-Baker et al. 2022).

Despite the growing trends of using vape products and engaging in heavy use of e-cigs among the public, healthcare professionals' knowledge and awareness of vape products remain insufficient (Hunter et al. 2021). This gap hampers healthcare professionals' ability to counsel patients and addresses the growing public health issue related to the use of vape products, particularly among younger users. As such, it is vital for healthcare providers to understand the implications of vaping and effectively communicate such risks to their patients, regardless of age. This book chapter aims to address (1) vape products and trends, (2) the background of vape and e-cigs, (3) regulatory landscape of vape usage, (4) healthcare providers' awareness and knowledge about vaping, and (5) future research directions to identify gaps related to vaping.

3.2 Trends of Vaping

Unlike conventional cigarettes, vape products are known for lacking lingering smoke odor, making them appealing for use in various settings (Keamy-Minor et al. 2019). These devices were originally conceptualized by Joseph Robinson in 1927 and later developed into the modern e-cigs, marketed as a safer smoking cessation product by a Chinese pharmacist in 2004 (Baldassarri 2020; Foster 2023). E-cigs are battery-operated devices that vaporize liquid nicotine into an aerosol. The Chinese pharmacist patented these devices in 2007. Since then, the market for e-cig has experienced rapid growth, with an estimated compound annual growth rate of 30.6% from 2023 to 2030 (Malaysian Vape Chamber of Commerce 2023). In Malaysia, e-cig retail value increased by approximately 53% from 2019 to 2023 (Malaysian Vape Chamber of Commerce 2023).

Globally, the number of e-cig users was estimated at 82 million in 2021 (Jerzyński and Stimson 2023). In the United States, 2.55 million middle and high school students were reported as current e-cig users in 2022 (Cooper et al. 2022). Despite a decline in e-cig use among middle and high school students in the United States reported from 2019 to 2021, there has been a concerning increase in the intensity of e-cig use. For example, the percentage of students using e-cigs within five minutes of waking rose from less than 1% in 2017 to 10.3% in 2021, underscoring the urgency of addressing youth addiction (Glantz et al. 2022). In Great Britain, the prevalence of current e-cig use among youths aged 11 to 17 grew by a whopping 137.5%, increasing from 3.2% in 2021 to 7.6% in 2023 (Action on Smoking and Health 2023). In Malaysia, there was a 27% increase in e-cig users from 2019 to 2022 (Malaysian Vape Chamber of Commerce 2023).

The addiction to vape products can be attributed to several factors, including increasing use of social media and exposure toward vape advertisements (Hopkinson et al. 2023), the perception that vaping helps smokers quit tobacco (Abu-Baker et al. 2022), the perception that vaping is less harmful than conventional smoking (Fong et al. 2019), and the enjoyment of various flavors available in vape products (Ahmed et al. 2021). Constant exposure to vape products on social media reinforces the belief that vaping is less harmful than conventional tobacco, influencing young adults to view vaping as a safer alternative. This perception increases the prevalence of vaping and normalizes this behavior.

3.3 The Evolution of Vape Products and E-cigs

The first commercial e-cig was developed in 2003 by a Chinese pharmacist, Hon Lik (DeAngelis 2014). Motivated by his father's death from smoking-related lung cancer and his own unsuccessful attempts to quit smoking using nicotine patches, Hon Lik sought to create a safer alternative. The inspiration for the e-cig came from a dream in which he was drowning in a sea that transformed into vapor (DeAngelis 2014). In 2003, he patented a nicotine delivery system that utilized piezoelectricity to vaporize liquid, which was first sold in China the following year (Foster 2023). Since their introduction to Europe and the United States in 2006, e-cigs have surged in popularity globally and evolved into different types of vape products (Foster 2023). Though vaping has provided an alternative to smoking, its widespread adoption has significantly impacted public health.

The first-generation e-cigs, termed "cig-alike," are disposable devices designed to mimic conventional cigarettes (Baldassarri 2020; Centers for Disease Control and Prevention 2019). However, their small e-liquid chamber often results in insufficient nicotine delivery, especially for heavy users. Second-generation e-cigs, characterized by their pen-like shapes, feature larger batteries to support increased usage frequency and stronger aerosol production. These devices come with prefilled or refillable cartridges offering a diverse selection of e-liquids, including varying nicotine strengths and flavors. Third-generation devices, known as modifiable ("mods"), include tank systems and larger batteries, enabling the production of denser clouds and intensified nicotine delivery. The fourth-generation, "pod style" devices like the well-known JUUL are compact and rectangular, resembling a thumb drive, and dominated the US market in 2020. These devices incorporate e-liquid mixtures containing benzoic acid, which lowers the pH of nicotine, reducing throat irritation while delivering nicotine to users (Baldassarri 2020; Centers for Disease Control and Prevention 2019).

With the evolution of vape products, their demand increased, particularly among young adults. In the United States, youth vaping has reached epidemic levels, driven by product availability, high nicotine levels, e-liquid flavorings, and social media posts promoting vaping culture (Brett et al. 2021). A notable trend among youths is the vape trick phenomenon, known as "cloud chasing," where users exhale e-cig aerosols to produce large and artistic vape clouds. By 2013, the evolution of vape products into more powerful and customizable devices allowed users to modify e-liquid mixtures and adjust the humectant increasing vegetable glycerin to produce larger and opaque clouds, thus propelling cloud chasing and vaping cultures. However, the complexity of the tank-based mods led to a decline in their popularity by 2017. The transition toward smaller and less powerful "pod-mod" devices increased the desire for subtle vaping and decreased interest in vape tricks. Regulatory policies and health crises, such as the e-cig or vaping product use-associated lung injury outbreaks in 2019 and 2020 in the towards, and the COVID-19 pandemic likely contributed to the shift in vaping culture and decline in cloud chasing (Brett et al. 2021).

In the past three years, a new form of disposable e-cig, resembling pod devices, has flooded the market under brand names such as "Puff Bar," "Elf Bar," and "Lost Mary" (Tattan-Birch et al. 2023, 2024). These devices effectively deliver nicotine using high-concentration nicotine salt e-liquids while being more affordable and convenient (Tattan-Birch et al. 2024). Among 18-year-old vapers in Great Britain, disposable e-cig use skyrocketed from 0.4% at the beginning of 2021 to 54.8% the following year (Tattan-Birch et al. 2023). Before this surge, research showed a stable or declining trend in inhaled nicotine use through smoking and vaping across all age groups in England before June 2021 (Tattan-Birch et al. 2024). However, with the advent of these new disposable e-cigs, trends reversed, showing the most significant increase among young adults. While smoking prevalence continues to decline among individuals aged 18–44, it has started to climb again among those over 45. Notably, there was a greater relative rise in vaping among people who had never regularly smoked, underscoring a pressing concern that demands urgent action (Tattan-Birch et al. 2024).

3.4 Regulatory Landscape of Vape Usage

The World Health Organization (WHO) has called for urgent policies to control e-cig access among children and non-smokers. According to statistics, 34 countries have banned e-cig sales, while 88 countries lack a minimum age requirement for purchasing e-cigs (World Health Organization 2023). Over the past decade, the United States Food and Drug Administration (FDA) has faced challenges in regulating e-cig products due to their rapid evolution and complex legal and political considerations, leading to delays in the limited effectiveness of policies (Thomas and Kaplan 2019). Despite the FDA's implementation of the Deeming Rule in 2016, which prohibits the sale of tobacco products, including e-cigs to minors (Sharpless 2019), the rule has not effectively restricted access, particularly through retail stores (Do et al. 2023). The inconsistent implementation of vaping policies globally has resulted in the failure to curb vape products use among young adults and minors.

In response to the growing prevalence of e-cig users, especially among youths, several countries have implemented or proposed new regulations to address public concerns. A significant factor driving youth e-cigs use is the availability of flavored e-liquids, especially fruit and sweet flavors (Notley et al. 2022; Sapru et al. 2020). On January 2, 2020, the FDA announced an enforcement policy targeting unauthorized e-cig products, encompassing flavored cartridge or pod-based e-cigs (excluding tobacco or menthol flavors), all other e-cig products that fail to prevent youth access, and any e-cig products likely to promote youth use (US Food and Drug Administration 2020a). To balance adult smokers transitioning to e-cigs, the FDA's enforcement policy does not prioritize tobacco and menthol e-cig flavors, which are less preferred by youths compared to mint and fruit flavors (US Food and Drug Administration 2020a). Using Twitter data, Xie et al. (2022) showed that the FDA flavor enforcement policy potentially influenced user vaping behavior, with an increase in females and young adults mentioning quitting vaping on X (previously known as Twitter).

However, because the policy did not include menthol and tobacco-flavored pod systems or other flavored e-cig products, such as the tank systems or disposable e-cigs, behavioral changes were identified, including switching to these other types of e-cig products, switching to conventional cigarettes, or quitting e-cig use (Li et al. 2022). Consequently, flavored disposable e-cigs, initially excluded from the policy, gained popularity, creating a loophole in youth access to flavored products (Boakye et al. 2022; Li et al. 2022). Given concerns about youth vaping and environmental impacts, regulations on disposable e-cig restrictions and disposal methods should be implemented (Jackson et al. 2024a, b; Pourchez et al. 2022).

In response, the FDA notified companies, including Puff Bar manufacturers, to remove their flavored disposable e-cigs from the market on 20th July 2020 (US Food and Drug Administration 2020b). This action led to a significant decline in online searches for Puff Bar, a popular disposable e-cig brand (Boakye et al. 2022). In Australia, disposable e-cig importation is prohibited from 1st January 2024, and the importation of non-therapeutic e-cigs is prohibited from 1st March 2024 (Therapeutic Goods Administration 2023). Similarly, the Prime Minister of the UK announced a ban on disposable vapes on the 29th of January 2024, aimed at protecting children's health and addressing excessive use of vape products (Department of Health and Social Care 2024).

Disposable e-cig use surged across all population subgroups in Great Britain between 2021 and 2023, with the most significant rise among younger users aged 16 to 24 (Jackson et al. 2024a, b). The implementation of banning disposable vapes by the UK government will impact approximately 2.6 million adults, particularly youngsters and those who have never regularly smoked. Conversely, the ban will also affect around 1.2 million current smokers and 744,000 ex-smokers who have switched to disposable e-cigs from conventional cigarettes (Jackson et al. 2024a, b). Therefore, it is crucial to carefully weigh the potential unintended consequences and consider various factors when implementing regulatory policies. Moreover, the Tobacco and Vapes Bill, passed during its first reading on 16th April 2024, targets restricting access to e-cig products among individuals under 18 to reduce youth vaping in the UK (Balogun 2024; Hui 2024).

In Malaysia, the issue of vaping among young adults and minors continues to pose challenges. E-liquids containing nicotine were exempted from the Poisons Act 1952 on 1st April 2023, leaving them unregulated and warranting serious consideration (The Sun 2024). Malaysia is one of the countries with the highest number of vapers after the United States (Abidin et al. 2018). The lower prices of vape products and accessories have led to an increase in vape users between 2011 and 2020 (Mat Salleh et al. 2023). Fortunately, strict regulations will be introduced to govern the use of e-cigs with the upcoming enforcement of the Control of Smoking Products for Public Health Act 2024 in June 2024 (Bernama 2024).

The debate over the potential risks and benefits of e-cigs compared to conventional cigarettes continues due to the uncertainties about long-term health effects and effectiveness in aiding cessation. These uncertainties have hindered government regulation and incentivization (Sanders-Jackson et al. 2016). A study by Fairchild et al. (2019) highlighted divergent views and priorities in e-cigs among major stakeholders, including the National Academies of Sciences, Engineering, and Medicine (NASEM) and Public Health England (PHE). NASEM focuses on protecting non-smokers, preventing smoking normalization, and emphasizing evidence demonstrating causality between e-cig use and harm. Conversely, PHE prioritizes reducing harm to health for current smokers and focuses on the immediate harms experienced by the current smokers. Interestingly, these two major public health organizations have opposing views on the impact of e-cigs on youth. While the NASEM concluded the risk posed by the use of e-cigs, the PHE reported a negligible impact on youth (Fairchild et al. 2019).

3.5 Healthcare Providers' Awareness and Knowledge

Healthcare professionals' knowledge and awareness of vape products' impact and effects have become a recent issue of concern. Studies (Hunter et al. 2021; Hwang et al. 2020; Kanchustambham et al. 2017) have highlighted the necessity for healthcare professionals to be aware and have knowledge of the impact of vape products and e-cigs to effectively counsel patients. Without current knowledge or awareness, patients' health could be at risk. Debates around vape products often focus on their effects compared to tobacco products or conventional smoking.

A recent study from the UK revealed significant misperception about e-cigs compared to combustible cigarettes over the last decades, emphasizing the need for better communication about the lower risk associated with e-cigs (Jackson et al. 2024a, b). The belief that e-cigs are less harmful than conventional cigarettes has decreased by 40%, potentially discouraging smokers from switching to vaping as a smoking cessation aid. This shift in perception poses challenges to the English government's "swap to stop" campaign launched in April 2023 to help smokers quit smoking by transitioning to vaping (Jackson et al. 2024a, b).

While e-cigs are generally less harmful than conventional cigarettes due to lower toxins in their contents, they are not entirely safe (Centers for Disease Control and Prevention 2023; Marques et al. 2021). One controversy is that heating and mixing primary active compounds with other chemicals in e-liquids can produce potentially harmful compounds, posing risks for respiratory illnesses such as EVALI (e-cig or vaping product use associated lung injury) and harmful second-hand exposure (Bhatt et al. 2020). In contrast, Alotaybi et al. (2022) as well as Gülşen and Uslu (2020) suggest that e-cigs are as harmful as conventional cigarettes or highlight the health consequences and complications. Yayan et al. (2024) argue that compared to the thousands of known chemical compounds in tobacco smoke, the aerosols produced by e-cigs may have fewer harmful effects, and their non-combustible nature may produce fewer carcinogenic substances. The diversity of e-cigs and e-liquid formulations, including nicotine concentrations and flavorings, may generate different toxic compounds and levels, resulting in varied health implications and safety risks (Yayan et al. 2024). Limited data support the use of e-cigs for smoking cessation in adults, including pregnant women (Centers for Disease Control and Prevention 2023). The uncertainty surrounding their long-term effects (Marques et al. 2021; Yayan et al. 2024) underscores the need for robust research. Furthermore, the absence of standardized safety assessments for e-cigs, variability among vape products, and reliance on self-reported data demand stronger evidence through better-designed studies (Yayan et al. 2024).

Healthcare providers play a crucial role in improving healthcare quality by communicating accurate information about the risks, benefits, and potential health impacts of e-cigs to their patients. Research indicates that patients who receive advice from healthcare providers tend to maintain longer smoking cessation durations, emphasizing the importance of healthcare providers in promoting quitting efforts (Sharma et al. 2024). Nevertheless, a systematic review by Selamoglu et al. (2022) revealed that most general practitioners had limited knowledge and confidence when discussing e-cig safety and their use as smoking cessation tools with patients. Due to limited evidence, concerns about health and safety risks, and insufficient dissemination of guidelines, healthcare providers feel limited in discussing vaping (Hunter et al. 2021).

Similarly, cross-sectional studies by Bharadwaj et al. (2020) among public health dentists, and by Thongsutt et al. (2023) among pharmacy students reported concerning trends. Despite high awareness, participants showed limited knowledge about e-cigs, with mean scores of 5.813 ± 1.83 out of 14 and 3.80 ± 1.71 out of 7, respectively (Bharadwaj et al. 2020; Thongsutt et al. 2023). Furthermore, many medical students, despite being aware of the dangers of vaping, had limited vaping education and knowledge, leading to a prevalence of vaping among them that surpassed the general population (Ruppel et al. 2021). This underscores the necessity of formal education on vaping and its harmful effects to promote health among dental, medical, and pharmacy students.

Despite formal training in respiratory medicine, many general practitioners remain unfamiliar with the harmful effects of e-cigs (25.5%), smoking (21.6%), nicotine replacement therapy (35.3%), and oral cessation medications (34.3%) (Zhong et al. 2023). Such knowledge deficiencies may lead to biased primary care or smoking cessation advice (Zhong et al. 2023). A study by Kulak et al. (2023) found that even with the possession of knowledge, 44% of college healthcare providers participated in e-cig training or educational activities, with some remaining misinformed. Most healthcare providers rely on peers and social media for e-cig information rather than accessing evidence-based sources such as medical journals or formal comprehensive education during their studies (Hwang et al. 2020; Mohammad et al. 2023).

A study by Koprivnikar et al. (2020) found that only a minority of healthcare professionals recommend e-cigs to patients. Those who did often exhibited a discrepancy between their actual and perceived knowledge they believed they had. Their actual understanding was lacking, potentially leading to suboptimal recommendations (Koprivnikar et al. 2020). Similarly, Kanchustambham et al. (2017) found that 53% of physicians were asked about e-cigs by patients wanting to quit smoking but significant gaps existed between their perceived and actual knowledge. Only 9% of physicians reported being very familiar with e-cigs while 25% were not familiar at all. Among those who reported being very or somewhat familiar, 28% were unaware of e-cig contents or the meaning of terms like "vape." Despite more than half of the physicians viewing a role for e-cigs in harm reduction and prescribing them for smoking cessation, some of them lacked familiarity and had safety concerns, and were compounded by conflicting evidence. Their perceptions of harm reduction and smoking cessation were not significantly associated with their level of training and degree of familiarity, emphasizing the need for better education and evidence-based guidelines to ensure they provide accurate advice based on reliable information rather than personal perceptions (Kanchustambham et al. 2017).

Notably, knowledge gaps among future and current healthcare professionals extend to regulations and policies related to e-cigs, second-hand exposure to vaping and toxins, strength and delivery of nicotine, prevalence of e-cig use and chemical components (Alfayoumi et al. 2022; Franks et al. 2017; Hwang et al. 2020; Kanchustambham et al. 2017). These alarming findings call for ongoing education and training on e-cigs and verifying the reliability and accuracy of information sources. Hence, conducting a timely systematic review to assess the current knowledge and awareness among future and present healthcare professionals from the existing literature is crucial for providing accurate and evidence-based advice to patients.

3.6 Future Direction and Needs Assessment

As vaping continues to grow in popularity, it is imperative to address the associated risks and misconceptions. There is an urgent need to enhance awareness and knowledge about the development of cape products and their potential health risks among healthcare providers. One critical area that healthcare providers must focus on is the knowledge of the long-term effects of vaping, which are often underestimated or misunderstood, or misperceived by many healthcare providers as in prior studies. Additionally, the general public's misconception that vaping is a safer alternative in comparison to smoking needs to be corrected. Given that vape products have been marketed as a safer option in comparison to conventional smoking in recent years, it is essential that healthcare providers improve their understanding of the components of e-cigs and identify those that pose health risks. By staying informed about emerging technologies and research, healthcare providers can better educate and protect their patients.

Furthermore, the rising prevalence of vaping among minors is a major concern for public health. Young adults are particularly susceptible to nicotine addiction, which can be detrimental to their health and have long-term consequences on their brain development and cognitive function. Targeted intervention strategies are needed among this age group, and healthcare providers are pivotal in these efforts. As the first point of contact for young patients and their families, healthcare providers can offer evidence-based information and resources to encourage and support young minors to quit e-cigs early.

Finally, the regulatory landscape on the use of vape products or e-cigs needs to be tightened. Currently, regulations vary widely across different countries due to inconsistent policy development worldwide. There is a pressing need for a comprehensive regulatory framework to address product safety, advertising restrictions, and the sale of such products to minors. Healthcare providers can advocate for stronger policies and contribute their expertise through clinical studies and patient interactions help to reduce the harm that vaping causes to young people.

3.7 Conclusion

The rise in vaping among the younger generation is concerning and presents significant public health concerns. Despite being marketed as a safe alternative compared to smoking, vaping poses its own risks. The rapid promotion and widespread efforts that are fueled by online advertisements necessitate the need for healthcare providers to stay abreast of the current and developing trends of vape and e-cigs. Knowledge of such products and their developments is crucial to effectively communicate the risks to patients, regardless of age, and to guide patients to healthier lifestyles.

The historical evolution of e-cigs, coupled with their increasing popularity, raises significant concerns and the need for more stringent regulatory frameworks that will promote educational initiatives and awareness among the public. Vape products and e-cigs have rapidly evolved from first-generation devices to containing various flavors that deliver high levels of nicotine. The evolution outpaces regulatory measures and calls for immediate attention among regulatory bodies, as well as from relevant stakeholders who can influence policy-setting. By addressing and raising awareness among the public, healthcare providers can better navigate the complexities of vaping and offer evidence-based counseling to their patients.

References

Abidin NZ, Abidin EZ, Zulkifli A, Ismail SNS, Karuppiah K, Nordin ASA, Musbah Z, Zulkipli NF, Praveena SM, Rasdi I (2018) Vaping topography and reasons of use among adults in Klang Valley, Malaysia. APJCP 19(2):457

Action on Smoking and Health (2023) Use of e-cigs (vapes) among young people in Great Britain. https://ash.org.uk/uploads/Use-of-vapes-among-young-people-GB-2023-v2.pdf?v=1697209531

Ahmed LA, Verlinden M, Alobeidli MA, Alahbabi RH, AlKatheeri R, Saddik B, Oulhaj A, Al-Rifai RH (2021) Patterns of tobacco smoking and nicotine vaping among university students in the united arab emirates: a cross-sectional study. Int J Environ Res Public Health 18(14). https://doi.org/10.3390/ijerph18147652

Alfayoumi I, Aqel O, Axon DR (2022) An assessment of student pharmacists' knowledge of electronic cigarettes or vapes – a cross sectional study at one college of pharmacy. Pharmacy 10(5):131. https://www.mdpi.com/2226-4787/10/5/131

Alotaybi M, Alzahrani SS, Algethmi AM et al (2022) E-cigs and vaping: a smoking cessation method or another smoking innovation? Cureus 14(12):e32435. https://doi.org/10.7759/cureus.32435

Baker KA, Campbell NJ, Noonan D, Thompson JA, Derouin A (2022) Vaping prevention in a middle school population using CATCH My Breath. J Pediatr Health Care. 36(2):90–98.

Baldassarri SR (2020) Electronic cigarettes: past, present, and future: what clinicians need to know. Clin Chest Med 41(4):797–807. https://doi.org/10.1016/j.ccm.2020.08.018

Balogun B (2024) Tobacco and vapes bill. UK Parliment. https://commonslibrary.parliament.uk/research-briefings/cbp-9992/

Bernama (2024) New act to strictly regulate use of electronic cigarettes – Dr Dzul. https://www.bernama.com/en/news.php?id=2305876

Bharadwaj AN, Vijayalakshmi B, Raju R, Gubbihal R, Kousalya PS (2020) Awareness regarding e-cigs among public health dentists in dental colleges of Bengaluru: a population study. J Indian Assoc Public Health Dent 18(4). https://doi.org/10.4103/jiaphd.jiaphd_42_20

Bhatt JM, Ramphul M, Bush A (2020) An update on controversies in e-cigs. Paediatr Respir Rev 36:75–86. https://doi.org/10.1016/j.prrv.2020.09.003

Boakye E, Dzaye O, Erhabor J, Osuji N, Obisesan O, Osei AD et al (2022) Impact of the Food and Drug Administration enforcement policy on flavoured e-cigs on the online popularity of disposable e-cigs: analyses of Google search query data. BMC Public Health 22(1):1937. https://doi.org/10.1186/s12889-022-14367-3

Brett E, Krissinger R, King A (2021) The rise and fall of e-cig cloud chasing appealing to youth. Prev Med Rep 24:101644. https://doi.org/10.1016/j.pmedr.2021.101644

Centers for Disease Control and Prevention (2019) E-cig, or vaping, products visual dictionary https://www.cdc.gov/tobacco/basic_information/e-cigs/pdfs/ecigarette-or-vaping-products-visual-dictionary-508.pdf

Centers for Disease Control and Prevention (2023) About electronic cigarettes (e-cigs). https://www.cdc.gov/tobacco/basic_information/e-cigs/about-e-cigs.html#:~:text=Top%20of%20Page-,Are%20e%2Dcigarettes%20less%20harmful%20than%20regular%20cigarettes%3F,in%20smoke%20from%20regular%20cigarettes

Cooper M, Park-Lee E, Ren C, Cornelius M, Jamal A, Cullen KA (2022) Notes from the field: E-cig use among middle and high school students – United States, 2022. Morb Mortal Wkly Rep 71(40):1283–1285. https://doi.org/10.15585/mmwr.mm7140a3

DeAngelis T (2014) Are e-cigs a game changer? Monit Psychol 45(3):48

Department of Health and Social Care (2024) Disposable vapes banned to protect children's health. GOV.UK. https://www.gov.uk/government/news/disposable-vapes-banned-to-protect-childrens-health

Do EK, Aarvig K, Donovan EM, Schillo BA, Vallone DM, Hair EC (2023) Underage youth continue to obtain e-cigs from retail sources in 2022: evidence from the truth continuous tracking survey. Int J Environ Res Public Health 20(2). https://doi.org/10.3390/ijerph20021399

Fairchild AL, Bayer R, Lee JS (2019) The e-cig debate: what counts as evidence? Am J Public Health Assoc 109(7):1000–1006. https://doi.org/10.2105/ajph.2019.305107
Fong GT, Elton-Marshall T, Driezen P, Kaufman AR, Cummings KM, Choi K, Kwan J, Koblitz A, Hyland A, Bansal-Travers M, Carusi C, Thompson ME (2019) U.S. adult perceptions of the harmfulness of tobacco products: descriptive findings from the 2013-14 baseline wave 1 of the path study. Addict Behav 91:180–187. https://doi.org/10.1016/j.addbeh.2018.11.023
Foster JA (2023) Consideration of vaping products as an alternative to adult smoking: a narrative review. Subst Abuse Treat Prev Policy 18(1):67. https://doi.org/10.1186/s13011-023-00571-w
Franks AM, Hawes WA, McCain KR, Payakachat N (2017) Electronic cigarette use, knowledge, and perceptions among health professional students. Curr Pharm Teach Learn 9(6):1003–1009. https://doi.org/10.1016/j.cptl.2017.07.023
Glantz S, Jeffers A, Winickoff JP (2022) Nicotine addiction and intensity of e-cig use by adolescents in the US, 2014 to 2021. JAMA Netw Open 5(11):e2240671. https://doi.org/10.1001/jamanetworkopen.2022.40671
Gülşen A, Uslu B (2020) Health hazards and complications associated with electronic cigarettes: a review. Turk Thorac J 21(3):201–208. https://doi.org/10.5152/TurkThoracJ.2019.180203
Hopkinson NS, Vrinten C, Parnham JC, Rado MK, Filippidis FT, Vamos E, Laverty AA (2023) Associations of media use with tobacco smoking and e-cigs: a national longitudinal study. https://doi.org/10.1101/2023.06.03.23290924
Hui S (2024) UK lawmakers back landmark bill to gradually phase out smoking for good. abc News. https://abcnews.go.com/International/wireStory/uk-lawmakers-vote-landmark-bill-aiming-create-countrys-109299538
Hunter A, Yargawa J, Notley C, Ussher M, Bobak A, Murray RL, Nath S, Cooper S (2021) Healthcare professionals' beliefs, attitudes, knowledge, and behavior around vaping in pregnancy and postpartum: a qualitative study. Nicotine Tob Res 23(3):471–478. https://doi.org/10.1093/ntr/ntaa126
Hwang J, Lee C, Mastrolonardo E, Frasso R (2020) Where there's smoke, there's fire: what current and future providers do and do not know about electronic cigarettes. BMC Public Health 20(1):1145. https://doi.org/10.1186/s12889-020-09265-5
Jackson SE, Tattan-Birch H, East K, Cox S, Shahab L, Brown J (2024a) Trends in harm perceptions of e-cigs vs cigarettes among adults who smoke in England, 2014-2023. JAMA Netw Open 7(2):e240582. https://doi.org/10.1001/jamanetworkopen.2024.0582
Jackson SE, Tattan-Birch H, Shahab L, Oldham M, Kale D, Brose L, Brown J (2024b) Who would be affected by a ban on disposable vapes? A population study in Great Britain. Public Health 227:291–298. https://doi.org/10.1016/j.puhe.2023.12.024
Jerzyński T, Stimson GV (2023) Estimation of the global number of vapers: 82 million worldwide in 2021. Drugs Habits Soc Policy 24(2):91–103. https://doi.org/10.1108/DHS-07-2022-0028
Kanchustambham V, Saladi S, Rodrigues J, Fernandes H, Patolia S, Santosh S (2017) The knowledge, concerns and healthcare practices among physicians regarding electronic cigarettes. J Community Hosp Intern Med Perspect 7(3):144–150. https://doi.org/10.1080/20009666.2017.1343076
Keamy-Minor E, McQuoid J, Ling PM (2019) Young adult perceptions of JUUL and other pod electronic cigarette devices in California: a qualitative study. BMJ Open 9(4):e026306. https://doi.org/10.1136/bmjopen-2018-026306
Koprivnikar H, Zupanič T, Farkas JL (2020) Beliefs and practices regarding electronic cigarettes in smoking cessation among healthcare professionals in Slovenia. Tob Prev Cessat 6(January):10.18332/tpc/115029
Kulak JA, Voit HE, Balfour CA, Homish GG (2023) College health providers' knowledge and confidence in addressing students' vaping: evidence from a pilot study in New York State. PEC Innov 2:100123. https://doi.org/10.1016/j.pecinn.2023.100123
Li D, Ossip DJ, Bansal-Travers M, Xie Z (2022) Impact of the FDA flavour enforcement policy on flavoured electronic cigarette use behaviour changes. Tob Control 31(Suppl 3):s176–s183. https://doi.org/10.1136/tc-2022-057492

Malaysian Vape Chamber of Commerce (2023) The Malaysian vape industry study 2023. https://dpvmalaysia.com/downloads/the_malaysian_vape_industry_study_2023.pdf

Marques P, Piqueras L, Sanz M-J (2021) An updated overview of e-cig impact on human health. Respir Res 22(1):151. https://doi.org/10.1186/s12931-021-01737-5

Mat Salleh R, Baharom N, Siau CS, Chan CMH, Amit N, Sia PY, Wee LH (2023) E-cig users' profiles and their association with identified impacts of COVID-19 on vaping among young adults in Malaysia. Healthcare 11:434. https://doi.org/10.3390/healthcare11030434

Mohammad H, Nassrawin S, Alhyasat A, AlMuhaissen S (2023) Physicians, pharmacists, and dentists' knowledge and attitudes toward the use of electronic cigarettes. Clin Epidemiol Global Health 24:101443. https://doi.org/10.1016/j.cegh.2023.101443

Notley C, Gentry S, Cox S, Dockrell M, Havill M, Attwood AS et al (2022) Youth use of e-liquid flavours-a systematic review exploring patterns of use of e-liquid flavours and associations with continued vaping, tobacco smoking uptake or cessation. Addiction 117(5):1258–1272. https://doi.org/10.1111/add.15723

Pourchez J, Mercier C, Forest V (2022) From smoking to vaping: a new environmental threat? Lancet Respir Med 10(7):e63–e64. https://doi.org/10.1016/s2213-2600(22)00187-4

Ruppel T, Alexander B, Mayrovitz HN (2021) Assessing vaping views, usage, and vaping-related education among medical students: a pilot study. Cureus 13(2):e13614. https://doi.org/10.7759/cureus.13614

Sanders-Jackson A, Tan AS, Bigman CA, Mello S, Niederdeppe J (2016) To regulate or not to regulate? Views on electronic cigarette regulations and beliefs about the reasons for and against regulation. PLoS One 11(8):e0161124. https://doi.org/10.1371/journal.pone.0161124

Sapru S, Vardhan M, Li Q, Guo Y, Li X, Saxena D (2020) E-cigs use in the United States: reasons for use, perceptions, and effects on health. BMC Public Health 20(1):1518. https://doi.org/10.1186/s12889-020-09572-x

Selamoglu M, Erbas B, Kasiviswanathan K, Barton C (2022) General practitioners' knowledge, attitudes, beliefs and practices surrounding the prescription of e-cigs for smoking cessation: a mixed-methods systematic review. BMC Public Health 22(1):2415. https://doi.org/10.1186/s12889-022-14696-3

Sharma A, King J, Krishnan-Sarin S, O'Malley SS, Morean M, Bold K (2024) How healthcare providers and the right information may play a critical role in quitting success among smokers interested in using e-cigs for quitting: results from a survey of U.S adults. PLoS One 19(5):e0303245. https://doi.org/10.1371/journal.pone.0303245

Sharpless N (2019) How FDA is regulating e-cigs. U.S. Food and Drug Administration. https://www.fda.gov/news-events/fda-voices/how-fda-regulating-e-cigs#:~:text=On%20the%20deeming%20effective%20date,to%20purchase%20a%20tobacco%20product

Tattan-Birch H, Jackson SE, Kock L, Dockrell M, Brown J (2023) Rapid growth in disposable e-cig vaping among young adults in Great Britain from 2021 to 2022: a repeat cross-sectional survey. Addiction 118(2):382–386. https://doi.org/10.1111/add.16044

Tattan-Birch H, Brown J, Shahab L, Beard E, Jackson SE (2024) Trends in vaping and smoking following the rise of disposable e-cigs: a repeat cross-sectional study in England between 2016 and 2023. Lancet Regional Health – Europe. https://doi.org/10.1016/j.lanepe.2024.100924

The debate over e-cigs demands stronger evidence of their value (2019) Nature 570(7762):415. https://doi.org/10.1038/d41586-019-01785-4

The Sun (2024) JTI Malaysia suggests government enact regulations for vaping products. https://thesun.my/local_news/jti-malaysia-suggests-government-enact-regulations-for-vaping-products-LG12146591

Therapeutic Goods Administration (2023) New regulation of vapes starting January 2024. https://www.tga.gov.au/news/media-releases/new-regulation-vapes-starting-january-2024

Thomas K, Kaplan S (2019) E-cigs went unchecked in 10 years of federal inaction. The New Tork Times. https://www.nytimes.com/2019/10/14/health/vaping-e-cigs-fda.html

Thongsutt T, Yusote C, Jubprang S, Sasisuwan A, Poonchuay N, Chanawong A et al (2023) Factors associated with knowledge and attitude towards e-cigs among undergraduate students in Thailand: a cross-sectional study. Asian Pac J Cancer Prev 24(2):559–567. https://doi.org/10.31557/apjcp.2023.24.2.559

US Food and Drug Administration (2020a) FDA finalizes enforcement policy on unauthorized flavoured cartridge-based e-cigs that appeal to children, including fruit and mint. https://www.fda.gov/news-events/press-announcements/fda-finalizes-enforcement-policy-unauthorized-flavoured-cartridge-based-e-cigs-appeal-children

US Food and Drug Administration (2020b) FDA notifies companies, including Puff Bar, to remove flavoured disposable e-cigs and youth-appealing e-liquids from market for not having required authorization. https://www.fda.gov/news-events/press-announcements/fda-notifies-companies-including-puff-bar-remove-flavoured-disposable-e-cigs-and-youth

World Health Organization (2023) Urgent action needed to protect children and prevent the uptake of e-cigs. https://www.who.int/news/item/14-12-2023-urgent-action-needed-to-protect-children-and-prevent-the-uptake-of-e-cigs

Xie Z, Ruan J, Jiang Y, Zhang B, Chen T, Luo J, Li D (2022) Potential impact of FDA flavour enforcement policy on vaping behavior on Twitter. Int J Environ Res Public Health 19(19). https://doi.org/10.3390/ijerph191912836

Yayan J, Franke K-J, Biancosino C, Rasche K (2024) Comparative systematic review on the safety of e-cigs and conventional cigarettes. Food Chem Toxicol 185:114507. https://doi.org/10.1016/j.fct.2024.114507

Zhong Q, An K, Wu Z, Zhang H, Li S, Zhang L, Li C, Li H, Ga QMZ, Zong DJY, An Z (2023) Knowledge and awareness of nicotine, nicotine replacement therapy, and electronic cigarettes among general practitioners with a special interest in respiratory medicine in China. Front Med 10:1236453. https://doi.org/10.3389/fmed.2023.1236453

4 Marketing Strategies and Tactics Used by E-cigarette Industry

Noor Afiza Abdul Rani and Yong Check Yoon

4.1 Introduction

Recent years have witnessed a substantial expansion of the e-cig industry, primarily driven by its diverse and multifaceted marketing approach. This rapid growth is attributed to the industry's strategic use both of traditional advertisements and digital platforms to promote e-cigs. The industry has extensively leveraged prominent social media platforms including TikTok, Instagram, and Facebook. These platforms offer cost-effective avenues for e-cig brands to reach broad audiences with tailored messages that leverage the distinctive features and user demographics platforms (Chu et al. 2015). Moreover, cross-site dissemination of marketing content is effective to maintain a consistent presence for any products across a wide range of social media platforms, which can enhance their reach and impact (Unnava and Aravindakshan 2021). This marketing approach also applies to e-cigs.

The e-cig industry often emphasizes health-related benefits and smoking cessation claims in their marketing messages. A systematic review found these themes prevalent in e-cig advertisements, highlighting supposed health benefits and portraying e-cigs as safer alternatives to conventional cigarettes (Lyu et al. 2022). The industry has aggressively targeted adolescents and young adults in their marketing strategies (Chu and Kong 2022). Research indicates that advertisements often employ themes, visuals, and imagery such as flavors, colors, and lifestyle depictions that are designed to appeal to this target audience's preferences (Struik et al. 2020). Besides that, the use of social media influencers and celebrities to promote e-cigs also further enhances their appeal to young people, increasing engagement and

N. A. A. Rani (✉)
Malaysia National Poison Centre, Universiti Sains Malaysia, Minden Penang, Malaysia
e-mail: noor_afiza@usm.my

Y. C. Yoon
Consumer Association Penang, Penang, Malaysia

Y.-H. Leong, A. S. Mohd Samin (eds.), *E-Cigarettes*,
https://doi.org/10.1007/978-981-96-5087-3_4

experimentation (Smith and Hilton 2023). However, the industry has consistently denied intentionally targeting adolescents and young adults by claiming their marketing is aimed at adult smokers seeking to quit conventional cigarettes.

The digital marketplace also presents unique challenges and opportunities for e-cig marketing. Many e-cig vendors have leveraged diverse promotional strategies and social networking platforms to boost product visibility and drive sales significantly (Mackey et al. 2015). This online environment allows vendors to circumvent traditional advertising restrictions, granting them greater flexibility to reach a broader audience, including vulnerable underaged users. While expanded access to digital media offers commercial advantages, it also raises important issues regarding the ethical and regulatory implications of e-cig marketing.

The objective of this chapter is to analyze the marketing strategies employed by the e-cig industry, particularly its use of digital platforms and social media to promote its products. It seeks to explore how these strategies have contributed to the growing acceptance and use of e-cigs, especially among adolescents and young adults. Additionally, the chapter highlights the problem that, despite the industry's claims that its marketing is aimed at adults, it continues to appeal to younger audiences, posing significant public health risks. The lack of comprehensive regulations to curb these marketing practices and the absence of effective educational campaigns to counter misleading messages necessitate urgent policy intervention.

Overall, the marketing strategies of the e-cig industry have significantly contributed to the widespread acceptance and use of these products, particularly among younger populations (Chu et al. 2015). Therefore, comprehensive regulation and education are essential to counteract these marketing and promotion strategies by the e-cig industry. Enforcing strict advertising guidelines and implementing age verification mechanisms can limit the reach of misleading advertisements, especially those targeting vulnerable populations like young adults (Soneji et al. 2016). Simultaneously, educational initiatives are needed to debunk myths and provide accurate information about the health risks of e-cig use, thereby reducing their appeal. By incorporating both regulatory and educational strategies, governments can minimize the influence of e-cig marketing, ensure that consumers make informed decisions, and protect public health.

4.2 Design Elements of E-cig Packaging

The e-cig industry leverages sophisticated marketing strategies to promote its products, with one key approach being the integration of strategic design elements into packaging. Branding and logo design play a crucial role in establishing a strong visual identity. Well-designed logos enhance brand recognition, making e-cigs easily identifiable and memorable for consumers (Gomes et al. 2024). Consistent branding across various marketing platforms helps these products stand out in a competitive market, allowing users to easily identify and recall them (Haardörfer et al. 2017).

Beyond logos, the physical appearance of the packaging, particularly the use of color, is a powerful marketing tool (Dutta 2023). Colors can evoke specific emotions and associations, aiding in targeting different consumer demographics. For example, calming colors like blue and green might be used to suggest relaxation and appeal to consumers seeking stress relief, while vibrant colors like red and orange are employed to convey excitement and energy, appealing to a more dynamic audience (Taylor et al. 2023). Visually striking packaging not only captures attention but also reinforces brand identity, helping products differentiate themselves in a crowded marketplace.

Moreover, e-cig packaging highlights technological advancements and unique product features, such as adjustable airflow and ergonomic designs (Brown and Cheng 2014). By emphasizing the sophisticated and customizable nature of vaping devices, the industry aims to attract a growing segment of users who are interested in the latest innovations in e-cig technology (Trumbo and Harper 2015).

E-cig boxes function as more than just packaging; they are a vital touchpoint between manufacturers and consumers. In addition to protecting products during shipping and storage, these boxes convey essential information—such as flavor, nicotine strength, and usage instructions—and play a crucial role in persuading users to choose one product over others.

4.3 E-liquid Flavor Variety and Labeling

The e-cig industry leverages a wide variety of flavors as a key strategy to expand its market appeal. By offering flavors ranging from conventional tobacco to an extensive selection of fruits, desserts, and beverages, the industry aims to cater to diverse tastes and preferences (Chaffee et al. 2022). This approach not only assists smokers in transitioning away from conventional cigs but also attracts non-smokers, particularly adolescents and young adults, who may be drawn to the novelty and enjoyment of flavored e-liquids (Mok et al. 2022). The variety of flavors helps create a more personalized and enjoyable vaping experience, making e-cigs more attractive to a broader audience (Gravely et al. 2020). This strategy enables the industry to tap into a larger market by addressing individual preferences and enhancing the overall attractiveness of their products (Ma et al. 2022a, b).

Flavored e-liquid labeling also plays a crucial role in this marketing strategy. Manufacturers often mimic popular foods, drinks, and sweets in their labeling to enhance product familiarity and appeal to younger buyers (McKelvey et al. 2019). This tactic can be misleading, as it might give a false impression that these e-liquids are harmless or pose less risk than they actually do. The use of bright colors, eye-catching images, and catchy names further heightens the appeal, drawing in potential users who might not have otherwise considered vaping (Soule et al. 2019).

Additionally, e-liquid packaging often resembles everyday consumer goods like cookies, juice boxes, or popular snacks, creating a misleading sense of safety and normalcy around vaping. This design strategy may downplay the potential risks of e-cigs, portraying them as harmless and enjoyable recreational items rather than nicotine delivery devices with potential harms (Fagan et al. 2018). This false impression can encourage experimentation and long-term use, particularly among younger populations who may perceive these products as benign and enjoyable rather than potentially harmful (Simonavičius et al. 2023).

Furthermore, e-cig marketing frequently includes misleading health or nutritional claims. For example, some brands assert that their e-liquid contains vitamin like B12 or essential oils, implying that inhaling these substances offers health benefits (Basáñez et al. 2019). However, there is no evidence that vitamins or nutrients can be effectively delivered through vaping, and inhalation of such substances may even pose health risks (Maggie O' Neill 2024). These misleading nutrition claims contribute to misconceptions about the safety and benefits of e-cigs, potentially influencing user behavior and undermining public health efforts to communicate the associated risks. The prevalence of these unsubstantiated claims highlights the need for stricter regulation and enforcement of virtual marketing practices for e-cigs.

The extensive range of e-liquid flavors, including fruit, dessert, and beverage options, can mislead users into thinking these products are harmless. However, users are at risk of developing E-cig or Vaping-Associated Lung Injury (EVALI), a severe inflammatory condition that can cause significant lung damage.

4.4 Promotion and Communication Strategies

The e-cig industry has significantly adapted its promotional strategies, with a strong emphasis on digital media, particularly social networking platforms to engage a wide audience. By strategically using influencer endorsements, celebrity involvement, and user-generated content, the industry creates appealing and engaging advertisements across various platforms like Instagram, Facebook, YouTube, and TikTok. These marketing techniques not only increase product visibility but also promote positive perceptions of vaping, especially among adolescents and young adults. Given the pervasively and loosely regulated online marketing of e-cigs, implementing effective advertising regulation and enforcement is crucial to protect vulnerable groups from potential harm.

4.4.1 Digital Platforms

The e-cig industry employs sophisticated strategies to promote its products across digital platforms, capitalizing on the vast reach and interactive potential of the internet. A key approach is the use of popular social media platforms such as Instagram, Facebook, and TikTok. These platforms enable the industry to share visually appealing content, including product images, promotional videos, and user testimonials that aim to attract a younger audience (Chu et al. 2015). By maintaining an active and engaging presence on social media, the e-cig industry fosters a sense of community around its products, strengthening brand loyalty and encouraging users to share their positive vaping experience (Smith and Hilton 2023).

User-generated content plays a key role in the e-cig industry's digital marketing strategy. The content typically involves individuals sharing personal experiences and opinions about using e-cigs on various online platforms such as forums, blogs, and social media. Often highlighting the perceived benefits and enjoyment associated with vaping, this user-generated content helps cultivate a positive narrative around e-cigs (Kwon and Park 2020). By actively encouraging and amplifying such user-generated content, the industry enhances the authenticity of its marketing efforts, making e-cigs seem more relatable and trustworthy to potential users (Murukutla et al. 2022).

The e-cig industry employs engaging and interactive digital marketing activities to further enhance its promotional efforts. These activities often include contests, giveaways, and challenges, which not only boost brand visibility and user engagement but also foster a sense of community and excitement around the products. By encouraging e-cig users to share their experiences and tag the brand on social media, the industry amplifies its marketing reach (Chu et al. 2015). Creating an interactive and participatory environment through these digital activities makes e-cigs particularly appealing and accessible to the younger generation, who are highly active on digital platforms (Smith and Hilton 2023).

The e-cig industry employs sophisticated digital strategies to promote its products, particularly targeting younger audiences through popular platforms like Instagram, Facebook, and TikTok. By sharing visually engaging content, such as product images, promotional videos, and user testimonials, the industry builds brand loyalty and fosters a community around e-cigs. A key tactic is leveraging user-generated content, where individuals share positive experiences with vaping across social media, blogs, and forums. This approach enhances authenticity, making e-cigs appear more relatable and trustworthy. Additionally, interactive marketing activities, such as contests, giveaways, and challenges, boost brand visibility and engagement, creating a participatory environment that appeals to a digitally active younger demographic.

4.4.2 Influencer Marketing and Content Creation

The e-cig industry relies heavily on influencer marketing as a key component of its digital strategy. By collaborating with social media influencers who have large followings, the industry leverages their status as trendsetters to promote its products. These influencers create content that portrays e-cigs in a positive light, often through personal endorsements, demonstrations of usage, and product reviews (Jancey et al. 2024). By harnessing the trust and reach of these influential figures, the industry effectively targets and influences the purchasing decisions of its followers, particularly adolescents and young adults who are susceptible to peer behavior and celebrity endorsements (Vassey et al. 2022).

In addition to influencer marketing, the e-cig industry relies on content creation as a crucial promotional strategy, using various digital platforms to reach and engage potential users. Video-sharing platforms like YouTube serve as significant avenues where both professional and user-generated content promote e-cigs, presenting them as cleaner, cheaper, and more convenient alternatives to conventional cigs (Sears et al. 2017). The industry also uses social media platforms like Facebook to run marketing campaigns, frequently featuring promotional giveaways and discounts to drive a high level of user engagement (Kong et al. 2021). Moreover, this trend extends to other popular social media platforms, such as X.com (formerly Twitter), where e-cig brands actively promote their products (Sun et al. 2020). The industry's extensive use of social media facilitates sophisticated marketing strategies that target specific demographics, contributing to the normalization and widespread acceptance of e-cigs (Chu et al. 2015).

The e-cig industry heavily utilizes influencer marketing and digital content creation to promote its products, targeting particularly impressionable demographics like adolescents and young adults. By partnering with social media influencers who act as trendsetters, the industry benefits from positive portrayals of e-cigs through personal endorsements, usage demonstrations, and product reviews, influencing followers' purchasing decisions. Additionally, video platforms like YouTube and social media sites such as Facebook and X (formerly Twitter) serve as key channels for distributing promotional content, often featuring giveaways and discounts to boost engagement. These strategies help normalize e-cig use and increase its acceptance among young audiences.

4.4.3 Role of Celebrities

Celebrities play a significant role in the marketing of e-cigs, influencing public perceptions and usage of these products. Their endorsements make e-cig use appear fashionable and widespread, particularly among younger audiences. By projecting an image of sophistication and modernity, celebrities inadvertently enhance the appeal of e-cigs. This influence is further amplified through social media, where celebrities frequently showcase their lifestyles, including e-cig use, which can sway the behaviors and choices of their followers. Studies have shown that celebrity endorsements have a more substantial impact on attitudes toward e-cigs compared to endorsements by non-celebrities (Phua et al. 2018).

Celebrities play a significant role in marketing e-cigs, shaping public perceptions and the usage of these products. Their endorsements make e-cig use appear fashionable and prevalent, particularly among younger audiences. By projecting an image of sophistication and modernity, celebrities enhance the appeal of e-cigs, which is further amplified on social media platforms where they often showcase their lifestyles, including e-cig use. Studies indicate that celebrity endorsements have a more substantial impact on attitudes toward e-cigs compared to those made by non-celebrities, influencing the behaviors and choices of their followers.

4.4.4 Movies and Television Dramas

The portrayal of e-cigs in movies and television dramas has raised significant public health concerns. Streaming platforms like Netflix have been criticized for incorporating e-cigs into storylines without highlighting the associated health risk (Allem et al. 2022). Popular films and television shows often depict vaping in a casual or rebellious light, making it appear appealing, especially to younger audiences. These portrayals can glamorize e-cig use, fostering the perception that vaping is socially acceptable or even sophisticated, which can lead to increased experimentation and use among adolescents (Escobedo et al. 2020a, b). Exposure to such depiction has been linked to heightened interest in vaping, particularly among impressionable adolescents (Hassanein et al. 2022). This trend allows the e-cig industry to bypass traditional regulations on tobacco advertising by exploiting the lack of stringent regulations governing product placement in entertainment media. As a result, e-cig brands capitalize on this opportunity to market their products, especially in regions where direct advertising is banned or heavily regulated.

The portrayal of e-cigs in movies and television dramas raises significant public health concerns, particularly as streaming platforms like Netflix have been criticized for featuring vaping without emphasizing its health risks. Popular films and TV shows often depict e-cig use in a casual or rebellious light, making it appealing to younger audiences and glamorizing vaping as socially acceptable or sophisticated. Such representations can lead to increased experimentation among adolescents, as exposure to these portrayals has been linked to heightened interest in vaping. Moreover, this trend allows the e-cig industry to circumvent traditional advertising regulations by exploiting the lack of strict rules governing product placement in entertainment media, enabling brands to market their products in areas where direct advertising is restricted.

Some movies and TV series that portrayed e-cig use:

- Drive Hard (2014. movie)
- The Tourist (2010. movie)
- Map to the Stars (2014. movie)
- House of Cards (2013. TV series)
- 2 Broke Girls (Season 4, Episode 11. TV series)
- HBO's True Detective (Season 2. TV series)
- Billions (2016. Episode 2)
- The Simpsons (2015. TV series)

4.5 In-store and Direct Marketing

The e-cig industry employs a variety of strategic marketing to promote its products and drive sales in retail settings. These strategies include eye-catching in-store displays, informative promotional materials, and direct user engagement through well-trained staff (Gaiha et al. 2024). Additionally, the industry leverages street marketing tactics such as product demonstrations and free samples to reach potential users in public spaces. To incentives purchases and foster brand loyalty, the e-cig industry implements discount programs, vouchers, and tiered loyalty schemes that offer rewards for repeat users, and bundling deals to create a sense of urgency and value, encouraging both new and existing users to make purchases (Ma et al. 2022a, b).

4.5.1 Promotional Material and Point-of-Sale Display

The e-cig industry uses several strategic tactics to promote its products through eye-catching in-store displays and promotional materials. A primary tactic is placing prominent displays that showcase new devices and e-liquid flavors (Eadie et al. 2015). These displays are strategically positioned to capture the attention of users as they enter retail spaces, making e-cig products more noticeable and appealing (Brocklebank et al. 2022). Additionally, cross-promotion with related accessories like chargers and carrying cases encourages users to explore the full range of complementary products (Wan et al. 2017).

The industry also frequently designs and positions promotional materials, such as signage, posters, and brochures, within stores to highlight special offers, new product launches, and unique features of e-cigs (Mantey et al. 2017). By placing these materials in high-traffic areas, visibility and impact are maximized, enhancing the likelihood of impulse purchases. This strategy is particularly effective in the rapidly evolving e-cig market, where retailers must find ways to capture users' attention and drive sales of new and innovative products.

4.5.2 Customer Engagement Strategies for In-store E-cig

The e-cig industry employs direct marketing tactics to engage users through face-to-face interactions in retail environments. Knowledgeable and approachable staff are employed to interact with users, address their questions, and provide recommendations (Cheney et al. 2015). This personalized approach fosters trust and loyalty, as consumers feel more confident and informed in their purchases (Begay et al. 2020). Moreover, product demonstrations and sampling sessions allow consumers to experience e-cigs firsthand, further influencing their purchasing decisions (Cheney et al. 2016). By combining appealing displays, informative promotional materials, and direct user engagement, the industry effectively promotes e-cigs and drives sales in retail settings.

4.5.3 Street Marketing

The e-cig industry uses street marketing as a hands-on strategy to engage potential users in public spaces through face-to-face interactions. This strategy involves hiring promoters to distribute free samples, flyers, and promotional materials in high-traffic areas such as shopping centers, busy streets, and events. By approaching potential users directly, promoters can generate immediate interest and provide first-hand information about e-cig products, creating a personal and impactful marketing experience.

Promoters often conduct demonstrations to showcase the features and benefits of e-cigs, allowing potential users to see the physical design and understand how the devices work (Chu et al. 2015). This approach allows promoters to highlight unique selling points such as ease of use, a variety of flavors, and advanced technology. The hands-on experience not only captures attention but also addresses any hesitation or skepticism, supporting a more informed purchasing decision.

Promoters may provide exclusive deals such as discounts, buy-one-get-one-free offers, and redeemable vouchers for local stores or online platforms, creating urgency and perceived value for potential users. These immediate incentives are designed to encourage trial and impulse purchases of e-cig products, particularly among price-conscious or curious individuals (Collins et al. 2019). Through these strategies, the e-cig industry seeks to drive initial adoption and promote long-term use of e-cigs among its target demographic (Dai and Hao 2017).

The industry strategically targets young adults and individual professionals who are more receptive to trying new products with street marketing. By placing promoters in locations frequented by these groups, the industry maximizes reach and marketing effectiveness (Giovenco et al. 2016). This targeted approach focuses on individuals most likely to become regular users, optimizing the return on investment for promotional activities. Through personalized interactions, immersive experiences, and attractive offers, the industry effectively promotes e-cigs to expand its user base.

4.6 Discount, Voucher, Loyalty, and Reward Programs

The e-cig industry employs a range of promotional strategies – such as discounts, vouchers, rewards, and loyalty programs – to attract new users and build brand loyalty. These tactics create a sense of value and urgency, incentivize purchases, enhance user retention, and support consistent sales. By leveraging these varied strategies, the industry effectively attracts new users, fosters brand loyalty, and maintains interest in a dynamic market landscape.

4.6.1 Discounts and Vouchers

The e-cig industry effectively promotes its products through strategic discounts and vouchers aimed at attracting both new and existing users. Promotions, such as percentage-based discounts or buy-one-get-one-free deals, create a sense of value and urgency (Ma et al. 2022a, b). These offers are frequently advertised via various channels, including industry websites, social media, and in-store displays (Chen-Sankey et al. 2024).

Meanwhile, vouchers offer additional purchase incentives for e-cigs. These vouchers are distributed through various channels, including direct mail, newsletters, and partnerships with retail outlets, and can be easily redeemed both in-store and online, making the offers highly accessible. By regularly issuing vouchers, the industry attracts new users but also encourages repeat purchases among existing users, thereby fostering brand loyalty (Chen-Sankey et al. 2024).

These promotional strategies effectively attract price-sensitive users by lowering the financial barrier to entry (Ma et al. 2022a, b). By making e-cigs more affordable, these tactics appeal to individuals who may hesitate to purchase at full price, thus driving higher purchase rates and increasing market penetration. Reduced costs also encourage repeat purchases and foster brand loyalty among cost-conscious users. This approach can be particularly effective in encouraging users to switch from conventional cigarettes to e-cigs, as the lower price point enhances product appeal. However, increased affordability through such promotions may inadvertently lead to greater experimentation and use among younger demographics who are more sensitive to price changes (Cantrell et al. 2020).

4.6.2 Loyalty and Reward Programs

The e-cig industry uses loyalty programs strategically to cultivate long-term relationships with users and enhance brand loyalty. These programs typically reward repeat users with points for each purchase, which can be redeemed for discounts, free products, or exclusive benefits (Davis 2020). By offering tangible incentives, loyalty programs encourage users to continue purchasing e-cig products from the same brand, thereby increasing user retention and supporting consistent sales (Komesarook 2023).

Similarly, the e-cig industry uses reward programs to promote products and build user loyalty. These programs often feature incentives like discounts, points for purchase, and exclusive access to new products or special events. By offering rewards for repeat purchases, reward programs encourage ongoing user engagement and foster a sense of loyalty among users. For instance, users may accumulate points with each purchase, which can later be redeemed for discounts or free products. This strategy not only incentivizes repeat business but also creates a sense of value and appreciation among users, leading to stronger brand loyalty and increased sales (Smith and Hilton 2023).

Both loyalty and reward programs are essential for creating a competitive edge by offering personalized incentives that are difficult for rivals to replicate, attracting and retaining users effectively. These programs enable brands to leverage user data to deliver tailored rewards, such as exclusive products or customized discounts, that resonate with individual preferences, establishing a distinctive value proposition. Research shows that personalized rewards, particularly those aligned with users' interests, enhance the perceived value of the program and foster long-term brand loyalty. Additionally, programs that incorporate rewards addressing emotional or psychological needs, rather than only financial incentives, can deepen user relationships, making them less likely to switch to competing brands. This not only enhances user retention but also establishes a long-term competitive advantage by fostering brand loyalty

4.6.3 Limited-Time Promotions and Special Offers

The e-cig industry leverages limited-time promotions and special offers as effective strategies to drive product sales. This strategy creates a sense of urgency and exclusivity that drives immediate action from the user (Collins et al. 2019). These promotions often include substantial discounts, buy-one-get-one-free deals, and exclusive bundle offers available for limited periods (Ma et al. 2022a, b). By offering time-sensitive deals, the industry prompts quick purchasing decisions from users eager to benefit from temporary cost savings, thereby boosting short-term sales.

These marketing strategies are often promoted through multiple channels, including social media, company websites, and email newsletters that ensure broad reach and maximum impact (Shroff and Sreeramareddy 2024). Another common marketing strategy is the use of flash sales, offering steep discounted prices for only a few hours or a single day (Begay et al. 2020). These flash sales generate excitement and attract a large number of potential users in a short period (Williams et al. 2018). The urgency and competitiveness of these deals create a buzz, encouraging users to share the offers with their peers, thereby expanding the promotion's reach through word-of-mouth and social sharing (Collins et al. 2019).

These promotional strategies are designed to drive immediate sales and user engagement with e-cig products (Wadsworth et al. 2018). By consistently offering new and exclusive deals, the product line remains dynamic and engaging, encouraging users to stay connected with the brand and anticipate future promotions. This sustained engagement helps cultivate a loyal user base, as individuals who take advantage of these limited-time offers are more likely to return for repeat purchases (McCausland et al. 2020). By leveraging the appeal of time-sensitive promotions and special offers, the e-cig industry effectively attracts and retains users, fostering ongoing interest and sales in a competitive market (Escobedo et al. 2020a, b).

4.6.4 Bundling and Package Deals

The e-cig industry employs bundling and package deals as effective strategies to promote e-cig products, enhancing the overall value proposition to users. Bundling offers multiple related products – such as e-cig devices, e-liquids, and accessories – at a discounted price when purchased together (Jiang et al. 2018). This approach encourages users to buy more items than they might individually, boosting the average transaction value and driving higher sales volumes (Jiang et al. 2018). By presenting a complete vaping solution in a single package, the industry offers convenience, making it easy for users to purchase everything they need and improving their shopping experience (Chu and Kong 2022).

Bundling and package deals often emphasize cost savings and convenience, making them attractive to both new and existing users. For new users, bundles provide a straightforward entry into vaping by offering complete starter kits with all the necessary components, reducing the complexity and decision-making process for first-time buyers and increasing the likelihood of choosing a bundled option (Collins et al. 2019). For existing users, these package deals can introduce them to new e-cigs or flavors, encouraging product exploration and brand loyalty.

Promotional campaigns frequently use bundling and package deals to create a sense of urgency and encourage immediate purchases. Limited-time offers and special bundle discounts are advertised through various channels, including online platforms, in-store promotions, and email marketing (Kim et al. 2015). These promotions boost short-term sales while keeping users engaged with the brand through regular attractive deals. By strategically using bundling and package deals, the e-cig industry can effectively increase product adoption, enhance user satisfaction, and maintain a competitive edge in the market (McQuilken et al. 2015).

4.7 Event Sponsorships and Participation

The e-cig industry bolsters its visibility and market position through event sponsorship, community engagement, and scholarship programs. These strategies enable direct interaction with the target audiences, showcase products, build trust, and promote a positive public image.

4.7.1 Sponsorship of Vaping-Related Events

The e-cig industry strategically promotes its products by sponsoring vaping-related events, significantly enhancing brand visibility and engagement within the vaping community. Through financial support and partnerships with events like vaping conventions, expos, and competitions, the industry gains direct access to its target audience (Taleb and Kalan 2018). These events provide a platform for brands to showcase products, offer demonstrations, and distribute promotional materials, creating a highly interactive environment that fosters consumer interest and product trials (Williams 2015).

Sponsoring these events enables e-cig brands to align themselves with the vaping culture and community, boosting their brand image and credibility (Howie 2020). Event sponsorships offer brands exclusive access to a highly engaged and relevant audience, which can drive brand loyalty and recognition. At these events, the companies often set up booths or kiosks to engage with attendees, offer samples, and provide detailed product information, building a stronger brand-user connection and repeated purchases (Williams 2015).

Furthermore, sponsoring vaping events offers valuable opportunities for the e-cig industry to introduce new products and gather user feedback (Taleb and Kalan 2018). Being present at these events allows brands to observe firsthand user reactions and preferences, helping them to refine their products and marketing strategies accordingly. The excitement and buzz generated by these events can also significantly boost brand awareness and attract media attention, amplifying the promotional impact (Hoek and Freeman 2019). Through strategic event sponsorships, the e-cig industry effectively promotes its products, drives consumer engagement, and strengthens its market position.

4.7.2 Community Engagement Through Local Events

The e-cig industry promotes its products by engaging with local communities at events, aiming to build strong connections and brand visibility (McCarthy 2014). By participating in and sponsoring events such as local fairs, festivals, and health and wellness activities, the e-cig industry can directly connect with a diverse audience in a community-oriented setting. These events provide brands opportunities to set up booths or stalls, where they can showcase their products, offer demonstrations, and distribute promotional materials, helping educate users about e-cigs, fostering a positive brand image, and encouraging product trials.

Local events offer a unique platform for the e-cig industry to personalize marketing efforts and build trust within the community. Companies often engage with event attendees through interactive activities like product sampling, vaping competitions, and educational workshops. By offering hands-on experiences and detailed information about the benefits and usage of e-cigs, they can effectively address consumer queries and concerns (Sussman et al. 2014). This personal interaction helps in creating a sense of community around the brand, fostering consumer connection and loyalty.

Through community engagement in local events, the e-cig industry positions itself as socially responsible and supportive of local initiatives. By sponsoring and participating in community-driven activities, companies can enhance their public image and demonstrate a commitment to the well-being of the community (Cheney et al. 2015). This approach not only builds goodwill but also helps counter negative perceptions associated with the e-cig industry. Through active participation in local events, the industry effectively promotes its products, strengthens community ties, and enhances brand loyalty and consumer trust.

4.7.3 Offering Scholarship

The e-cig industry strategically offers scholarships to improve its public image and foster goodwill within communities. By providing financial support for education, the industry positions itself as socially responsible and dedicated to community well-being. Scholarships often target students from diverse fields of study, aligning with the industry's interests and potential future workforce needs (Baler et al. 2020). This approach benefits recipients by alleviating financial burdens and creates a positive association between the brand and education.

Offering scholarships allows the e-cig industry to engage directly with educational institutions and students, establishing a presence within academic communities. These scholarship programs often include brand-sponsored events, workshops, and seminars, providing additional opportunities for the industry to promote products and convey brand values (Baler et al. 2020). By aligning themselves with educational initiatives, the e-cig industry can reach vulnerable groups, potentially shaping perceptions and attitudes toward vaping from an early age (de Andrade et al. 2016). This engagement helps to build a positive brand narrative, presenting the industry as an advocate for education and personal development growth.

Furthermore, scholarships serve as a valuable public relations tool that counters negative perceptions and criticisms of the e-cig industry. By investing in education, the industry can demonstrate a commitment to giving back to society and supporting future generations. This strategy enhances its reputation and fosters brand loyalty among scholarship recipients and their communities. Through scholarship offerings, the e-cig industry promotes its products, builds community goodwill, and strengthens its image as a socially responsible entity (Williams and Knight 2015).

4.7.4 Sports Events

The promotion of e-cigs through sports events associates these products with health and vitality, potentially normalizing e-cigs as a "healthy" choice. This association is especially misleading when e-cig companies sponsor sports events that symbolize fitness and well-being, aligning their products with an image of physical health (Milicic et al. 2017). Such marketing strategies strongly impact public perception, particularly among youth, by suggesting that e-cigs are a healthier alternative to conventional cigs and aligning them with athletes and healthy lifestyles.

Promoting e-cigs through sports events often targets youth, leading to increased experimentation and uptake (McCarthy 2014). The e-cig industry leverages sponsorships and advertisements during widely viewed sporting events to create associations between their products and the excitement of sports. This approach can glamorize e-cigs and make them appear as trendy, socially acceptable behavior. Exposure to e-cig marketing, especially on high-visibility platforms like sports, significantly increases the likelihood of experimentation among youth. Studies indicated that youth exposed to e-cig marketing are more likely to experiment with e-cigs, underscoring the effectiveness of these promotional strategies in driving youth uptake (Chen-Sankey et al. 2024).

4.8 Marketing and Promotion E-cig Regulation

Research indicates that as the volume and intensity of marketing communications increase, e-cig use among adolescents may become more normalized. Through widespread marketing, young people may perceive e-cig use as desirable, socially acceptable, and common, which could encourage both the initiation and continued use of e-cigs (Krugman 2016).

The marketing and promotion of e-cigs navigate a complex regulatory landscape aimed at protecting public health, particularly among vulnerable populations like adolescents. E-cig regulations vary significantly across countries, reflecting diverse approaches to managing the promotion and sale of these products. While some nations, like Brunei and Singapore, have implemented complete bans on e-cigs, others such as the United States and the European Union enforce strict advertising restrictions to limit exposure and appeal to minors (Kennedy et al. 2017). These regulations typically include bans on television and radio advertising, stringent packaging requirements, and restrictions on certain promotional strategies, including sponsorships and social media campaigns (Wadsworth et al. 2018).

To navigate these regulations, e-cig industries have adapted their marketing strategies, emphasizing less regulated channels and innovative promotional techniques. For example, the industry may utilize social media influencers and digital marketing campaigns to reach users indirectly, often blurring the lines of regulatory compliance. Influencers can create engaging content that promotes e-cigs subtly, allowing companies to bypass traditional advertising restrictions. Additionally, companies may focus on point-of-sale promotions and in-store displays, which are subject to different regulatory standards, to maintain a strong market presence (McCausland et al. 2019).

However, the effectiveness of e-cig regulations is often challenged by industry resistance and the rapid evolution of marketing practices. E-cig companies continually seek loopholes and exploit ambiguities in the law to continue promoting their products. This ongoing tension between regulators and the industry highlights the need for comprehensive and adaptable regulatory frameworks. Effective regulation requires robust enforcement mechanisms and continuous monitoring to keep pace with the industry's dynamic marketing strategies. Strengthening regulations and closing loopholes are crucial steps in reducing public health risks associated with e-cig use, especially among youth (Lindblom 2015).

4.9 Conclusion

The marketing strategies and tactics employed by the e-cig industry reveal a complex and adaptive approach to user engagement, with a particular focus on younger demographics. Through sophisticated digital marketing campaigns including the use of social media influencers and celebrities, the industry effectively promotes its products while circumventing traditional advertising restrictions. The design and variety of e-liquid flavors are strategically used to appeal to a broad customer base, enhancing the attractiveness of e-cigs. In addition, direct and in-store marketing techniques create personalized, engaging experiences for consumers, further strengthening appeal. These strategies are complemented by community engagement efforts, such as event sponsorships and participation, which boost brand visibility and credibility within targeted communities. The use of discounts, loyalty programs, and limited-time promotions further incentivize repeat purchases and foster long-term customer relationships.

However, the aggressive marketing of e-cigs raises significant public health concerns, especially given their popularity among adolescents and young adults. Regulatory bodies face ongoing challenges in implementing effective controls to mitigate the impact of these marketing tactics. The dynamic nature of marketing strategies employed by the e-cig industry necessitates robust and adaptable regulatory frameworks to protect vulnerable populations from the risks associated with nicotine addiction. Understanding the full scope of these marketing strategies is crucial for public health officials and policymakers as it enables the development of more effective interventions and regulations that can keep pace with technological advancements and marketing innovations. Only through comprehensive and enforced regulatory measures can the public health risks posed by e-cigs be effectively managed and mitigated, ensuring that consumer safety and well-being take precedence over industry profits.

References

Allem J-P, Van Valkenburgh SP, Donaldson SI, Dormanesh A, Kelley TC, Rosenthal EL (2022) E-cigarette imagery in Netflix scripted television and movies popular among young adults: a content analysis. Addict Behav Rep 16:100444. https://doi.org/10.1016/j.abrep.2022.100444

Baler G, Paci K, Kowitt SD, Goldstein AO (2020) Vaping industry-funded academic scholarships. Tob Control 29(e1):e181–e182. https://doi.org/10.1136/tobaccocontrol-2019-055158

Basáñez T, Majmundar A, Cruz TB, Allem J-P, Unger JB (2019) E-cigarettes are being marketed as "vitamin delivery" devices. Am J Public Health 109(2):194–196. https://doi.org/10.2105/AJPH.2018.304804

Begay C, Soto C, Baezconde-Garbanati L, Barahona R, Rodriguez YL, Unger JB, Smiley SL (2020) Cigarette and E-cigarette retail marketing on and near California tribal lands. Health Promot Pract 21(1_suppl):18S–26S. https://doi.org/10.1177/1524839919883254

Brocklebank LA, Blackwell AKM, Marteau TM, De-loyde K, Morris RW, Burgoine T, Hobson A, Ventsel M, Munafo MR (2022) Electronic cigarette and smoking paraphernalia point of sale displays: an observational study in England. Tob Control 31(e2):e201–e206. https://doi.org/10.1136/tobaccocontrol-2020-056314

Brown CJ, Cheng JM (2014) Electronic cigarettes: product characterisation and design considerations. Tob Control 23(Suppl 2):ii4–ii10. https://doi.org/10.1136/tobaccocontrol-2013-051476

Cantrell J, Huang J, Greenberg MS, Xiao H, Hair EC, Vallone D (2020) Impact of e-cigarette and cigarette prices on youth and young adult e-cigarette and cigarette behaviour: evidence from a national longitudinal cohort. Tob Control 29(4):374–380. https://doi.org/10.1136/tobaccocontrol-2018-054764

Chaffee BW, Halpern-Felsher B, Croker JA, Werts M, Couch ET, Cheng J (2022) Preferences, use, and perceived access to flavored e-cigarettes among United States adolescents and young adults. Drug Alcohol Depend Rep 3. https://doi.org/10.1016/j.dadr.2022.100068

Cheney M, Gowin M, Wann TF (2015) Marketing practices of vapor store owners. Am J Public Health 105(6):e16–e21. https://doi.org/10.2105/AJPH.2015.302610

Cheney MK, Gowin M, Wann TF (2016) Vapor store owner beliefs and messages to customers. Nicotine Tob Res 18(5). https://doi.org/10.1093/ntr/ntv129

Chen-Sankey J, Jeong M, Wackowski OA, Unger JB, Niederdeppe J, Bernat E, Bansal-Travers M, Moran M, Kennedy RD, Broun A, Hacker K, Choi K (2024) Noticing people, discounts and non-tobacco flavours in e-cigarette ads may increase e-cigarette product appeal among non-tobacco-using young adults. Tob Control 33(1):30–37. https://doi.org/10.1136/tobaccocontrol-2022-057269

Chu S, Kong J (2022) Electronic cigarette devices targeting youth in Korea. Tob Prev Cessat 8:30. https://doi.org/10.18332/tpc/151549

Chu K-H, Sidhu AK, Valente TW (2015) Electronic cigarette marketing online: a multi-site, multi-product comparison. JMIR Public Health Surveill 1(2):e11. https://doi.org/10.2196/publichealth.4777

Collins L, Glasser AM, Abudayyeh H, Pearson JL, Villanti AC (2019) E-cigarette marketing and communication: how E-cigarette companies market E-cigarettes and the public engages with E-cigarette information. Nicotine Tob Res 21(1):14–24. https://doi.org/10.1093/ntr/ntx284

Dai H, Hao J (2017) Direct marketing promotion and electronic cigarette use among US adults, national adult tobacco survey, 2013-2014. Prev Chronic Dis 14:E84. https://doi.org/10.5888/pcd14.170073

Davis B (2020) Building a loyalty program for a vape or E-cigarette store. https://www.stampme.com/blog/vape-loyalty-program

de Andrade M, Angus K, Hastings G (2016) Teenage perceptions of electronic cigarettes in Scottish tobacco-education school interventions: co-production and innovative engagement through a pop-up radio project. Perspect Public Health 136(5):288–293. https://doi.org/10.1177/1757913915612109

Dutta SS (2023) Colorful packaging, hidden dangers: the push to regulate vape aesthetics. https://www.google.com/search?q=Colorful+Packaging%2C+Hidden+Dangers%3A+The+Push+to+Regulate+Vape+Aesthetics&safe=active&sca_esv=25dd1b6fd40bfe61&rlz=1C1GCEU_enIN1080IN1080&ei=jgLhZ-G9HbrG4-EPktTQoQc&ved=0ahUKEwjhvIqKkqKMAxU64zgGHRIqNHQQ4dUDCBA&uact=5&oq=Colorful+Packaging%2C+Hidden+Dangers%3A+The+Push+to+Regulate+Vape+Aesthetics&gs_lp=Egxnd3Mtd2l6LXNlcnAiSENvbG9yZnVsIFBhY2thZ2luZywgSGlkZGVuIERhbmdlcnM6IFRoZSBQdXNoIHRvIFJlZ3VsYXRlIFZhcGUgQWVzdGhldGljc0jhCVCuB1iuB3ABeAGQAQCYAbIBoAGyAaoBAzAuMbgBA8gBAPgBAvgBAZgCAaACBsICChAAGLADGNYEGEfCAg0QLhiABBiwAxhDGIoF-wgINEAAYgAQYsAMYQxiKBcICDhAAGLADGOQCGNYE2AEBwgITEC4YgAQY-sAMYQxjIAxiKBdgBAZgDAIgGAZAGE7oGBggBEAEYCZIHATGgB9UCsgcAuAcA&sclient=gws-wiz-serp

Eadie D, Stead M, MacKintosh AM, MacDonald L, Purves R, Pearce J, Tisch C, van der Sluijis W, Amos A, MacGregor A, Haw S (2015) E-cigarette marketing in UK stores: an observational audit and retailers views. BMJ Open 5(9). https://doi.org/10.1136/bmjopen-2015-008547

Escobedo P, Garcia R, Soto C, Rodriguez Y, Barahona R, Baezconde-Garbanati L (2020a) Comparison of e-cigarette marketing and availability in tobacco retail outlets among diverse low-income communities in California. Tob Control 29(4). https://doi.org/10.1136/tobaccocontrol-2019-054985

Escobedo P, Rosenthal EL, Saucier CJ, Unger JB, Cruz TB, Kirkpatrick M, Allem J-P (2020b) Electronic cigarette product placement and imagery in popular music videos. Nicotine Tob Res 23(8):1367–1372. https://doi.org/10.1093/ntr/ntaa273

Fagan P, Pokhrel P, Herzog TA, Guy MC, Sakuma K-LK, Trinidad DR, Cassel KD, Jorgensen D, Lynch T, Felicitas-Perkins JQ, Palafox S, Hamamura FD, Maloney S, Degree K, Sterling K, Moolchan ET, Clanton MS, Eissenberg TE (2018) Warning statements and safety practices among manufacturers and distributors of electronic cigarette liquids in the United States. Nicotine Tob Res 20(8):970–976. https://doi.org/10.1093/ntr/ntx101

Gaiha SM, Lempert LK, Lung H, Vescia F, Halpern-Felsher B (2024) Appealing characteristics of E-cigarette marketing in the retail environment among adolescents. Prev Med Rep 43:102769. https://doi.org/10.1016/j.pmedr.2024

Giovenco DP, Casseus M, Duncan DT, Coups EJ, Lewis MJ, Delnevo CD (2016) Association between electronic cigarette marketing near schools and E-cigarette use among youth. J Adolesc Health 59(6):627–634. https://doi.org/10.1016/j.jadohealth.2016.08.007

Gomes MN, Reid JL, Hammond D (2024) The effect of branded versus standardized e-cigarette packaging and device designs: an experimental study of youth interest in vaping products. Public Health 230:223–230

Gravely S, Cummings KM, Hammond D, Lindblom E, Smith DM, Martin N, Loewen R, Borland R, Hyland A, Thompson ME, Boudreau C, Kasza K, Ouimet J, Quah ACK, O'Connor RJ, Fong GT (2020) The association of E-cigarette flavors with satisfaction, enjoyment, and trying to quit or stay abstinent from smoking among regular adult vapers from Canada and the United States: findings from the 2018 ITC four country smoking and vaping survey. Nicotine Tob Res 22(10):1831–1841. https://doi.org/10.1093/ntr/ntaa095

Haardörfer R, Cahn Z, Lewis M, Kothari S, Sarmah R, Getachew B, Berg CJ (2017) The advertising strategies of early E-cigarette brand leaders in the United States. Tob Regul Sci 3(2):222–231. https://doi.org/10.18001/TRS.3.2.10

Hassanein ZM, Barker AB, Murray RL, Britton J, Agrawal S, Leonardi-Bee J (2022) Impact of smoking and vaping in films on smoking and vaping uptake in adolescents: systematic review and meta-analysis. Health Educ Behav 49(6):1004–1013. https://doi.org/10.1177/10901981221086944

Hoek J, Freeman B (2019) BAT(NZ) draws on cigarette marketing tactics to launch Vype in New Zealand. Tob Control 28(e2):e162–e163. https://doi.org/10.1136/tobaccocontrol-2019-054967

Howie L (2020) Vape companies target festi. https://tewahanui.nz/health/vape-companies-target-festival-goers

Jancey J, Carey RN, Freeman B, Leaver T, Wolf K, Bromberg M, Chai K, Bialous S, Adams P, Mcleod M, McCausland K (2024) E-cigarettes on instagram: exploring vape content via an Australian vaping influencer. Tob Induc Dis 22(January):1–11. https://doi.org/10.18332/tid/175619

Jiang Y, Liu Y, Wang H, Shang J, Ding S (2018) Online pricing with bundling and coupon discounts. Int J Prod Res 56(5):1773–1788. https://doi.org/10.1080/00207543.2015.1112443

Kennedy RD, Awopegba A, De León E, Cohen JE (2017) Global approaches to regulating electronic cigarettes. Tob Control 26(4):440–445. https://doi.org/10.1136/tobaccocontrol-2016-053179

Kim AE, Hopper T, Simpson S, Nonnemaker J, Lieberman AJ, Hansen H, Guillory J, Porter L (2015) Using twitter data to gain insights into E-cigarette marketing and locations of use: an infoveillance study. J Med Internet Res 17(11):e251. https://doi.org/10.2196/jmir.4466

Komesarook C (2023) Why your vape shop can benefit from a digital loyalty card. https://www.stampme.com/blog/vape-shop-digital-loyalty-card

Kong G, Kuguru KE, Bhatti H, Sen I, Morean ME (2021) Marketing content on E-cigarette brand-sponsored Facebook profile pages. Subst Use Misuse 56(4):442–448. https://doi.org/10.1080/10826084.2021.1878223

Krugman DM (2016) Understanding the impact that marketing, advertising, and promotion have on adolescent E-cigarette behavior. J Adolesc Health 59(2016):609e611

Kwon M, Park E (2020) Perceptions and sentiments about electronic cigarettes on social media platforms: systematic review. JMIR Public Health Surveill 6(1):e13673. https://doi.org/10.2196/13673

Lindblom EN (2015) Effectively regulating E-cigarettes and their advertising – and the first amendment. Food and Drug Law Journal 70(1):55–92

Lyu JC, Huang P, Jiang N, Ling PM (2022) A systematic review of E-cigarette marketing communication: messages, communication channels, and strategies. Int J Environ Res Public Health 19(15). https://doi.org/10.3390/ijerph19159263

Ma S, Jiang S, Ling M, Chen J, Shang C (2022a) Price promotions of E-liquid products sold in online stores. Int J Environ Res Public Health 19(14). https://doi.org/10.3390/ijerph19148870

Ma S, Qiu Z, Yang Q, Bridges JFP, Chen J, Shang C (2022b) Expanding the E-liquid flavor wheel: classification of emerging e-liquid flavors in online vape shops. Int J Environ Res Public Health 19(21). https://doi.org/10.3390/ijerph192113953

Mackey TK, Miner A, Cuomo RE (2015) Exploring the e-cigarette e-commerce marketplace: Identifying Internet e-cigarette marketing characteristics and regulatory gaps. Drug Alcohol Depend 156:97–103. https://doi.org/10.1016/j.drugalcdep.2015.08.032

Maggie O' Neill (2024) Vitamin vape is not the health trend you want to follow. https://www.google.com/search?q=Vitamin+Vape+is+not+the+Health+Trend+you+want+to+Follow&rlz=1C1GCEU_enIN1080IN1080&oq=Vitamin+Vape+is+not+the+Health+Trend+you+want+to+Follow&gs_lcrp=EgZjaHJvbWUyBggAEEUYOTIHCAEQIRigATIGCAIQIRgVMgcIAxAhGI8C0gEHMzI3ajBqOagCALACAA&sourceid=chrome&ie=UTF-8

Mantey DS, Pasch KE, Loukas A, Perry CL (2017) Exposure to point-of-sale marketing of cigarettes and E-cigarettes as predictors of smoking cessation behaviors. Nicotine Tob Res 21(2):212–219. https://doi.org/10.1093/ntr/ntx246

McCarthy M (2014) E-cigarette companies target youth, US congressional study finds. BMJ 348:g2871. https://doi.org/10.1136/bmj.g2871

McCausland K, Maycock B, Leaver T, Jancey J (2019) The messages presented in electronic cigarette-related social media promotions and discussion: scoping review. J Med Internet Res 21(2):e11953. https://doi.org/10.2196/11953

McCausland K, Maycock B, Leaver T, Wolf K, Freeman B, Thomson K, Jancey J (2020) E-cigarette promotion on twitter in Australia: content analysis of tweets. JMIR Public Health Surveill 6(4):e15577. https://doi.org/10.2196/15577

McKelvey K, Baiocchi M, Ramamurthi D, McLaughlin S, Halpern-Felsher B (2019) Youth say ads for flavored e-liquids are for them. Addict Behav 91:164–170. https://doi.org/10.1016/j.addbeh.2018.08.029

McQuilken L, Robertson N, Polonsky M, Harrison P (2015) Consumer perceptions of bundles and time-limited promotion deals: do contracts matter? J Consum Behav 14(3):145–157. https://doi.org/10.1002/cb.1513

Milicic S, Piérard E, DeCicca P, Leatherdale ST (2017) Examining the association between physical activity, sedentary behavior and sport participation with E-cigarette use and smoking status in a large sample of Canadian youth. Nicotine Tob Res 21(3):285–292. https://doi.org/10.1093/ntr/ntx238

Mok Y, Jeon J, Levy DT, Meza R (2022) Associations between E-cigarette Use and e-cigarette flavors with cigarette smoking quit attempts and quit success: evidence from a U.S. large, nationally representative 2018–2019 survey. Nicotine Tob Res 25(3):541–552. https://doi.org/10.1093/ntr/ntac241

Murukutla N, Magsumbol MS, Raskin H, Kuganesan S, Dini S, Martinez-Mejia C, Rachfiansyah, & Aguilar, B. G. R. (2022) A content analysis of e-cigarette marketing on social media: Findings from the Tobacco Enforcement and Reporting Movement (TERM) in India, Indonesia and Mexico. Front Public Health 10:1012727. https://doi.org/10.3389/fpubh.2022.1012727

Phua J, Jin SV, Hahm JM (2018) Celebrity-endorsed e-cigarette brand Instagram advertisements: Effects on young adults' attitudes towards e-cigarettes and smoking intentions. J Health Psychol 23(4). https://doi.org/10.1177/1359105317693912

Sears CG, Walker KL, Hart JL, Lee AS, Siu A, Smith C (2017) Clean, cheap, convenient: promotion of Electronic cigarettes on YouTube. Tob Prev Cessat 3. https://doi.org/10.18332/tpc/69393

Shroff SM, Sreeramareddy CT (2024) Marketing claims, promotional strategies, and product information on Malaysian e-cigarette retailer websites-a content analysis. Subst Abuse Treat Prev Policy 19(1):11. https://doi.org/10.1186/s13011-024-00592-z

Simonavičius E, East K, Taylor E, Nottage M, Reid JL, Arnott D, Bunce L, McNeill A, Hammond D (2023) Impact of E-liquid packaging on vaping product perceptions among youth in England, Canada, and the United States: a randomized online experiment. Nicotine Tob Res 26(3):370–379. https://doi.org/10.1093/ntr/ntad144

Smith M, Hilton S (2023) Youth's exposure to and engagement with e-cigarette marketing on social media: a UK focus group study. BMJ Open 13:e071270. https://doi.org/10.1136/bmjopen-2022-071270

Soneji S, Gerling M, Yang J, Sargent J (2016) Online electronic cigarette marketing – violation of self-regulated standards by tobacco companies. JAMA Pediatr 170(5):511–512. https://doi.org/10.1001/jamapediatrics.2015.4501

Soule EK, Sakuma K-LK, Palafox S, Pokhrel P, Herzog TA, Thompson N, Fagan P (2019) Content analysis of internet marketing strategies used to promote flavored electronic cigarettes. Addict Behav 91:128–135. https://doi.org/10.1016/j.addbeh.2018.11.012

Struik LL, Dow-Fleisner S, Belliveau M, Thompson D, Janke R (2020) Tactics for drawing youth to vaping: content analysis of electronic cigarette advertisements. J Med Internet Res 22(8):e18943. https://doi.org/10.2196/18943

Sun L, Tao C, Xie Z, Li D (2020) Promotion of disposable electronic cigarette flavors and topics on Twitter. Int J Environ Res Public Health 17(24). https://doi.org/10.3390/ijerph17249221

Sussman S, Garcia R, Cruz TB, Baezconde-Garbanati L, Pentz MA, Unger JB (2014) Consumers' perceptions of vape shops in Southern California: an analysis of online Yelp reviews. Tob Induc Dis 12(1):22. https://doi.org/10.1186/s12971-014-0022-7

Taleb ZB, Kalan ME (2018) World vapor expo 2017: e-cigarette marketing tactics. Tob Control 27

Taylor E, Arnott D, Cheeseman H, Hammond D, Reid JL, McNeill A, Driezen P, East K (2023) Association of fully branded and standardized e-cigarette packaging with interest in trying products among youths and adults in Great Britain. JAMA Netw Open 6(3):e231799. https://doi.org/10.1001/jamanetworkopen.2023.1799

Trumbo CW, Harper R (2015) Perceived characteristics of e-cigarettes as an innovation by young adults. Health Behav Policy Rev 2(2):154–162. https://doi.org/10.14485/HBPR.2.2.7

Unnava V, Aravindakshan A (2021) How does consumer engagement evolve when brands post across multiple social media? J Acad Mark Sci 49(5):864–881. https://doi.org/10.1007/s11747-021-00785-z

Vassey J, Valente T, Barker J, Stanton C, Li D, Laestadius L, Cruz TB, Unger JB (2022) E-cigarette brands and social media influencers on Instagram: a social network analysis. Tob Control. https://doi.org/10.1136/tobaccocontrol-2021-057053

Wadsworth E, McNeill A, Li L, Hammond D, Thrasher JF, Yong H-H, Cummings KM, Fong GT, Hitchman SC (2018) Reported exposure to E-cigarette advertising and promotion in different regulatory environments: findings from the International Tobacco Control Four Country (ITC-4C) Survey. Prev Med 112:130–137. https://doi.org/10.1016/j.ypmed.2018.04.022

Wan N, Siahpush M, Shaikh RA, McCarthy M, Ramos A, Correa A (2017) Point-of-sale e-cigarette advertising among tobacco stores. J Community Health 42(6):1179–1186. https://doi.org/10.1007/s10900-017-0368-x

Williams RS (2015) VapeCons: E-cigarette user conventions. J Public Health Policy 36(4):440–451. https://doi.org/10.1057/jphp.2015.31

Williams RJ, Knight R (2015) Insights in public health: Electronic cigarettes: marketing to Hawai'i's adolescents. Hawai'i J Med Public Health 74(2):66–70

Williams RS, Derrick J, Liebman AK, LaFleur K (2018) Content analysis of e-cigarette products, promotions, prices and claims on Internet tobacco vendor websites, 2013-2014. Tob Control 27(e1):e34–e40. https://doi.org/10.1136/tobaccocontrol-2017-053762

Overview of Legislation and Regulation on E-cigarettes

5

Yin-Hui Leong and Eunice Ting Shi Ying

Abstract

This chapter offers a comprehensive overview of the current legislative and regulatory landscape for electronic cigarettes (e-cigs) globally. It explores the various approaches taken by different countries and jurisdictions to regulate these products, including sales restrictions, product standards, marketing limitations, taxation, and use in public spaces. The chapter also analyzes key policy trends, challenges in implementation and enforcement, and emerging regulatory issues as the e-cig market continues to evolve rapidly. By synthesizing data from multiple sources, it provides insights into the complex and dynamic nature of e-cig regulation worldwide.

Keywords

E-cigarette regulation · Tobacco policy · Vaping legislation · Public health law

5.1 Introduction

The surge in popularity of electronic cigarettes (e-cigs) in the past decade has led governments and regulatory agencies worldwide to grapple with how to regulate these new nicotine delivery devices that do not fit neatly into existing tobacco control frameworks. Governments typically classify e-cigs as tobacco products, imitation tobacco, medicinal or pharmaceutical products, consumer products, poisons, or electronic nicotine delivery systems (ENDS). Currently, there is no worldwide standard for regulating e-cigs, and the regulations vary from one country or region to

Y.-H. Leong (✉) · E. T. S. Ying
Malaysia National Poison Centre, Universiti Sains Malaysia, Penang, Malaysia
e-mail: yhleong@usm.my

Y.-H. Leong, A. S. Mohd Samin (eds.), *E-Cigarettes*,
https://doi.org/10.1007/978-981-96-5087-3_5

another (Bianco et al. 2021). As a result, different international approaches to regulating e-cigs have been adopted. E-cigs are often promoted as a less harmful alternative to conventional cigarettes. However, the long-term health effects of e-cig use remain uncertain, and there are concerns about their appeal to youth and potentially becoming a gateway to conventional cigarette smoking (Stratton et al. 2018). In response to these public health concerns, various policies govern the sale, marketing, product standards, packaging, and labeling requirements, and use of e-cigs at international, national, and local levels.

The most common approach to regulation has been to use existing tobacco control laws or create new legislation to regulate e-cigs similarly to conventional tobacco products. At least 88 countries permit the sale of e-cigs and regulate them in line with the World Health Organization Framework Convention on Tobacco Control (WHO FCTC) (Tobaccofreekids 2023a). The regulation of e-cigs is still evolving as policymakers aim to weigh the potential benefits and risks of these products based on emerging scientific evidence (Stratton et al. 2018; Yan et al. 2023). The recent revelations regarding the health hazards linked to vaping, and the absence of manufacturing standards have spurred increased advocacy for tighter regulations on the use of e-cigs (Kennedy et al. 2017; Shih and Etter 2019; USC 2024). On the other hand, some argue highlighting the potential of e-cigs to reduce smoking-related harm in adults who are unable or unwilling to quit (Stratton et al. 2018).

This chapter aims to provide a comprehensive overview of the current global landscape of e-cig legislation and regulation. It examines the range of policy approaches that have been implemented, from outright bans to more permissive regulatory frameworks. Key areas of focus include product standards and safety requirements, restrictions on sales and marketing, taxation policies, regulations on use in public spaces, and packaging and labeling requirements. The regulatory environment for e-cigs remains highly dynamic, with policies continuing to evolve as new evidence emerges about the public health impacts of these products. This chapter also analyzes major trends in e-cig regulation, ongoing debates and controversies, and emerging issues that policymakers are grappling with. Furthermore, some of the key challenges in regulating e-cigs and future regulatory directions are discussed. By synthesizing data from multiple sources, including government policies, academic research, and public health reports, this chapter provides readers with a thorough overview of regulatory issues surrounding e-cigs. This can serve as a foundation for further exploration of specific policy approaches and their impacts in subsequent chapters.

5.2 Current Regulatory Landscape

The regulation of e-cigs differs widely across countries and jurisdictions, reflecting various approaches to balancing potential public health benefits and risks. The regulatory landscape from different international, national, and state/local levels is highlighted in this section.

The WHO FCTC is the first global public health treaty aimed at addressing the global tobacco epidemic by creating an international policy consensus around tobacco control. It has expressed concerns about e-cigs and stressed the importance of strict regulations to safeguard public health, especially among young people. Recently, the WHO urged governments to enforce measures similar to those for tobacco products, such as prohibiting all e-cig flavors, imposing high taxes, and banning public use (WHO 2024a). However, despite the global adoption of FCTC policies aimed at addressing cigarettes, countries have adopted wildly disparate approaches to regulating this evolving product. The lack of scientific consensus about the long-term health effects and potential benefits/risks of e-cigs, corporate interests (especially from the tobacco and e-cig industries), different safety evidence standards required by different policymakers in various countries have led to varying regulatory approaches; and various governmental, cultural, economic, political, and statutory differences between countries have also contributed to policy divergence (Feldman 2018; McKee 2019; Newman and Nurfaiza 2022). While the WHO's recommendations are not legally binding, they often influence national policies. Please refer to the WHO FCTC's chronology and progress in Appendix 1.

National regulations on e-cigs differ significantly across major markets. Several countries—the United States, UK, and China—are selected for discussion and comparison in this section. North America is the world's largest e-cig market (Tobacco Tactics 2023), China is the world's largest producer of e-cigs (Reuters 2024), and the UK is the only country to have announced support for the medicinal licensing of e-cigs (Hopkinson et al. 2022). In the United States, the Food and Drug Administration (FDA) has the authority to regulate e-cigs as tobacco products (FDA 2023) The US FDA regulates the sale, advertising, promotion, distribution, manufacture, import, packaging, and labeling of e-cigs classified as tobacco products based on the laws set forth in the Tobacco Control Act and the Food, Drug, and Cosmetic Act (FD&C Act) (Institute for Global Tobacco Control 2024). Since 2016, the FDA has granted extensive powers to reduce their availability to young people and requires manufacturers to obtain premarketing approval based on safety grounds (McKee 2020). E-cig manufacturers are required to undergo the FDA's Premarket Tobacco Product Application (PMTA) process to demonstrate that their products are "appropriate for the protection of public health" in order to obtain marketing authorization (FDA 2024c). In January 2020, the FDA prioritized enforcement against certain flavored cartridge-based e-cigs, except for tobacco and menthol flavors. As of May 2024, the FDA authorized 34 tobacco- and menthol-flavored e-cig products and devices after rigorous scientific review and toxicologic assessments, meeting the statutory public health standard (FDA 2024a).

In the UK, e-cigs are primarily regulated under the Tobacco and Related Products Regulations (TRPR) 2016 (the regulation) and based on the European Union (EU) Tobacco Products Directive (TPD) with additional national measures (Rough 2024). The TRPR 2016 was amended by the Tobacco Products and Nicotine Inhaling Products (Amendment Etc.) (EU Exit) Regulations 2019 and 2020 to enable tobacco and e-cig regulation to continue to function following the UK's withdrawal from the EU (Rough 2024). However, the regulations still implement the majority of the EU's directives in the UK. The UK Government endorses the use of e-cig products as part of a harm reduction strategy. On October 29, 2021, the UK Medicines and Healthcare Products Regulatory Agency (MHRA) issued guidelines on how to grant licenses to e-cigs and other nicotine-containing products as medicines. This move could position England as the first country to prescribe e-cigs licensed as medical products to help reduce smoking rates (GOV.UK 2021). As of January 2024, no e-cig products have been licensed as medicine (Rough 2024). Plans to ban disposable vapes from April 2025, regulate flavors and packaging, and extend age restrictions to non-nicotine vaping products have been announced in the UK (GOV.UK 2024a, b). These measures aim to balance the use of e-cigs as a smoking cessation tool while addressing public health and environmental concerns.

Conversely, China has prohibited the sale of flavored e-cigs within the country but allows their manufacturing for export (Reuters 2024). Since e-cigs were first introduced in 2004, regulations have evolved from being lenient to strict. Initially, there was limited international regulatory experience for classifying e-cigs. E-cig companies promoted their products as a safer alternative to tobacco and a tool to quit smoking. As evidence of health effects was lacking, government agencies started regulating e-cigs more strictly (Xu et al. 2016). In 2018, the State Administration for Market Regulation (SAMR) and the State Tobacco Monopoly Administration (STMA) banned e-cig sales to minors and online sales in 2019 (Zhao et al. 2020). In November 2021, China's State Council updated the Regulations for the Implementation of the Tobacco Monopoly Law to include e-cigs and related products under the same rules as tobacco. In March 2022, STMA introduced new regulations for the production, marketing, and sales of e-cigs. Then, in April 2022, the SAMR approved a standard banning the sale of any e-cig flavors other than tobacco. These measures signify stricter regulation of e-cigs in mainland China since their introduction (China Briefing 2022).

5.3 Key Areas of E-cig Regulation

E-cigs are regulated in several key areas: usage, marketing, retail, pricing, and product standards. Each area has multiple regulations, so it is crucial to have a clear regulatory stance within each domain. This section explores various legislation based on these key areas, different approaches to regulating e-cigs, and analyzes their effectiveness with examples from specific countries.

5.3.1 Product Standards and Safety

Regulations on product standards and safety focus on the nicotine volume or concentration, harmful ingredients (excluding nicotine) or flavors in e-liquid, and the quality of e-liquid content (require safety and quality evaluation).

5.3.1.1 Device and e-liquid Regulations

At least 39 counties/jurisdictions regulate the concentration/volume of nicotine in e-cig products (Institute for Global Tobacco Control 2024). The EU's TPD (2014/40/EU) specifies that e-liquids must not exceed a nicotine concentration of 20 mg/ml, and the same amount of nicotine should be delivered when puffed at the same strength and duration (European Union 2014). It also sets size limits for e-liquids, with dedicated refill containers limited to 10 ml and e-cig cartridges and tanks limited to 2 ml. Additionally, the directive requires safety mechanisms, such as child-proof fastenings and opening mechanisms, for e-liquid containers, cartridges, and tanks (European Union 2014). These standards have been effective in harmonizing product requirements across EU member states. In 2022, China regulated the total amount of nicotine in each e-cig to 200 mg and the nicotine emissions of each puff to 0.2 mg (Tobacco Control Law 2024b).

According to Jenkins et al. 2024, at least seven countries have prohibited characteristic flavors in e-cig products, with the exception of tobacco flavor as of April 2024. Finland was the first to implement the ban in 2016, followed by Hungary, Lithuania, China, Ukraine, the Netherlands, and Latvia. Denmark, Estonia, and Slovenia also have similar bans, except for menthol flavors. Furthermore, certain Canadian provinces and US states have banned characteristic flavors. This is important because there is evidence that certain chemicals used to flavor e-liquids, such as diacetyl, are linked to lung disease (Allen et al. 2016).

Several countries have put in place regulations regarding the ingredients in e-cigs to prevent potential harm or false health claims. For instance, the EU TPD requires member states to ban the use of colorings, vitamins, caffeine, taurine, and other stimulants, as well as any additives with carcinogenic, mutagenic, or reprotoxic (CMR) properties in their unburnt form (European Union 2014). New Zealand, Canada, and China also have restrictions on various additives, stimulants, and substances with health risks in e-cigs.

5.3.1.2 Testing and Reporting Requirements

E-cigs, being a relatively new product category, have required the development of specific testing protocols and reporting requirements to ensure product safety, quality, and regulatory compliance. Regulatory bodies such as the US FDA and the MHRA in the UK have implemented premarket notification systems requiring manufacturers to submit detailed product information and test results before introducing new e-cig products to the market (FDA 2024c; European Union 2014). This typically includes reporting on product design, ingredients, emissions data, and manufacturing processes. Many jurisdictions have also mandated ongoing monitoring and periodic reporting requirements for manufacturers, as seen in Australia and the United States (FDA 2024d; TGA 2024). This may include submitting sales data, adverse event reports, and updates on any product modifications.

There has been some progress in developing e-cig-specific testing standards, but there is still a need for further standardization and harmonization of testing methods across different regulatory frameworks. Organizations such as the International Organization for Standardization (ISO), the Cooperation Centre for Scientific

Research Relative to Tobacco (CORESTA), and the European Committee for Standardization (CEN) have been working to establish consensus standards for e-cig testing. A study found that the tobacco industry plays a significant role in developing standards related to tobacco products, particularly through CORESTA and its influence on ISO processes (Bialous and Yach 2001). Product testing is used to generate data necessary to support the disclosure requirements to both regulators and the public. It is important to note that there are concerns about the tobacco industry's influence on these standards and generated data. Tobacco companies are driven to sell their products and find new users, which conflicts with public health policy (WHO FCTC 2013). They introduce products claiming to reduce harm while distorting evidence of their harmful effects (WHO 2021a).

The WHO Tobacco Laboratory Network (TobLabNet), a global network of tobacco industry independent laboratories of WHO, is responsible for developing and standardizing testing methods for e-cigs and other tobacco products. The network aims to establish global tobacco testing and research capacity to assist with regulatory compliance and policy decisions. TobLabNet creates standard operating procedures (SOPs) for analyzing various aspects of e-cigs and e-liquids, such as nicotine content, glycerol, and propylene glycol levels. These SOPs are designed to be Global Public Health Goods, supporting countries in strengthening nicotine and tobacco product regulation and implementing Articles 9 and 10 of the WHO FCTC (WHO FCTC 2013). By developing harmonized reporting standards, TobLabNet facilitates the comparison of test results globally, providing valuable data to inform regulatory efforts.

5.4 Sale and Marketing Restrictions

Sale and marketing restrictions are a vital component of e-cig regulations in many countries. Sale regulations usually involve age restrictions, licensing requirements for retailers, and product standards. Marketing regulations cover aspects such as advertising, promotions, packaging, and sponsorship. These restrictions are designed to reduce the appeal and availability of e-cigs, especially to young people.

5.4.1 Sale and Use Restrictions

According to the Institute for Global Tobacco Control 2024, the sale of all types of e-cigs is prohibited in 31 countries. In some countries, the bans on sales pertain to nicotine e-cigs only, while in others, the bans cover both nicotine and non-nicotine e-cigs. Additionally, the use of e-cigs is banned in at least 7 countries, including Cambodia, Iran, Nicaragua, Syria, Singapore, Taiwan, and Venezuela. In Singapore, the Tobacco Control Act bans the sale, distribution, and importation of e-cigs. Additionally, four countries (Jamaica, Japan, Mexico, and Switzerland) prohibit the sale of nicotine-containing e-cigs (Institute for Global Tobacco Control 2024). As of September 2024, 56 countries/jurisdictions have minimum age requirements for the purchase and use of e-cigs (Institute for Global Tobacco Control 2024). Many countries have the same age requirement for purchasing both e-cigs and conventional cigarettes, while others have different age limits for these products. Most countries that regulate age restrictions set the minimum age for buying e-cigs at 18. For example, the UK, Canada, New Zealand, and many EU countries have set the minimum age at 18 (Conway 2024). In Canada, e-cigs are regulated under the Tobacco & Vaping Products Act (TVPA), which prohibits the sale of e-cigs to individuals under the age of 18. It also restricts the sale of e-cigs that may attract youth through their appearance or functionality (Campus et al. 2021). Additionally, Jordan, the Republic of Korea, and Turkey have set the minimum age of purchase at 19 years (Institute for Global Tobacco Control 2024). In the United States, the federal minimum age for purchasing e-cigs is 21 under legislation known as Tobacco 21 or T21 (CDC 2024).

The effectiveness of age restrictions has been questioned due to the increase in youth vaping in many countries. In the UK, for instance, a survey revealed that the percentage of children trying vaping increased by 50% annually from 2022 to 2023 (Conway 2024). A study investigating how underage youth obtain e-cigs in countries with minimum age sales restrictions discovered that young people often acquire e-cigs through social sources (friends, family) or commercial sources, bypassing age restrictions (Graham-DeMello et al. 2023). A recent study found that 72% of online vendors failed to comply with age verification measures for e-cig purchases (Reiter et al. 2024). Children aged 13–15 years are using e-cigs at higher rates than adults in all WHO regions, indicating that age restrictions alone may not be enough to prevent youth access (WHO 2023).

5.4.2 Licensing Requirement for Retailers

Retailers play a crucial role in enforcing e-cig sales regulations. The specific responsibilities and restrictions placed on retailers vary widely between countries. Many countries restrict how retailers can display and advertise e-cig products, enforce age restrictions, require special licenses or certifications to sell e-cigs, and limit the types of e-cig flavors that retailers can sell. In the United States, retailers are required to verify the ages of customers and can face penalties for non-compliance (FDA 2024b). Moreover, retailers are prohibited from selling tobacco products, including e-cigs, and from using vending machines in facilities where individuals under 21 are present or are permitted to enter at any time. Berg et al. 2022 have documented the reshaping of the e-cig retail environment from the perspective of evolution and public health concerns in the United States. On the other hand, New Zealand's Smokefree Environments and Regulated Products Act 1990 provides some specific exemptions for vaping products compared to tobacco products, including allowing all retailers to display vaping products in their retail and online stores (including in vending machines) and allowing specialist vape retailers to give advice and recommendations about vaping products to customers, and to permit vaping within their approved vaping premises (Smokefree Environments and Regulated Products

(Vaping) Amendment Act 2020). Public health campaigns, subject to approval by the Director-General of Health, may also promote vaping as an alternative for cigarette smokers (Bateman et al. 2019). However, ENDS retailers must meet the requirement to sell these products by obtaining a licence in New Zealand (Bateman et al. 2019).

5.4.3 Advertising and Promotion Restrictions

Many countries either prohibit entirely or heavily regulate the advertising and promotion of e-cigs. For example, countries such as Argentina, Brazil, Brunei Darussalam, Cambodia, Colombia, Egypt, The Gambia, India, Iran, Kuwait, Lao People's Democratic Republic, Lebanon, Mauritius, Mexico, Nepal, Nicaragua, Oman, Panama, Qatar, Seychelles, Singapore, Sri Lanka, Suriname, Syria, Thailand, Timor-Leste, Turkey, Turkmenistan, and Uganda have regulations in place (Campus et al. 2021). Moreover, countries like China, Jordan, Moldova, and Guyana have explicitly prohibited the online advertising and promotion of e-cigs (Jenkins et al. 2024). Bans or restrictions on e-cig advertising are selectively implemented across various media platforms, including television, radio, print media, online, social media, and outdoor advertising, by different countries. In the EU, the TPD prohibits most advertising and sponsorship associated with e-cigs (European Union 2014). Canada has prohibited lifestyle advertising, advertisements that easily attract and appeal to youth, and the use of affidavits and advocacies in promoting e-cigs, as part of its regulations under TVPA (Institute for Global Tobacco Control 2024).

The impact of advertising restrictions on reducing vaping product use is still uncertain. Scottish Government (2024) found conflicting results and insufficient evidence to support the effectiveness of advertising restrictions. However, in Canada, provinces with comprehensive restrictions on e-cig marketing had lower levels of exposure to marketing and lower prevalence of e-cig use among youth compared to provinces with fewer restrictions (Hammond et al. 2020). Another study by Ylitörmänen et al. 2024 found that bans on internet advertising were associated with a lower prevalence of use after controlling for individual-level factors and country income level. A study comparing four countries (Australia, Canada, England, and the United States) found that reported exposure to e-cig advertising generally reflected each country's regulations. For example, exposure to point-of-sale advertising was higher in England and the United States, which have fewer restrictions, compared to Australia and Canada (Cho et al. 2019).

5.5 Taxation

Taxation has become an essential tool for regulating e-cigs, with many countries implementing excise taxes on these products. The approach to e-cig taxation often differs from conventional tobacco taxes due to the diverse nature of the e-cig market, which includes variations in nicotine concentration, e-liquid volume, and device types. Thirty-nine countries have imposed excise taxes on e-cigs, with Italy being the first to do so in 2014 (Nian et al. 2023). Among these countries, 82.1% solely levy a specific tax, 12.8% impose only an ad valorem tax, and 5.1% use a combination of both.

5.5.1 Comparison to Conventional Tobacco Taxes

The taxation of e-cigs is a relatively new development, whereas conventional cigarette taxes have been in place for a long time and are generally higher. In the United States, the tax rate for conventional cigarettes per pack was $ 0.991 while the tax rate for e-cigs per ml of nicotine was $ 0.262 in 2018 (Pesko et al. 2020), which indicated significant differences between them. Besides, a standard JUUL (US's leading e-cig manufacturer) pod containing 0.7 ml of fluid would, therefore, be assessed a tax of less than $0.04 for Kansas and Louisiana, and $1.27 for Minnesota. For comparison, in 2018 the weighted federal and state conventional cigarette excise tax was $2.51 per pack (Pesko et al. 2020), which the tax rate was more expensive than e-cigs. In 2024, the authority with the highest tax rate on cigarettes is currently New York at $5.35 for a pack of 20 (Roy 2024). Washington DC is the second highest at $5.03 per pack, while Kansas ($1.29 per pack), Louisiana ($1.08 per pack), and Minnesota ($3.04 per pack) (Roy 2024) have their cigarettes and tobacco taxes increase when compared with taxes of 2018.

The WHO recommends that excise taxes on tobacco products should be at least 75% of the retail price (WHO 2024b, 2024c). However, among ASEAN countries, only Thailand meets this recommendation with an excise tax rate of approximately 81.3% (Tan and Dorotheo 2024). In Malaysia, an ad valorem excise tax of 10% is applied to all smoking devices, including e-cigs, heated tobacco products, and conventional tobacco devices like hookahs and smoking pipes (Tan and Dorotheo 2024). Additionally, in Malaysia, a 10% ad valorem sales tax and a specific excise duty of MYR 0.40 (USD 0.08) per ml (Tan and Dorotheo 2024) were imposed on non-nicotine e-liquid and came into effect on 1 January 2021. From April 2023 and onwards, the taxation applies to nicotine e-liquids as well after the government removed liquid nicotine used in e-cigs and vape from the Poisons List of controlled substances (Code Blue 2023).

In 2018, Indonesia imposed a 57% tax on the retail prices of e-liquid, which was higher than the average 40% tax on cigarettes (Kowitt et al. 2022). According to Kowitt et al. (2022), this difference in taxation may have unintended consequences, as economic studies suggest that e-cigs and conventional cigarettes can be substitutes for each other. Research shows that increasing taxes on conventional cigarettes can lead to a decrease in adult cigarette use and an increase in e-cig use. On the other hand, higher taxes on e-cigs can have the opposite effect (Pesko et al. 2020). This demonstrates the importance of taking into account the entire range of tobacco products when creating tax policies.

5.5.2 Variations in Tax Approaches

A study conducted by Nian et al. (2023) revealed that 76.5% of countries with specific taxes utilize a fixed rate determined by factors such as liquid volume, cartridge units, device units, or nicotine concentration. For example, Portugal, Sweden, Ukraine, and Uzbekistan tax only nicotine-containing liquids, while Italy imposes different tax rates for nicotine and non-nicotine liquids. Certain countries have variable tax rates based on factors such as nicotine concentration, device type, or type of nicotine. Sweden, for instance, nearly doubles the tax rate for e-cigs with nicotine concentrations ≥15 mg/ml compared to those with lower concentrations. Israel and Indonesia have a tiered specific excise tax based on the type of ENDS product, with closed systems (disposable devices and pods) incurring significantly higher tax rates compared to open systems. The Philippines employs tiered tax rates based on the type of nicotine (free-base or salt-base) (Tobaccofreekids 2023b).

Studies on vaping regulations suggest that increasing taxes or prices on e-cigs generally leads to a decrease in vaping use, with no corresponding rise in tobacco use. The research indicates that higher e-cig taxes result in a more significant reduction in vaping compared to areas with less stringent regulations (Reiter et al. 2024). As the e-cig market continues to evolve, policymakers face the challenge of balancing public health goals with the potential unintended consequences of differential taxation between e-cigs and conventional tobacco products.

5.6 Use Restrictions

The use of e-cigs is being more strictly regulated, with many countries implementing bans or restrictions on vaping in public spaces such as schools, public transport, and workplaces. These measures aim to protect non-users from potential secondhand exposure and to maintain smoke-free environments.

According to the Institute for Global Tobacco Control 2024, 85 countries/jurisdictions prohibit or restrict e-cig use in public places. In the EU, countries that have entirely banned e-cigs in enclosed public places, workplaces and offices include Croatia, Cyprus, the Czech Republic, Estonia, Finland, Georgia, Greece, Hungary, Lithuania, Malta, Montenegro, Poland, Slovenia, Turkey, and Ukraine (Amalia et al. 2022). Regulation of e-cigs not banned in workplaces and offices includes Italy, Denmark, Portugal, the Republic of Moldova, Tajikistan, Luxembourg, and Spain (Amalia et al. 2022). In Cyrus, the use of e-cigs in private vehicles with a pregnant woman or a minor under 18 years present is prohibited (Institute for Global Tobacco Control 2024). Similarly, since 11 November 2020, New Zealand's Smokefree Environments and Regulated Products (Vaping) Amendment Act 2020 prohibits vaping in indoor workplaces, schools, early childhood education and care centers. From 28 November 2021, the Act prohibits smoking and vaping in cars carrying children, with specific exemptions for single occupants and vehicles used as a dwelling (Smokefree Environments and Regulated Products (Vaping) Amendment Act 2020 2020). In China, e-cigs are banned in public places, indoor workplaces, public transportation, and outdoor areas such as schools, kindergartens, and hospitals. Entertainment venues were given a grace period to comply with these regulations and set up designated smoking rooms if they chose to. Violations of this ban are subject to fines (Rachel 2019). In the United States, there is a nationwide ban on the use of e-cigs in restaurants and workplaces (Reiter et al. 2024).

5.7 Packaging and Labeling Requirements

Packaging and labeling requirements for e-cigs vary among countries, but they generally aim to provide consumers with information about potential health risks and product contents. The placement of health warnings on e-cig packaging is required in at least 51 countries (Institute for Global Tobacco Control 2024). Israel, Denmark, and Finland require plain packaging on all e-cigs and refill containers, and Uruguay prohibits brands/patents for e-cigs (Tobacco Control Law 2024a). Plain packaging is designed to remove all marketing elements, such as brand logos and colors for health warnings.

In the EU, health warnings are required on e-cig products to inform consumers about nicotine content and the fact that they are not for non-smokers. Health warnings must cover 30% of the front and back of the package, stating: "This product contains nicotine which is a highly addictive substance." There is ongoing debate about whether e-cigs are a viable alternative to conventional smoking, given that smoking remains a significant public health issue in the EU, with nearly 700,000 cigarette-related deaths annually (Rachel 2019). The EU TPD and its UK counterpart, the TRPR, mandate that nicotine-containing vaping products include specific health warnings and information (Nottage et al. 2024). These requirements include a clear nicotine warning, a list of ingredients, instructions for safe use and storage, information on nicotine content, and promotional elements that are prohibited on e-cig packaging (European Union 2014). In addition, Canada has implemented strict labeling requirements, including health warnings that must be displayed on the main display panel, nicotine concentration must be stated, a list of ingredients is required, and child-resistant packaging is mandatory (Minister of Justice Canada 2019).

According to the US FDA, for e-liquids and aerosol products pre-loaded with tobacco-derived substances containing nicotine, product packages are required to display health warnings. The warnings must clearly show the risks associated with nicotine, such as its addictive potential and toxicity (Nottage et al. 2024). The warnings should be prominently placed to ensure visibility, highlighting that nicotine is a harmful substance, especially for young people and non-smokers, and that it should be kept out of reach of children (Snell et al. 2021). Furthermore, the FDA has taken regulatory action against e-cig products with packaging that strongly resembles candies, cookies, and other snacks (FDA 2018).

In food and medicine, ingredients are listed in descending order of quantity, with the first ingredient being the most abundant. According to a study by Buonocore et al. (2017), 13% of products did not have an ingredient list. Additionally, the brand

Apollo used the vague label "may contain" for its ingredients. Blue Star used terms like "approved flavors" and "approved colorings," while others referred to "food-grade flavoring" or "in-house flavoring agents". The study also found that in 40% of cases, manufacturers listed flavors before nicotine, suggesting that flavors might be present in greater quantities than nicotine.

5.8 Challenges in Regulating E-cigs

Regulating e-cigs presents significant challenges because of their complex interplay with public health, economics, and sociopolitical factors. One of the primary difficulties is the rapid evolution of e-cig products as well as marketing strategies designed to target youths and young adults. These have led to widespread use and nicotine addiction among these age groups (CDC 2024; Truth Initiative 2024). Furthermore, the rapidly evolving nature of e-cig technology makes it even more challenging to establish comprehensive regulations to keep pace with new products and delivery systems. The lack of comprehensive regulation has enabled many e-cig products to remain in the market without undergoing a full review of their public health impact thus contributing to the ongoing crisis of youth vaping (Truth Initiative 2024). Regulations enforcement is often slow and inconsistent, with many popular e-cigs for youths remaining available despite regulations such as FDA denials of marketing permissions (Truth Initiative 2024). Moreover, the global regulatory landscape is fragmented, with different countries employing various legal mechanisms to regulate e-cigs, including bans, market authorizations, and amendments to existing legislation (Kennedy et al. 2017). The challenge is further complicated by the influence of tobacco and e-cig industries, which have aggressively marketed their products, undermining scientific evidence to promote their interests (Banks et al. 2023). Effective regulation requires a balanced approach that protects public health while considering the potential benefits of e-cigs for smoking cessation, emphasizing the need for evidence-based and thoughtfully designed policies (Sindelar 2020).

5.9 Future Regulatory Directions

Future regulations for e-cigs are expected to be influenced by ongoing research and evolving public health concerns. Policymakers may move toward implementing more stringent product standards, including limitations on nicotine concentrations and restrictions on certain e-liquid ingredients or flavors. There is a growing emphasis on developing comprehensive age verification systems for both online and in-person sales to curb youth access. Regulators may also prioritize the standardization of product labeling and packaging to improve consumer awareness of potential risks. As research on the long-term health effects of e-cig use continues, future regulations may incorporate new findings to refine safety guidelines and usage recommendations (Wackowski et al. 2017). Additionally, there is likely to be an increased focus on harmonizing regulations across different jurisdictions to create a more consistent global approach to e-cig control. The future regulatory landscape may also see a greater emphasis on post-market surveillance and the implementation of track-and-trace systems to monitor the supply chain and combat illicit trade (WHO 2021b).

5.10 Conclusion

In conclusion, regulating e-cigs is a complex and contentious issue in public health policy. As the market evolves and new evidence emerges, policymakers must stay alert and flexible in their approaches to effectively balance public health concerns with the potential benefits of these products for adult smokers seeking alternatives to conventional cigarettes.

Appendix 1: ENDS and the WHO FCTC[1]

The WHO Framework Convention on Tobacco Control (WHO FCTC) is the first international treaty negotiated under the auspices of the World Health Organization. It was adopted by the World Health Assembly on May 21, 2003, and came into force on February 27, 2005. Since its adoption, it has become one of the most rapidly and widely endorsed treaties in United Nations history (https://fctc.who.int).

This treaty, which is legally binding for the 182 countries that have ratified it, represents a coordinated international effort to implement evidence-based strategies and policies to reduce tobacco use, protect public health, and promote global well-being.

ENDS were introduced to the market in 2003 and have since become widely available worldwide. This extensive availability has raised concerns among Parties to the World Health Organization Framework Convention on Tobacco Control (WHO FCTC) regarding the potential impact of these products on individual and public health. In 2008, the Working Group on Articles 9 and 10 of the WHO FCTC recommended that the WHO identify best practices for reporting to regulators on the contents, emissions, and characteristics of products, including electronic systems. This recommendation was incorporated into decision FCTC/COP3(9) at the Third session of the Conference of the Parties (COP3) (https://apps.who.int/gb/fctc/PDF/cop3/FCTC_COP3_DIV3-en.pdf).

At the Fourth Session of the Conference of the Parties (COP4), in 2010, the Convention Secretariat presented a report (FCTC/COP/4/12) (https://apps.who.int/gb/fctc/PDF/cop4/FCTC_COP4_DIV6-en.pdf) on the Control and Prevention of Smokeless Tobacco Products and Electronic Cigarettes. The report provided information on ENDS, along with an overview of recommendations made by the WHO

[1] Contributed by Dr. Cheah Nuan Ping/Chair of WHO TobLabNet. Information up to date as at 9 July 2024.

Study Group on Tobacco Product Regulation (TobReg) and the outcome of a regulatory consultation convened by WHO. It noted growing global concerns about the quality, safety, and "regulatory gap" of these emerging products as they continued to penetrate new markets.

At the same session, through decision FCTC/COP4(14) (https://apps.who.int/gb/fctc/PDF/cop4/FCTC_COP4_DIV6-en.pdf), Parties requested that the Convention Secretariat, in collaboration with WHO, prepare a comprehensive report based on the experiences of the Parties regarding smokeless tobacco products and nicotine delivery systems, including electronic cigarettes, for consideration at the Fifth session of the Conference of the Parties (COP5).

In 2012, at the Fifth Session of the COP, the Convention Secretariat presented document FCTC/COP/5/13 (https://apps.who.int/gb/fctc/PDF/cop5/FCTC_COP5_13-en.pdf), which included results from a questionnaire sent to all Parties in November 2011. The survey covered topics such as availability, regulatory framework, sales volume, and scientific studies on ENDS. The report also outlined regulatory strategies undertaken by the Parties and highlighted policy areas that needed consideration.

COP5 decision FCTC/COP/5/13 tasked the Convention Secretariat with identifying options for the prevention and control of ENDS and examining emerging evidence on the health impacts of their use.

In 2014, WHO presented a report FCTC/COP/6/10 Rev.1 (https://apps.who.int/gb/fctc/PDF/cop6/FCTC_COP6_10Rev1-en.pdf) on ENDS, to the Sixth session of the Conference of the Parties (COP6). This report included scientific recommendations from TobReg and analysis from a WHO survey on tobacco products, which had responses from 90 WHO Member States, 86 of which were Parties to the WHO FCTC. The survey revealed that over 50% of the Parties did not regulate ENDS, prompting the same report to propose a regulatory framework for these products.

COP6 decision FCTC/COP6(9) (https://apps.who.int/gb/fctc/PDF/cop6/FCTC_COP6(9)-en.pdf) requested the Convention Secretariat to invite WHO to prepare a report on ENDS and electronic non-nicotine delivery systems (ENNDS) for the Seventh session of the Conference of the Parties (COP7). This decision also invited Parties to take measures to address the challenges posed by ENDS/ENNDS and consider prohibiting or regulating them as tobacco products, medicinal products, consumer products, or other categories, as appropriate, ensuring a high level of protection for human health.

In 2016, WHO presented a report FCTC/COP/7/11 (https://fctc.who.int/publications/i/item/fctc-cop-7-11-electronic-nicotine-delivery-systems-and-electronic-non-nicotine-delivery-systems-(-ends-ennds)-of-the-who-framework-convention-on-tobacco-control-of-the-seventh-session-of-the-conference-of-the-parties-(cop7)) to the Seventh Session of the COP, providing updates on the health impact of ENDS/ENNDS, their potential role in tobacco cessation, their effect on tobacco control efforts, and an assessment of regulatory options.

In response to growing concerns about e-cigs, the seventh session of the FCTC Conference of the Parties (COP) adopted a decision urging Parties to consider implementing regulatory measures to "prohibit or restrict the manufacture, importation, distribution, presentation, sale, and use of ENDS/ENNDS* in accordance with their national laws and public health objectives.

The decision also requested the Convention Secretariat to invite Parties to monitor and report on scientific, regulatory, and market developments related to initiation, cessation, advertising, and promotion. Additionally, it requested WHO to report on the development of methods by regional and international standards-development organizations for testing and measuring the contents and emissions of these products, either at the eighth or ninth session of the COP, as applicable, with a report to be prepared and presented at the ninth session.

Upon the request of the Parties, at the Seventh Session of the COP, the COP7 Bureau asked the Convention Secretariat to present a report at COP8 on the progress of ENDS and ENNDS regulatory measures undertaken by the Parties and the current situation (https://fctc.who.int/publications/i/item/fctc-cop7(9)-electronic-nicotine-delivery-systems-and-electronic-non-nicotine-delivery-systems-parties-to-the-who-framework-conventionon-tobacco-control-seventh-session).

A report presented in FCTC/COP/8/10 was prepared by the Convention Secretariat where the ENDS and ENNDS situation based on information from the Parties and other sources, were compiled. Based on the information provided, many countries began by establishing precise definitions and categorizations for ENDS within their existing legal frameworks. Countries have reported classifying ENDS into seven categories: (1) tobacco products; (2) products simulating tobacco; (3) medicinal products; (4) pharmaceutical products; (5) consumer products; (6) poisons; or (7) as ENDS themselves.

At the Tenth Session of the COP of the WHO FCTC, (COP10) regarding smoking cessation options, WHO reported that there is currently insufficient evidence to support the use of electronic cigarettes as a cessation device. Heated tobacco products (HTPs) are not considered cessation options, and current evidence does not support the classification of electronic nicotine delivery systems (ENDS), including electronic cigarettes, as cessation products at the population level (https://iris.who.int/bitstream/handle/10665/377610/fctc-cop10-26-en.pdf?sequence=1).

The WHO also shared that, in response to data indicating that far more children than adults were starting to use electronic nicotine delivery systems (ENDS), some from a very early age, they issued a "Call To Action" in December 2023, presenting options to address this public health issue (https://www.who.int/publications/m/item/electronic-cigarettes%2D%2D-call-to-action; https://www.who.int/publications/m/item/technical-note-on-call-to-action-on-electronic-cigarettes).

For the 35 countries that have banned the sale and importation of these products, WHO recommended strong and effective enforcement, surveillance, and monitoring measures. For countries that allow their commercialization, stringent regulatory measures were advised to restrict children's and young people's access to and use of these products. Additionally, WHO recommended banning flavours and flavouring agents across all tobacco and nicotine products to make them less appealing to children and young people. WHO also suggested that all countries, regardless of their regulatory environments and policy goals, implement measures to help smokers quit and ensure that smoking cessation products are licensed as medicines for adult use only. Furthermore, WHO encouraged the adoption of smoke-free regulations, particularly in response to tobacco industry efforts to undermine legislative action in this area.

Despite ongoing discussions in the COP over the past decade about potential approaches to regulate ENDS, many Parties have yet to implement regulations for these products. This regulatory gap has several potential consequences, including increased uptake among young people, negative impacts on existing tobacco control measures, the proliferation of misleading health claims and deceptive marketing strategies, and ultimately, a lack of proper information available to consumers. It is imperative and timely for Parties to develop regulations that encompass the policy domains outlined in decisions FCTC/COP6(9) and FCTC/COP7(9), using appropriate regulatory mechanisms tailored to each Party's legal framework.

References

Allen JG, Flanigan SS, LeBlanc M, Vallarino J, MacNaughton P, Stewart JH, Christiani DC (2016) Flavoring chemicals in E-cigarettes: diacetyl, 2,3-Pentanedione, and acetoin in a sample of 51 products, including fruit-, candy-, and cocktail-flavored E-cigarettes. Environ Health Perspect 124(6):733–739. https://doi.org/10.1289/ehp.1510185

Amalia B, Fu M, Feliu A, Tigova O, Fayokun R, Mauer-Stender K, Fernández E (2022) Regulation of electronic cigarette use in public and private areas in 48 countries within the WHO European region: a survey to in-country informants. J Epidemiol 32(3):JE20200332. https://doi.org/10.2188/jea.JE20200332

Banks E, Yazidjoglou A, Joshy G (2023) Electronic cigarettes and health outcomes: epidemiological and public health challenges. Int J Epidemiol 52(4):984–992. https://doi.org/10.1093/ije/dyad059

Bateman J, Robertson L, Marsh L, Thornley L, Hoek J (2019) New Zealand tobacco retailers' understandings of and attitudes towards selling electronic nicotine delivery systems: a qualitative exploration. Tobacco Control:tobaccocontrol-2019-055173. https://doi.org/10.1136/tobaccocontrol-2019-055173

Berg CJ, Melena A, Wittman FD, Robles T, Henriksen L (2022) The reshaping of the E-cigarette retail environment: its evolution and public health concerns. Int J Environ Res Public Health 19(14):8518. https://doi.org/10.3390/ijerph19148518

Bialous SA, Yach D (2001) Whose standard is it, anyway? How the tobacco industry determines the International Organization for Standardization (ISO) standards for tobacco and tobacco products. Tob Control 10(2):96–104. https://doi.org/10.1136/tc.10.2.96

Bianco E, Skipalskyi A, Goma F, Odeh H, Hasegawa K, Al Zawawi M, Stoklosa M, Dalmau R, Dorotheo EU, Berteletti F, Mwangi J, Wang Y (2021) E-cigarettes: a new threat to cardiovascular health—a world heart federation policy brief. Global Heart 16(1). https://doi.org/10.5334/gh.1076

Buonocore F, Gomes ACNM, Nabhani-Gebara S, Barton SJ, Calabrese G (2017) Labelling of electronic cigarettes: regulations and current practice. Tob Control 26(1):46–52. https://doi.org/10.1136/tobaccocontrol-2015-052683

Campus B, Fafard P, Pierre JS, Hoffman SJ (2021) Comparing the regulation and incentivization of e-cigarettes across 97 countries. Soc Sci Med 291:114187. https://doi.org/10.1016/j.socscimed.2021.114187

CDC (2024) STATE system E-Cigarette fact sheet. https://www.cdc.gov/statesystem/factsheets/ecigarette/ECigarette.html#:~:text=As%20of%20March%2031%2C%202024%20all%2050%20states%2C%20the%20District,e%2Dcigarettes%20to%20underage%20persons

China Briefing (2022) China's vape industry: consumption tax imposed. https://www.china-briefing.com/news/china-vape-industry-how-will-new-administrative-measures-impact-industry/#regulatingchinasvapeindustryHeader

Cho YJ, Thrasher J, Cummings M, Yong HH, Hitchman SC, McNeill A, Fong GT, Hammond D, Hardin J, Li L, Lindblom EN (2019) Cross-country comparison of cigarette and vaping product marketing exposure and use: findings from 2016 ITC four country smoking and vaping survey. Tobacco Control:tobaccocontrol-2018-054650. https://doi.org/10.1136/tobaccocontrol-2018-054650

Code Blue (2023) Malaysia shockingly Legalises vape with nicotine for tax—without existing E-cigarette regulations. https://codeblue.galencentre.org/2023/03/malaysia-shockingly-legalises-vape-with-nicotine-for-tax-without-existing-e-cigarette-regulations/

Conway L (2024) Advertising, marketing and promotion of vaping products. Commons Library Research Briefing. https://researchbriefings.files.parliament.uk/documents/SN07001/SN07001.pdf

European Union (2014) Directive 2014/40/EU of the European Parliament and of the council of 3 April 2014 on the approximation of the laws, regulations and administrative provisions of the member states concerning the manufacture, presentation and sale of tobacco and related products and repealing directive 2001/37/EC (Article 20). In Official journal of the European Union, 2014. https://eur-lex.europa.eu/legal-content/EN/TXT/?uri=CELEX%3A02014L0040-20150106

FDA (2018) FDA warns company for selling e-liquids that resemble kid-friendly foods as part of the agency's ongoing Youth Tobacco Prevention Plan. https://www.fda.gov/news-events/press-announcements/fda-warns-company-selling-e-liquids-resemble-kid-friendly-foods-part-agencys-ongoing-youth-tobacco

FDA (2023) Facts about E-Cigarettes. https://www.fda.gov/news-events/rumor-control/facts-about-e-cigarettes

FDA (2024a) E-Cigarettes, vapes, and other Electronic Nicotine Delivery Systems (ENDS). https://www.fda.gov/tobacco-products/products-ingredients-components/e-cigarettes-vapes-and-other-electronic-nicotine-delivery-systems-ends#:~:text=To%20date%2C%20FDA%20has%20authorized,About%20E%2DCigarettes%E2%80%9D%20webpage

FDA (2024b) FDA issues final rule increasing the minimum age for certain restrictions on tobacco sales. https://www.fda.gov/news-events/press-announcements/fda-issues-final-rule-increasing-minimum-age-certain-restrictions-tobacco-sales

FDA (2024c) Premarket tobacco product applications. https://www.fda.gov/tobacco-products/market-and-distribute-tobacco-product/premarket-tobacco-product-applications

FDA (2024d) Tobacco product problem reports. https://www.fda.gov/tobacco-products/tobacco-science-research/tobacco-product-problem-reports

Feldman EA (2018) Regulating E-cigarettes: why policies diverge. In: Faculty scholarship at penn law. https://papers.ssrn.com/sol3/papers.cfm?abstract_id=3236410

GOV.UK (2021) E-cigarettes could be prescribed on the NHS in world first. https://www.gov.uk/government/news/e-cigarettes-could-be-prescribed-on-the-nhs-in-world-first#:~:text=This%20could%20mean%20England%20becomes,to%20help%20them%20quit%20smoking

GOV.UK (2024a) Creating a smokefree generation and tackling youth vaping: what you need to know. https://healthmedia.blog.gov.uk/2024/04/15/creating-a-smokefree-generation-and-tackling-youth-vaping-what-you-need-to-know/

GOV.UK (2024b) Tobacco and vapes bill 2024. https://www.gov.uk/government/collections/tobacco-and-vapes-bill-2024

Graham-DeMello A, Hoek J, Drew J (2023) How do underage youth access e-cigarettes in settings with minimum age sales restriction laws? A scoping review. BMC Public Health 23(1):1809. https://doi.org/10.1186/s12889-023-16755-9

Hammond D, Reid JL, Burkhalter R, Rynard VL (2020) E-cigarette marketing regulations and youth vaping: cross-sectional surveys, 2017–2019. Pediatrics 146(1). https://doi.org/10.1542/peds.2019-4020

Hopkinson NS, Vestbo J, Bush A, Grigg J (2022) Should e-cigarettes be licensed as medicines? BMJ n2912. https://doi.org/10.1136/bmj.n2912

Institute for Global Tobacco Control (2024) Country laws regulating E-cigarettes. https://www.globaltobaccocontrol.org/en/policy-scan/e-cigarettes

Jenkins S, Flintoff AG, Greenhalgh EM, Grace C, Scollo MM (2024) 18.14 international regulatory overview. In: Tobacco in Australia: facts and issues. Cancer Council Victoria, Melbourne. https://www.tobaccoinaustralia.org.au/chapter-18-e-cigarettes/18-14-regulatory-overview

Kennedy RD, Awopegba A, De León E, Cohen JE (2017) Global approaches to regulating electronic cigarettes. Tob Control 26(4):440–445. https://doi.org/10.1136/tobaccocontrol-2016-053179

Kowitt SD, Anshari D, Orlan EN, Kim K, Ranney LM, Goldstein AO, Byron MJ (2022) Impact of an e-cigarette tax on cigarette and e-cigarette use in a middle-income country: a study from Indonesia using a pre–post design. BMJ Open 12(5):e055483. https://doi.org/10.1136/bmjopen-2021-055483

McKee M (2019) Evidence and E-cigarettes: explaining English exceptionalism. Am J Public Health 109(7):965–966. https://doi.org/10.2105/AJPH.2019.305132

McKee M (2020) Will E-cigarette regulation evolve in pace with new products? Am J Public Health 110(6):782–783. https://doi.org/10.2105/AJPH.2020.305668

Minister of Justice Canada (2019) Vaping products labelling and packaging regulations. https://laws-lois.justice.gc.ca/PDF/SOR-2019-353.pdf

Newman J, Nurfaiza MW (2022) Policy design, non-design, and anti-design: the regulation of e-cigarettes in Indonesia. Policy Studies 43(2):226–243. https://doi.org/10.1080/01442872.2019.1708887

Nian Q, Welding K, Dai Z (2023) An overview of national-level excise taxes on e-cigarettes across the world. Tob Induc Dis 21(October):1–3. https://doi.org/10.18332/tid/171353

Nottage MK, Taylor EV, East KA, McNeill A, Thrasher JF, Reid JL, Hammond D, Simonavičius E (2024) Packaging of disposable vaping products and e-liquids in England. A content analysis. Addiction, Canada and the United States. https://doi.org/10.1111/add.16611

Pesko MF, Courtemanche CJ, Maclean JC (2020) The effects of traditional cigarette and e-cigarette tax rates on adult tobacco product use. J Risk Uncertain 60(3):229–258. https://doi.org/10.1007/s11166-020-09330-9

Rachel EZ (2019) Puff puff pass the legislation: a comparison of E-Cigarette regulations across borders. J National Assoc Admin Law Judiciary

Reiter A, Hébert-Losier A, Mylocopos G, Filion KB, Windle SB, O'Loughlin JL, Grad R, Eisenberg MJ (2024) Regulatory strategies for preventing and reducing nicotine vaping among youth: a systematic review. Am J Prev Med 66(1):169–181. https://doi.org/10.1016/j.amepre.2023.08.002

Reuters (2024) E-cigarettes around the world. https://www.reuters.com/business/healthcare-pharmaceuticals/e-cigarettes-around-world-2023-12-19/

Rough E (2024) The regulation of E-cigarettes. UK House of Commons Library https://research-briefings.files.parliament.uk/documents/CBP-8114/CBP-8114.pdf

Roy C (2024) Tobacco & cigarette state excise tax for 2024. In: ComplyIQ

Scottish Government (2024) Vaping—effectiveness of advertising restrictions and role of advertising and free distribution in uptake. https://www.gov.scot/publications/vaping-effectiveness-advertising-restrictions-role-advertising-free-distribution-uptake/pages/6/

Shih C-S, Etter J-F (2019) Stakeholders' views on E-cigarette legislation: a qualitative study in Taiwan. Front Public Health 7. https://doi.org/10.3389/fpubh.2019.00354

Sindelar JL (2020) Regulating vaping — policies, possibilities, and perils. N Engl J Med 382(20). https://doi.org/10.1056/NEJMp1917065

Smokefree Environments and Regulated Products (Vaping) Amendment Act 2020 (2020). https://www.globaltobaccocontrol.org/sites/default/files/2021-12/New%20Zealand%20-%20Smokefree%20Environments%20and%20Regulated%20Products%20Vaping%20Amendment%20Act%202020.pdf

Snell LM, Nicksic N, Panteli D, Burke S, Eissenberg T, Fattore G, Gauci C, Koprivnikar H, Murauskiene L, Reinap M, Barnes AJ (2021) Emerging electronic cigarette policies in European member states, Canada, and the United States. Health Policy 125(4):425–435. https://doi.org/10.1016/j.healthpol.2021.02.003

Stratton K, Kwan LY, Eaton DL (2018) Public health consequences of E-cigarettes. National Academies Press. https://doi.org/10.17226/24952

Tan YL, Dorotheo U (2024) ASEAN tobacco control atlas, 6th edn. Southeast Asia tobacco control Alliance (SEATCA

TGA (2024) Cost recovery implementation statement 2024–2025. https://www.tga.gov.au/resources/publication/publications/cost-recovery-implementation-statement-2024-2025

Tobacco Control Law (2024a) Legislation. https://www.tobaccocontrollaws.org/legislation

Tobacco Control Law (2024b) Legislation by country/jurisdiction-China. https://www.tobaccocontrollaws.org/legislation/china/laws

Tobacco Tactics (2023) E-cigarettes. https://tobaccotactics.org/wiki/e-cigarettes/

Tobaccofreekids (2023a) E-cigarettes: global regulation and recommended measures. https://assets.tobaccofreekids.org/emerging-products/resources/E-Cigarettes-Global-Regulation-2023.pdf

Tobaccofreekids (2023b) Global taxation of ENDS and ENNDS. https://assets.tobaccofreekids.org/content/what_we_do/international_issues/Electronic-Cigarettes/Final_ENDSandENNDS_7.7.23.pdf

Truth Initiative (2024) E-cigarettes: facts, stats and regulations. https://truthinitiative.org/research-resources/emerging-tobacco-products/e-cigarettes-facts-stats-and-regulations

USC (2024) Vaping: a growing public health crisis. https://mphdegree.usc.edu/blog/vaping-a-growing-public-health-crisis

Wackowski O, Hammond D, O'Connor R, Strasser A, Delnevo C (2017) Considerations and future research directions for E-cigarette warnings—findings from expert interviews. Int J Environ Res Public Health 14(7):781. https://doi.org/10.3390/ijerph14070781

WHO (2021a) WHO report on the global tobacco epidemic 2021: addressing new and emerging products. https://iris.who.int/bitstream/handle/10665/343287/9789240032095-eng.pdf?sequence=1

WHO (2021b) WHO study group on tobacco product regulation. Report on the scientific basis of tobacco product regulation: eighth report of a WHO study group https://www.who.int/publications/i/item/9789240022720

WHO (2023) Urgent action needed to protect children and prevent the uptake of e-cigarettes. https://www.who.int/news/item/14-12-2023-urgent-action-needed-to-protect-children-and-prevent-the-uptake-of-e-cigarettes

WHO (2024a) New Report: WHO urges urgent government actions to protect youth and prevent the uptake of e-cigarettes. https://www.who.int/laos/news/detail/14-02-2024-new-report–who-urges-urgent-government-actions-to-protect-youth-and-prevent-the-uptake-of-e-cigarettes

WHO (2024b) Promoting taxation on tobacco products. https://www.who.int/europe/activities/promoting-taxation-on-tobacco-products#:~:text=The%20World%20Health%20Organization%20recommends,53%20countries%20in%20the%20Region

WHO (2024c) Regulating nicotine and tobacco products. https://www.who.int/activities/regulating-nicotine-and-tobacco-products/who-tobacco-laboratory-network

WHO FCTC (2013) Guidelines for implementation: article 5.3, article 8, articles 9 and 10, article 11, article 12, article 13, article 14. World Health Organization (WHO), Geneva. https://apps.who.int/iris/bitstream/10665/80510/1/9789241505185_eng.pdf

Xu X, Wang X, Zhang X, Liu Y, He H, Mackay J (2016) The debate on regulation of e-cigarettes in China. Lancet Respir Med 4(11):856–858. https://doi.org/10.1016/S2213-2600(16)30313-7

Yan D, Wang Z, Laestadius L, Mosalpuria K, Wilson FA, Yan A, Lv X, Zhang X, Bhuyan SS, Wang Y (2023) A systematic review for the impacts of global approaches to regulating electronic nicotine products. J Glob Health 13:4076. https://doi.org/10.7189/jogh.13.04076

Ylitörmänen T, Tarasenko Y, Hiilamo H, Ruokolainen O, Puska P, Ollila H (2024) Cross-sectional study of the associations between the implementation of the WHO FCTC tobacco advertising, promotion and sponsorship bans and current e-cigarette use among youth from countries with different income levels. Tobacco Control:tc-2023-058160. https://doi.org/10.1136/tc-2023-058160

Zhao Z, Zhang M, Wu J, Xu X, Yin P, Huang Z, Zhang X, Zhou Y, Zhang X, Li C, Wang L, Gao GF, Wang L, Li X, Zhou M (2020) E-cigarette use among adults in China: findings from repeated cross-sectional surveys in 2015–16 and 2018–19. Lancet Public Health 5(12):e639–e649. https://doi.org/10.1016/S2468-2667(20)30145-6

E-cigarette Injuries and Poisoning Crisis and Mitigation Strategies

6

Sulastri Samsudin

Abstract

The widespread use of electronic cigarettes (e-cigs) has raised significant public health concerns, particularly with the increasing number of poisoning cases among children and adolescents. This chapter offers an overview of epidemiology, clinical manifestations, and mitigation strategies related to e-cig poisoning and injuries. The prevalence and epidemiology section delves into the high number of e-cig poisoning cases as a result of accidental ingestion and exposure to e-liquids. Geographical variations in e-cig use and poisoning incidents are discussed, underscoring the global nature of the issue. Clinical manifestations of e-cig poisoning are detailed, including the range of symptoms from gastrointestinal disturbances to severe neurological effects attributed to nicotine toxicity. The risks of substance abuse and injuries caused by explosions are discussed to emphasize the need for stringent regulation and monitoring of e-cig devices. The chapter concludes with a discussion on mitigation strategies that focus on regulatory measures, public health interventions, and research efforts to reduce the incidence of e-cig poisoning and injuries. Collaborative efforts across various sectors are needed to effectively address this emerging public health challenge.

Keywords

E-cigarette poisoning · Vaping · Nicotine toxicity · Burn injuries · Prevention strategies

---

S. Samsudin (✉)
Malaysia National Poison Centre, Universiti Sains Malaysia, Minden, Penang, Malaysia
e-mail: sulastri@usm.my

Y.-H. Leong, A. S. Mohd Samin (eds.), *E-Cigarettes*,
https://doi.org/10.1007/978-981-96-5087-3_6

6.1 Introduction

Electronic cigarettes (e-cigs) have rapidly evolved over the past two decades and are now available in attractive designs and a variety of flavors and e-liquid formulations (Wang et al. 2022). E-liquid typically contains propylene glycol (PG), vegetable glycerin (VG), nicotine, and flavorings. Ongoing research indicates the presence of additional chemicals, including newly identified compounds generated from heating during vaping, whose toxic effects are still not yet known. As a result, e-cig users inhale a diverse blend of chemicals, which include nicotine, solvent carriers, flavorings, tobacco-specific nitrosamines, volatile organic compounds, aldehydes, free radicals, and metals, many of which are linked to negative health outcomes (Das 2022; Dinardo and Rome 2019; Gray et al. 2022; Marques et al. 2021; Rahman Ahmed 2022; Sun et al. 2024; Tzortzi et al. 2020).

While often perceived as safer than combustible cigarettes, e-cigs are not without risks. Short-term use can cause respiratory irritation and inflammation, and long-term health effects remain largely unknown (Dinardo and Rome 2019; Marques et al. 2021). Apart from typical health risks associated with vaping, related injuries and poisoning are a significant concern. Such incidents encompass cases of ingestion among young children, inhalation effects among adolescents, and burns resulting from device explosions.

A rising prevalence of e-cig use has been reported in several countries. A systematic review of global cases of e-cig vaping revealed an overall 23% lifetime prevalence and 11% current prevalence. Europe had the highest prevalence rates, with a lifetime prevalence of 26% and a current prevalence of 14%. The overall trend in e-cig use from 2011 to 2019 indicated a consistent upward trajectory (Tehrani et al. 2022).

A study conducted in 2020 reported that percentage of e-cig use among Malaysian adults increased significantly, with 11.4% of them used vape monthly and 5.4% daily since 2016 (Driezen et al. 2022). The growing popularity of e-cig use has led to a notable increase in the number of individuals experiencing unintended effects from ingesting, inhaling, or other forms of exposure to e-cigs and e-liquids. The instances of e-cig exposure have significantly surged in the general population and the pediatric population globally. Current data on e-cig and vaping substance poisoning are limited. This chapter will delve into the prevalence of e-cig poisoning incidents, explore the characteristics of exposure, examine the clinical manifestations, and identify common substances associated with these symptoms. It concludes by emphasizing various mitigation strategies and their effect on reducing the burden of e-cig poisoning.

6.2 Prevalence and Epidemiology of E-cigs Poisoning Exposure and Injuries

6.2.1 Poisoning Trends Associated with the Prevalence of E-cig Users

Since e-cigs have become available on the market, health issues and poisoning incidents directly associated with their use have been reported. The significant annual percentage change in e-cig poisoning is attributed to the increased marketing exposure of these devices. As the popularity and availability of e-cigs have increased across countries, monitoring the current profile of exposure to e-cig liquid is important.

The increasing trend of vape poisoning and injuries can be attributed to several other key factors. Many e-cigs contain high nicotine concentrations, which increase the risk of poisoning even with small amounts. Appealing flavors and attractive packaging make these products enticing, especially to children, leading to accidental ingestion. The lack of consistent safety standards in the rapidly evolving e-cig industry contributes to device malfunctions and contaminated e-liquids. Misuse and modification of devices along with the use of black-market products further elevate the risks. In general, awareness about the dangers of e-cigs is lacking, leading to improper use and inadequate storage and disposal practices, which play significant roles in the rising cases of vape-related poisoning and injuries.

Since e-cigs were introduced to the US market in 2007, the number of calls to poison control centers related to this product has increased (Kamboj et al. 2016). Data from the National Poison Data System shows that the total annual count of e-cig exposure cases escalated from 57 in 2010 to 2901 in 2018. From 2010 to 2014, the annual number of e-cig exposure cases significantly increased, reaching a peak of 3742 in 2014, a total of 6465% increase compared with the value in 2010.

Subsequently, there was a consistent decline of e-cig exposure cases each year from 2015 to 2017 with 38% reduction. However, between 2017 and 2018, an increase of 25% was recorded, with the overall number increasing from 2320 to 2901 cases (Wang et al. 2020). The first national data on the prevalence of e-cig use revealed that 4.5% of US adults, projected to be 10.8 million in the total US adult population, were current e-cig users in 2016 (Mirbolouk et al. 2018). Between 2017 and 2018, the overall use of e-cigs increased from 4.3% in 2017 to 5.4% in 2018 (Parekh and Desai 2020) and the use among high school students increased by more than 10-folds from 1.5% in 2011 to 19.6% in 2020 (Wang et al. 2018).

In Europe, e-cigs were introduced in 2006 and consumption continuously increased over the years (Prieger et al. 2019). Vardavas et al. (2017) analyzed data reported by 10 poison control centers in the EU from 2012 to 2014. Their findings revealed that the number of reported cases of acute exposure to e-cigs increased from 10 cases in 2012 to 140 cases in 2014. Similar trends were reported by the Czech Toxicological Information Centre, and the cases peaked in 2014. A subsequent increase was noted between 2016 and 2018 (Obertova et al. 2020). A recent study conducted in France documented 919 cases of exposure to e-liquids between July 2019 and December 2020 (Franchitto et al. 2023). The UK National Poison Information Service (NPIS) observed a considerable increase in telephone inquiries linked to e-cig exposure, surging from six calls in 2007 to over 75 inquiries. This upward trend tripled in 2014, reaching 241 calls (National Poisons Information Service 2014), which could be associated with the increasing number of e-cig users from 700,000 in 2012 to 2.1 million in 2014 (Use of E-Cigs (Vapes) among Adults in Great Britain, 2021).

In Malaysia, vaping was commercially introduced in 2015. The Malaysia National Poison Centre has received 111 cases of vape-related poisoning from January 2015 to December 2023 (Dermawan 2024). Cases of e-cig exposure have not been reported before 2015. The number of exposures significantly increased by 22.7% in 2020 and 31.8% in 2021 during the Covid-19 pandemic ("Abstract Book APAMT" (Asia Pacific Association of Medical Toxicology) 2022). In 2022 and 2023, the number of e-cig exposure calls increased following the Movement Control Order (MCO). Despite a slight decrease in 2022 to 18%, in 2023, a significant increase to 30.6% was recorded, marking a five-time increase compared with the number in 2019 (Nambiar 2024). The Centre also reported a shift in the trend of nicotine poisoning cases, with an increasing number of teenagers aged 15–19 years experiencing poisoning from deliberate use of e-cigs. This trend is possibly linked to the increasing prevalence of e-cigs and vaping among teenagers in Malaysia. The country documented an escalation in the number of e-cig users from 3.2% in 2016 to 5.4% in 2020, with the highest prevalence seen among individuals aged 20 to 24 years (Ab Rahman et al. 2019; Driezen et al. 2022) (Table 6.1).

In other Asian countries, no published reports on e-cig poisoning exposure were found. The absence of reports does not necessarily imply the absence of poisoning incidents in the region, but it suggests potential underreporting or a lack of data available for publication. Countries, such as Thailand and Singapore, which banned e-cigs (Jin and Jiang 2024), may not have reported any cases of poisoning or may have an insignificant number of cases due to restricted access to e-cigs.

Table 6.1 Trends of e-cigs poisoning exposure in different countries

Countries	E-cigs introduced in the market (year)	Initial e-cig poisoning data (number/year)	Peak	Total cases reported	Reporting period	References
United States of America	2006	57 cases in 2010	3742 cases in 2014	17,358 cases	2010–2018	Wang et al. (2020)
Europe union MS	2006	10 cases in 2012	140 cases in 2014	223 cases	Aug 2018–Dec 2019	Vardavas et al. (2017)
United Kingdom	2006	6 cases in 2007	241 cases in 2014	278 cases	April 2008—March 2016	Ang et al. (2018)
Malaysia	2012	5 cases in 2015	34 cases in 2023	111 cases	2015–2023	Malaysia National Poison Centre [dataset]

6.2.2 Poisoning Characteristics from E-cigs and Vaping Substances

The profile of incident characteristics remained largely consistent across studies conducted in various countries. Specific trends and risk factors, such as age category, reasons and route of exposure, and poisoning symptoms, were identified. Poisoning can occur due to exposure to various substances found in e-cigs and vaping products. Nicotine and tetrahydrocannabinol (THC) are the primary substances of concern, but additives, such as propylene glycol, vegetable glycerin, flavorings, and vitamin E acetate, also pose significant health risks. Recent studies identified new chemicals generated through heating that may have a detrimental effect on the lungs or even with completely unknown or enhanced toxicity (Sleiman et al. 2016; Gray et al. 2022).

Unintentional exposure is the most frequent type, with ingestion as the primary route of exposure. This can occur when e-liquid is accidentally swallowed, a significant concern among young children (Franchitto et al. 2023; Vardavas et al. 2017, 2021; Chang et al. 2019; Kamboj et al. 2016; Obertova et al. 2020; McFaull et al. 2020).

Obertova et al. (2020) revealed that the most common reason for calls to poison control centers was unintentional oral exposure of neonates and young children to e-liquids. This was primarily due to the appealing flavors and attractive design of the products made them particularly enticing to children. Young children are susceptible to ingestion because of their curiosity and oral exploration tendencies, thereby explaining the high incidence of acute exposure of young children to e-cigs and e-liquids at home.

In recent years, the trend has shifted, with more adolescents being exposed intentionally to e-cig poisoning, while vaping increased due to the increased prevalence of vaping habits among the age group between 18 and 24 years old (Mirbolouk et al. 2018; National Institutes of Health 2020; Vardavas et al. 2021). The evolution of new e-cigs with attractive designs and flavors has led to an increase in the number of vape users among teenagers. The use of high-concentration nicotine salt formulations in the latest vaping products increases the risk of nicotine poisoning. Additionally, vape devices that can be modified to include other substances can increase the risk of introducing prohibited substances into vape liquids, leading to poisoning or addiction comparable with that of nicotine. Dermal and ocular contact are a few common routes of exposure, accounting for 6.5% and 2.1% of cases, respectively, in a California study (2021). Similar findings were reported in European countries where these routes accounted for less than 10% of all exposures (Vardavas et al. 2021). The ocular route of exposure could be due to mishandling of e-cig products and patients mistakenly using the e-liquid as eye drops or accidentally spraying it in the eye (Wang et al. 2020; Driller et al. 2021). Intravenous administration, a rare method of exposure, accounted for 2% of cases in the Czech study (Obertova et al. 2020). Additionally, certain cases involved multiple routes of exposure.

6.2.3 Clinical Manifestations of E-cigs and Vaping Substances Poisoning

Incidents of e-cig poisoning with intentional intent were associated with a higher frequency of symptoms compared with unintentional incidents (Vardavas et al. 2017). In general, individuals exposed to e-cigs either showed no symptoms or experienced mild effects; many poisoning cases are managed on site or treated and discharged from the hospital (Chang et al. 2019; Kamboj et al. 2016). Severe clinical effects and life-threatening symptoms were less frequent, although fatalities were reported (Chen et al. 2015; Kamboj et al. 2016; Morley et al. 2017).

One of the fatal cases involved a 1-year-old child who died from cardiac arrest after ingesting nicotine liquid from an open refill container (Kamboj et al. 2016). In another case, a 24-year-old woman intentionally consumed up to 3000 mg of liquid nicotine for e-cigs. She was found in a state of pulseless electrical activity but regained spontaneous circulation after about 10 min of CPR. Despite intensive supportive care, she succumbed to multiple acute infarcts indicating severe anoxic brain injury. Toxicology tests shortly after ROSC showed plasma nicotine and cotinine levels exceeding 1000 ng/mL, highlighting the severe toxicity risk of suicidal liquid nicotine ingestion (Chen et al. 2015). Additionally, a 32-year-old man ingested 1440 mg of nicotine liquid while intoxicated, resulting in cardiac arrest and brain hypoxia. Despite receiving intensive care, he died 3 days later. His blood alcohol level was 246 mg/100 mL, and no other drugs were detected. The autopsy revealed no other anatomical cause of death (Morley et al. 2017).

Common clinical manifestations include gastrointestinal symptoms, with vomiting as the most prevalent. In the case of children exposed to e-cigs, the most common symptoms were vomiting, coughing, and drowsiness. Children exposed to e-cigs exhibited a significantly higher likelihood of healthcare admission (5.2 times) and more severe medical outcomes (2.6 times) than those exposed to conventional cigarettes (Kamboj et al. 2016).

Most patients exhibit symptoms of nicotine poisoning that can be associated with the presence of nicotine as the primary ingredient in e-liquid. Nicotine, derived from *Nicotiana* spp. plants, has extensive absorption across various routes, though its oral bioavailability is limited due to significant first-pass metabolism. It readily crosses physiological barriers, such as the blood-brain barrier, placenta, and breast milk distribution. Nicotine is metabolized primarily in the liver, and undergoes rapid renal elimination with a short half-life (Benowitz et al. 2009). The ionized state affects membrane permeability, with e-cig aerosols presenting varying pH levels compared with conventional tobacco smoke. Nicotine intoxication manifests in stimulatory and inhibitory phases, affecting cardiovascular, gastrointestinal, and central nervous system functions, potentially leading to severe complications, such as convulsions and respiratory failure (Henstra et al. 2022). The ingestion of even small doses of nicotine can have toxic effects (Driller et al. 2021). Toxbase Information estimates the toxic nicotine dose as 0.2 mg/kg for children and adults. The approximate lethal dose of nicotine is 1 mg/kg, equating to an estimated 30–60 mg in adults and 10 mg in children (Kim and Baum 2015).

The situation is exacerbated by the presence of high nicotine concentrations in the formulation of e-liquid, a consequence of inadequate regulation in certain countries. Studies also indicate deception in labeling regarding nicotine concentrations in e-cig fluid (Jackson et al. 2020). Nicotine salt formulations have been introduced, and these formulations are increasingly becoming the choice for manufacturers, allowing them to be formulated with high concentrations reaching 60–70 mg/mL, while reducing the throat hit sensation for users (Pennings et al. 2023). This factor may contribute to an increased risk of nicotine overdose poisoning. However, other chemical compounds in e-liquids, or their combinations, might also cause or exacerbate health issues. Therefore, ongoing monitoring of e-liquid ingredients is essential as more poisoning data becomes available.

Based on cases with reported medical outcomes, the severity of symptoms and outcomes can vary depending on factors, such as the concentration of nicotine in the e-liquid and the amount of e-liquid exposed to (Wang et al. 2020; Obertova et al. 2020). In many studies, calculating the exact nicotine dose was challenging due to missing information on the device, its nicotine content, or body weight. This phenomenon complicates prognosis and medical recommendations. Improved reporting of nicotine concentration and ingested e-liquid can enhance understanding of nicotine health risks in children and inform regulatory actions (Obertova et al. 2020).

6.2.4 Substance Abuse via E-cigs

Illicit drug use with e-cigs poses challenges for healthcare, public health, and law enforcement. The misuse of substances through such an easy method can increase access to users, including new users, escalating the risk of toxicity to both users and children. At the same time, e-cigs are promoted as harm reduction tools to replace conventional cigarettes, this may create a similar perception in drug abuse, where users might attempt to consume illicit drugs through vaping, perceiving it as safer. Furthermore, the modern design of e-cigs and appealing flavorings can conceal drug use even in public settings.

While inhalation has gained popularity as a method for administering illicit drugs because of its rapid effects, research on the utilization of e-cigs for this purpose remains limited. Many e-cig users also use cannabis (Chen et al. 2023). As e-cig use rises, manufacturers are making products for both nicotine and cannabis. A national study found that nearly 40% of high school e-cig users included THC in their devices, and another study showed up to 27% vaped cannabis (Morean et al. 2015). E-cig users are three times more likely to use cannabis than non-users (Miech et al. 2016). A study of college students found that nearly 7% reported using an e-cig to vaporize and inhale a substance other than nicotine, with the majority reporting the use of cannabis or some derivative of cannabis (Kenne et al. 2017).

Another survey conducted on e-cigs (Blundell et al. 2018) showed nearly 10% of the respondents reported using e-cigs for recreational drugs, with cannabis being the most common drug vaped (65%) and there was a high overall reported lifetime vaping across all drug classes. These findings were further supported by Breitbarth et al. (2018) who reviewed evidence of e-cig use for illicit drugs, such as cannabis, synthetic cannabinoids (SCs), synthetic cathinones, benzoylmethylecgonine (cocaine), gamma-hydroxybutyric acid (GHB), heroin, fentanyl, 3,4-methylenedioxyamphetamine (MDA), 3,4-methylenedioxymethamphetamine (MDMA), and methamphetamine through scientific literature and illicit drug use internet forums, highlighting the potential risk of e-cigs as a newer illicit drug delivery system. Adolescent e-cig use might serve as an intermediary step that could lead to more serious substance use (Kristjansson et al. 2015). Therefore, further research is necessary to determine if e-cigs genuinely act as a gateway to other forms of substance use, particularly conventional cigarette smoking, among adolescents who might not have otherwise engaged in such behaviors.

6.2.5 Injuries from E-cigs Explosion

Fires and explosions caused by e-cigs have led to significant injuries. This situation is caused by several factors, such as short circuits or changes in the chemical reactions within the e-cig battery, ultimately resulting in excessive heating that leads to explosions or fires that capable of causing harm to the user (Corey et al. 2018). These explosions can occur in various contexts, including while a person is changing or replacing the device's battery (Rossheim et al. 2020). Most often, these incidents occurred when the e-cig device or batteries were in the user's pockets, implying that they were unintentional (Corey et al. 2018).

Although e-cig explosions and burn injuries are considered rare, surveillance methods are likely to underestimate the actual occurrence frequency. At present, no national system exists in any country for monitoring explosions and burn injuries caused by e-cigs. Such incidents might not be well documented as case reports, covered by media, or reported to regulatory authorities.

According to initial data, 92 explosions or fires were reported in the United States between 2009 and September 2015 through literature review, reports from federal agencies, and new media. In 2016, the US Fire Administration identified 195 incidents involving e-cig explosions or fires (Rudy and Durmowicz 2017). Rossheim et al. (2019) used data from the National Electronic Injury Surveillance System (NEISS) to estimate the incidents of e-cig explosion and burn injuries presenting to US hospital emergency departments. From 2015 to 2017, the number of cases was 2035 cases and the total increase was 2693 after a year (Rossheim et al. 2019, 2020). In most cases, burns from e-cigs were thermal and affected the upper leg and lower trunk (Corey et al. 2018), causing second-degree (35%), or combination of second-degree and third-degree burn (20%) (Seitz and Kabir 2018). Thermal burns are injuries to the skin or other organic tissues caused by exposure to heat sources. The severity of a thermal burn depends on the temperature of the heat source and the duration of exposure. In 65% of cases, e-cigs exploded while being carried in pockets.

The initial fatality resulting from an e-cig explosion was documented in 2018 (Seitz and Kabir 2018). A German study analyzed retrospective data from 46 consecutive patients with injuries from the exploding Electronic Nicotine Delivery System (ENDS). The majority of patients were male (98.2%), with a median age of 36 years. All patients suffered at least partial thickness burns, with 85% experiencing deep partial thickness burns and 17% having full thickness burns. Common injury sites included the hand, lower thigh, and upper thigh, with treatment involving wound debridement and thorough disinfection (Kaltenborn et al. 2023).

These safety concerns emphasize the need for improved regulation and monitoring of e-cig devices, including their batteries (Rossheim et al. 2019). Industry needs to address technical issues that could lead to these explosions (Rossheim et al. 2020).

6.3 Mitigation Strategies to Reduce E-cigs Poisoning Incidents

The rising popularity of e-cigs seems correlated with the increasing number of poisoning cases reported to poison centers globally. Therefore, preventive measures should be implemented to alleviate the risk associated with exposure to e-cig liquid. Prevention strategies for e-cig poisoning are multifaceted and involve regulatory and public health approaches (Department of Health and Human Services 2017).

6.3.1 Regulation and Legislation

Measures and legislation are proposed to address safety and accessibility concerns associated with e-cigs. Public policy interventions that have demonstrated effectiveness in reducing the incidence of e-cig poisoning exposure among children have been documented.

A significant regulation is the introduction of childproof packaging for e-liquids to prevent unintentional ingestion by children. This initiative decreased the instances of exposure to e-cigs in the United States between 2015 and 2017, coinciding with the enforcement of childproof container regulations by several states starting in 2014. Furthermore, in 2016, the enactment of the Child Nicotine Poisoning Prevention Act mandated the adoption of childproof packaging nationwide (Wang et al. 2020).

In addition to unsafe containers, the enticing aroma of e-liquids emitted from unsecured devices or cartridges has been identified as a leading cause of unintentional oral exposure in young children. Therefore, there is a suggestion to consider prohibiting flavored e-liquids because they are known for their appeal to young individuals (Obertova et al. 2020; Driller et al. 2021). In 2015, the American Academy of Pediatrics and other health organizations called on the US Food and Drug Administration (FDA) to ban the use of flavors and attractive labeling for liquid nicotine (Farber et al. 2015). FDA finally banned the flavored cartridge/pod-based products in 2020, whereas 2 countries in EU, Hungary has banned all types of flavors and Finland prohibited candy and fruit flavors (Yingst et al. 2021; Prieger et al. 2019).

Another critical element involves the establishment of standards for the manufacturing and sale of e-cigs and e-liquids, ensuring that these products are less susceptible to malfunctions and the release of harmful substances. Govindarajan et al. (2018) suggested extra measures, such as implementing child-resistant e-cig devices and incorporating flow restrictors on liquid nicotine containers. These measures could further decrease exposure occurrences and mitigate the potential for serious medical outcomes.

Other regulatory measures include clear warning signs for parents on packaging and devices (vape pen, cartridge, tank, and e-liquid refill) and introduce small-volume cartridges and tanks with low nicotine content per ml and use of a bitter additive to deter ingestion (Obertova et al. 2020). The European Union established regulations stipulating a maximum nicotine content of 20 mg/mL in e-liquid. Additionally, for single-use e-cigs, the liquid volume is limited to 2 mL; for refill containers, it is capped at 10 mL (Prieger et al. 2019).

Stringent regulations on e-cig batteries should be imposed given the ongoing e-cig explosion issue. In 2016, the FDA granted the Center for Tobacco Products (CTP) to regulate all e-cig components, including batteries (Rossheim et al. 2020).

The recommendation extends to enforcing age restrictions on e-cig sales to limit access among minors. These regulatory measures collectively strive to improve the safety, accessibility, and allure of e-cigs and significantly change the e-cig prevalence in the population (Jun and Kim 2021).

6.3.2 Clinical and Community Intervention

Addressing challenges posed by e-cigs requires comprehensive interventions at clinical and community levels. Proposed measures include the implementation of public education campaigns aimed at raising awareness among users and the public about the risks of e-cigs poisoning, with a strong emphasis on proper storage and usage. These initiatives can be carried out through community-based programs and include community health workers and parent education initiatives. The programs can be designed to provide culturally appropriate prevention messaging specifically tailored for populations, such as the Latinx community (Herrmann et al. 2023). Their goals include educating parents on the hazards of nicotine and e-cigs and executing evidence-based vaping prevention initiatives targeting youth and parents to increase awareness of the dangers associated with e-cig use. Furthermore, an essential preventive step involves training healthcare providers to discuss e-cig risks, particularly with vulnerable groups, such as parents and youth.

In the case of e-cig explosion or burn, consumer education on safe practices for carrying and operating e-cig devices and batteries may prevent battery failure (Corey et al. 2018). The Control of Tobacco Product also advises consumers to use recommended batteries and avoid extreme temperatures to prevent explosions (Rossheim et al. 2019).

6.4 Research and Surveillance

Emerging yet restricted data are available on e-cig poisoning, and obtaining additional comprehensive data is essential. Data in the present study primarily originated from poison centers, where exposure calls are voluntary. Consequently, the outcomes might underestimate the actual incidence of e-cig exposure and may not fully depict the entire range of poisoning cases (Obertova et al. 2020; Govindarajan et al. 2018; McFaull et al. 2020; Vardavas et al. 2021).

Implementing surveillance systems to track e-cig use and incidents of poisoning is vital to enable swift responses by public health officials to emerging trends. These systems not only improve the efficiency of response efforts but also offer valuable data to shape the development of prevention strategies and regulatory policies (Ravara et al. 2023). In addition, enhanced surveillance of e-cig burns and injuries is crucial, and implementing a unique e-cig code would enhance the classification system and improve the monitoring of these injuries (Rossheim et al. 2019).

By combining these strategies, public health officials, healthcare providers, educators, and policymakers can work together to reduce the incidence of e-cig poisoning and injuries and mitigate the associated health risks.

6.5 Conclusion

The increasing prevalence of e-cig use has been accompanied by a significant increase in reported poisoning cases, particularly among young children who accidentally ingest e-liquids and adolescents who intentionally inhale them. Nicotine poisoning from e-cigs can manifest with symptoms, ranging from mild effects, such as vomiting, to severe, life-threatening conditions, such as seizures and respiratory failure, depending on the dosage. While infrequent, e-cig explosions that cause burn injuries highlight the need for stringent device regulation and accountability from manufacturers. The vaping industry must address the technical issues contributing to these explosions to improve product safety. Additionally, the misuse of e-cigs for illicit substance abuse, such as vaping cannabis or other drugs, poses new challenges for healthcare and law enforcement.

Continuous monitoring of e-cig poisoning incidents and injuries is crucial to enable swift public health responses and inform prevention strategies. A comprehensive approach spanning regulatory measures, clinical interventions, community outreach, and research efforts is necessary to mitigate the escalating risks effectively. Regulatory actions, such as implementing childproof packaging, flavor bans, and manufacturing standards, can prevent unintentional exposure and curb youth access. Public education campaigns and healthcare provider training can raise awareness about proper storage, usage, and risks. Robust surveillance systems tracking e-cig use patterns and poisoning cases are vital for timely intervention.

Through collaborative implementation of these multi-pronged strategies, stakeholders from various sectors can work toward reducing the incidence of e-cig poisoning and associated injuries, thereby mitigating the public health burden posed by this emerging issue. Continuous vigilance and proactive measures are essential to address the evolving challenges surrounding e-cig use.

References

Ab Rahman J, Mohd Yusoff MF, Nik Mohamed MH, Mahadir Naidu B, Lim KH, Tee GH, Mohamad MS, Kartiwi M, Draman S, Ab Rahman NS, Aris T (2019) The prevalence of E-cigarette use among adults in Malaysia. Asia Pac J Public Health 31(7_suppl):9S–21S. https://doi.org/10.1177/1010539519834735

abstract book APAMT 2022 (2022)

Ang E, Tuthill D, Thompson J (2018) Accidental issue in children E-cigarette liquid ingestion: a fast growing. http://adc.bmj.com/content/early/2018/02/15/archdischild-2018-31488

Benowitz NL, Hukkanen J, Jacob P (2009) Nicotine chemistry, metabolism, kinetics and biomarkers. In: Handbook of experimental pharmacology, vol 192, pp 29–60. https://doi.org/10.1007/978-3-540-69248-5_2

Blundell M, Dargan P, Wood D (2018) A cloud on the horizon-a survey into the use of electronic vaping devices for recreational drug and new psychoactive substance (NPS) administration. QJM. 111(1):9–14. https://doi.org/10.1093/qjmed/hcx178. PMID: 29025078.

Breitbarth AK, Morgan J, Jones AL (2018) E-cigarettes—an unintended illicit drug delivery system. In: Drug and alcohol dependence, vol 192. Elsevier Ireland Ltd., pp 98–111. https://doi.org/10.1016/j.drugalcdep.2018.07.031

Chang JT, Wang B, Chang CM, Ambrose BK (2019) National estimates of poisoning events related to liquid nicotine in young children treated in US hospital emergency departments, 2013-2017. Injury Epidemiol 6(1). https://doi.org/10.1186/s40621-019-0188-9

Chen BC, Bright SB, Trivedi AR, Valento M (2015) Death following intentional ingestion of e-liquid. Clin Toxicol 53(9):914–916. https://doi.org/10.3109/15563650.2015.1090579

Chen G, Rahman S, Lutfy K (2023) E-cigarettes may serve as a gateway to conventional cigarettes and other addictive drugs. Adv Drug Alcohol Res 3. https://doi.org/10.3389/adar.2023.11345

Corey CG, Chang JT, Rostron BL (2018) Electronic nicotine delivery system (ENDS) battery-related burns presenting to US emergency departments, 2016. Injury Epidemiol 5(1). https://doi.org/10.1186/s40621-018-0135-1

Das M (2022) Increase in vaping and e-cigarette use by children in Malaysia. Lancet Oncol 23(11):1360. https://doi.org/10.1016/S1470-2045(22)00596-4

Department of Health and Human Services (2017) E-cigarette use among youth and young adults: a report of the surgeon general. Department of Health and Human Services, Centers for Disease Control and Prevention, National Center for Chronic Disease Prevention and Health Promotion, Office on Smoking and Health, Atlanta.

Dermawan (2024) NPC urges govt to address rising cases of electronic cigarette poisoning. https://api.nst.com.my/news/nation/2024/03/1031234/npc-urges-govt-address-rising-cases-electronic-cigarette-poisoning, New Straits Times Press (M) Bhd.

Dinardo P, Rome ES (2019) Vaping: the new wave of nicotine addiction. Clev Clin J Med 86(12):789–798. https://doi.org/10.3949/ccjm.86a.19118

Driezen P, Nordin ASA, Hairi FM, Yee A, Tajuddin NAA, Hasan SI, Danaee M, Kamaludin IS, Kaai SC, Yan M, Grey M, Quah ACK, Thompson ME, Fong GT (2022) E-cigarette prevalence among Malaysian adults and types and flavors of e-cigarette products used by cigarette smokers who vape: findings from the 2020 ITC Malaysia survey. Tob Induc Dis 20. https://doi.org/10.18332/TID/146363

Driller G, Plasencia E, Apollonio DE (2021) Retrospective review of nicotine exposures in California from 2012 to 2018 and analysis of the impacts of e-cigarette regulations. BMJ Open 11(3). https://doi.org/10.1136/bmjopen-2020-043133

Farber HJ, Nelson KE, Groner JA, Walley SC (2015) Public policy to protect children from tobacco, nicotine, and tobacco smoke. Pediatrics 136(5). https://doi.org/10.1542/peds.2015-3109

Franchitto PN, Bloch PJ, Solal Pharm DC, Pélissier F (2023) Self-poisoning by E-cigarette and E-liquids: national reports to French Poison Control Centers from July 2019 to December 2020: VIGIlance and VAPE: the VIGIVAPE study. Nicot Tob Res. https://doi.org/10.1093/ntr/ntad116

Govindarajan P, Spiller HA, Casavant MJ, Chounthirath T, Smith GA (2018) E-cigarette and liquid nicotine exposures among young children. Pediatrics 141(5):20173361. http://publications.aap.org/pediatrics/article-pdf/141/5/e20173361/1065781/peds_20173361.pdf

Gray N, Halstead M, Valentin-Blasini L, Watson C, Pappas RS (2022) Toxic metals in liquid and aerosol from pod-type electronic cigarettes. J Anal Toxicol 46(1):69–75. https://doi.org/10.1093/jat/bkaa185

Henstra C, Dekkers BGJ, Olgers TJ, ter Maaten JC, Touw DJ (2022) Managing intoxications with nicotine-containing e-liquids. Expert Opin Drug Metab Toxicol 18(2):115–121. https://doi.org/10.1080/17425255.2022.2058930

Herrmann AK, Ferullo SL, Hernandez M, Barriga VA, Leggis B, Ruiz J, Glenn BA, Bastani R (2023) Adapting E-cigarette prevention programming to reach the latinx community. Cancer Causes Control. https://doi.org/10.1007/s10552-023-01796-7

Jackson R, Huskey M, Brown S (2020) Labelling accuracy in low nicotine e-cigarette liquids from a sampling of US manufacturers. Int J Pharm Pract 28(3):290–294. https://doi.org/10.1111/ijpp.12596

Jin P, Jiang JY (2024) E-cigarette in ten Southeast Asian countries: a comparison of national regulations. Glob Health J 1(3):1–10

Jun J, Kim JK (2021) Do state regulations on e-cigarettes have impacts on the e-cigarette prevalence? Tob Control 30(2):221–226. https://doi.org/10.1136/tobaccocontrol-2019-055287

Kaltenborn A, Dastagir K, Bingoel AS, Vogt PM, Krezdorn N (2023) E-cigarette explosions: patient profiles, injury patterns, clinical management, and outcome. JPRAS Open 37:34–41. https://doi.org/10.1016/j.jpra.2023.05.001

Kamboj A, Spiller HA, Casavant MJ, Chounthirath T, Smith GA (2016) Pediatric exposure to e-cigarettes, nicotine, and tobacco products in the United States. Pediatrics 137(6). https://doi.org/10.1542/peds.2016-0041

Kenne DR, Fischbein RL, Tan ASL, Banks M (2017) The use of substances other than nicotine in electronic cigarettes among college students. Subst Abuse Res Treat 11. https://doi.org/10.1177/1178221817733736

Kim JW, Baum CR (2015). Liquid nicotine toxicity. www.pec-online.com

Kristjansson AL, Mann MJ, Sigfusdottir ID (2015) Licit and illicit substance use by adolescent E-cigarette users compared with conventional cigarette smokers, dual users, and nonusers. J Adolesc Health 57(5):562–564. https://doi.org/10.1016/j.jadohealth.2015.07.014

Marques P, Piqueras L, Sanz MJ (2021) An updated overview of e-cigarette impact on human health. Respir Res 22(1). https://doi.org/10.1186/s12931-021-01737-5

McFaull SR, Do MT, Champagne A, Bang F (2020) Injuries and poisonings associated with e-cigarettes and vaping substances, electronic Canadian hospitals injury reporting and prevention program, 2011–2019. Health Promot Chronic Dis Prev Can 40(7–8):250–254. https://doi.org/10.24095/hpcdp.40.7/8.05

Miech RA, O'Malley PM, Johnston LD, Patrick ME (2016) E-cigarettes and the drug use patterns of adolescents. Nicot Tob Res 18(5):654–659. https://doi.org/10.1093/ntr/ntv217

Mirbolouk M, Charkhchi P, Kianoush S, Uddin SMI, Orimoloye OA, Jaber R, Bhatnagar A, Benjamin EJ, Hall ME, DeFilippis AP, Maziak W, Nasir K, Blaha MJ (2018) Prevalence and distribution of E-cigarette use among U.S. adults: behavioral risk factor surveillance system, 2016. Ann Intern Med 169(7):429–438. https://doi.org/10.7326/M17-3440

Morean ME, Kong G, Camenga DR, Cavallo DA, Krishnan-Sarin S (2015) High school students' use of electronic cigarettes to vaporize cannabis. Pediatrics 136(4):611–616. https://doi.org/10.1542/peds.2015-1727

Morley S, Slaughter J, Smith PR (2017) Death from ingestion of E-liquid. J Emerg Med 53(6):862–864. https://doi.org/10.1016/j.jemermed.2017.06.029

Nambiar P (2024) Vape, e-cigarette emergencies surge five-fold over 4 years, says poison centre. https://www.freemalaysiatoday.com/category/nation/2024/03/27/vape-e-cigarette-emergencies-surge-five-fold-over-4-years-says-poison-centre/

National Institutes of Health (2020) National health and morbidity survey 2019: NCDs. Ministry of Health Malaysia: Vol I (Issue September)

National Poisons Information Service (2014) Use of e-cigarettes (vapes) among adults in Great Britain (2021), XXXIV International Congress of the European Association of Poisons Centres and Clinical Toxicologists (EAPCCT) 27–30 May 2014, Brussels, Belgium

Obertova N, Navratil T, Zak I, Zakharov S (2020) Acute exposures to e-cigarettes and heat-not-burn products reported to the Czech toxicological information Centre over a 7-year period (2012–2018). Basic Clin Pharmacol Toxicol 127(1):39–46. https://doi.org/10.1111/bcpt.13393

Parekh T, Desai R (2020) A comparative analysis of E-cigarette users and state-specific prevalence change in the United States between 2017 and 2018. Cureus. https://doi.org/10.7759/cureus.12079

Pennings JLA, Havermans A, Pauwels CGGM, Krüsemann EJZ, Visser WF, Talhout R (2023) Comprehensive Dutch market data analysis shows that e-liquids with nicotine salts have both higher nicotine and flavour concentrations than those with free-base nicotine. Tob Control 32(e1):e78–e82. https://doi.org/10.1136/tobaccocontrol-2021-056952

Prieger JE, Kleiman M, Kulick J, Aziani A, Levi M, Hampsher S, Manning C, Hahn R (2019) The impact of E-cigarette regulation on the illicit trade in tobacco products in the European Union. SSRN Electron J. https://doi.org/10.2139/ssrn.3435177

Rahman Ahmed A (2022) A review of electronic cigarettes and liquid nicotine poisoning exposure cases in the United States. In J Pharm Pharm Sci 25. www.cspsCanada.org

Ravara S, Corrêa PC, Calheiros J, Pisinger C (2023) The public health impact of e-cigarette use: revisiting Geoffrey Rose's prevention strategies. Pulmonology 29(6):452–454. https://doi.org/10.1016/j.pulmoe.2023.04.008

Rossheim ME, Livingston MD, Soule EK, Zeraye HA, Thombs DL (2019) Electronic cigarette explosion and burn injuries, US emergency departments 2015–2017. Tob Control 28(4):472–474. https://doi.org/10.1136/tobaccocontrol-2018-054518

Rossheim ME, McDonald KK, Soule EK, Gimm GW, Livingston MD, Barnett TE, Jernigan DH, Thombs DL (2020) Electronic cigarette explosion/burn and poisoning related emergency department visits, 2018–2019. Am J Emerg Med 38(12):2637–2640. https://doi.org/10.1016/j.ajem.2020.08.017

Rudy SF, Durmowicz EL (2017) Electronic nicotine delivery systems: overheating, fires and explosions. Tob Control 26(1):10–18. https://doi.org/10.1136/tobaccocontrol-2015-052626

Seitz CM, Kabir Z (2018) Burn injuries caused by e-cigarette explosions: a systematic review of published cases. Tob Prev Cessat 4. https://doi.org/10.18332/tpc/94664

Sleiman M, Logue JM, Montesinos VN, Russell ML, Litter MI, Gundel LA, Destaillats H (2016) Emissions from electronic cigarettes: key parameters affecting the release of harmful chemicals. Environ Sci Technol 50(17):9644–9651. https://doi.org/10.1021/acs.est.6b01741

Sun Y, Prabhu P, Li D, McIntosh S, Rahman I (2024) Vaping: public health, social media, and toxicity. Online J Public Health Inf 16:e53245. https://doi.org/10.2196/53245

Tehrani H, Rajabi A, Ghelichi-Ghojogh M, Nejatian M, Jafari A (2022) The prevalence of electronic cigarettes vaping globally: a systematic review and meta-analysis. Arch Public Health 80(1). https://doi.org/10.1186/s13690-022-00998-w

Tzortzi A, Kapetanstrataki M, Evangelopoulou V, Beghrakis P (2020) A systematic literature review of E-cigarette-related illness and injury: not just for the respirologist. Int J Environ Res Public Health 17(7). https://doi.org/10.3390/ijerph17072248

Vardavas CI, Girvalaki C, Filippidis FT, Oder M, Kastanje R, De Vries I, Scholtens L, Annas A, Plackova S, Turk R, Gruzdyte L, Rato F, Genser D, Schiel H, Balázs A, Donohoe E, Vardavas AI, Tzatzarakis MN, Tsatsakis AM, Behrakis PK (2017) Characteristics and outcomes of e-cigarette exposure incidents reported to 10 European poison centers: a retrospective data analysis. Tob Induc Dis 15(1):36. https://doi.org/10.1186/s12971-017-0141-z

Vardavas CI, Girvalaki C, Odani S, Nikitara K, de Vries I, van Riel A, van Sommeren-de Potter I, Grass JN, Grassi MC, Deim S, Balázs A, Fosztó S, Schiel H, Arif T, Eronen AK, Alonso AA, Menor JLC, Arrieta RM, Babić Ž et al (2021) Profile of incidental exposures to e-cigarette liquids in Europe, 2018–2019. Hum Exp Toxicol 40(6). https://doi.org/10.1177/0960327120975828

Wang TW, Gentzke A, Sharapova S, Cullen KA, Ambrose BK, Jamal A (2018) Tobacco product use among middle and high school students—United States, 2011–2017. MMWR Morb Mortal Wkly Rep 67(22). https://doi.org/10.15585/mmwr.mm6722a3

Wang B, Liu S, Persoskie A (2020) Poisoning exposure cases involving e-cigarettes and e-liquid in the United States, 2010–2018. Clin Toxicol 58(6). https://doi.org/10.1080/15563650.2019.1661426

Wang L, Wang Y, Chen J, Liu P, Li M (2022) A review of toxicity mechanism studies of electronic cigarettes on respiratory system. Int J Mol Sci 23(9). https://doi.org/10.3390/ijms23095030

XXXIV international congress of the European Association of Poisons Centres and Clinical Toxicologists (EAPCCT) 27–30 May 2014, Brussels, Belgium (2014) Clin Toxicol 52(4):295–443. https://doi.org/10.3109/15563650.2014.906213

Yingst JM, Bordner CR, Hobkirk AL, Hoglen B, Allen SI, Krebs NM, Houser KR, Livelsberger C, Foulds J (2021) Response to flavored cartridge/pod-based product ban among adult juul users: "you get nicotine however you can get it". Int J Environ Res Public Health 18(1):1–13. https://doi.org/10.3390/ijerph18010207

Exploring E-cigarette Metabolomics: Insights from In Vitro, In Vivo, and Clinical Investigations

7

Salaudeen Abdulwasiu Olawale,
Firli Rahmah Primula Dewi, and Vuanghao Lim

Abstract

Globally, the use of electronic cigarettes (e-cigs) has skyrocketed in the past several years. The impacts of e-cigs on the health of both humans and animals must be clarified. Thus, this chapter aims to provide a brief summary of findings from in vitro, in vivo, and clinical research that has drawn attention to the possible genotoxic and cytotoxic effects of e-cig vapor. The mechanism of action of e-cig vapor exposure is unclear, although its adverse effect on health involves the generation of trace metals and oxygen radicals and increased inflammation due to the propylene glycol (PG) and vegetable glycerine (VG) contents of e-liquid, both of which have the potential to be cytotoxic and genotoxic. Few studies on the possible consequences of e-cig usage that combine clinical, in vivo, and in vitro methods for long-term inhalation toxicity evaluations are available; hence, this chapter is imperative. The findings from this research revealed the harmful effects of e-cig aerosols on human health while also suggesting potential benefits. Exposure to e-cig vapor in a human model has been shown to alter pulmonary and vascular functions, interfere with lung homeostasis, increase the risk of myocardial infarction, and exacerbate inflammation and oxidative stress.

S. A. Olawale
Advanced Medical and Dental Institute, Universiti Sains Malaysia, Bertam, Penang, Malaysia

Chemistry Unit of Applied Mathematics Programme, National Mathematical Centre, Abuja, Nigeria

F. R. P. Dewi
Department of Biology, Faculty of Science and Technology, Universitas Airlangga, Surabaya, Indonesia

V. Lim (✉)
Advanced Medical and Dental Institute, Universiti Sains Malaysia, Bertam, Kepala Batas, Penang, Malaysia
e-mail: vlim@usm.my

Y.-H. Leong, A. S. Mohd Samin (eds.), *E-Cigarettes*,
https://doi.org/10.1007/978-981-96-5087-3_7

Finally, e-cig use alters gene expression and molecular pathway regulation. According to the data acquired, e-liquid exposure may be hazardous to cells in vitro and to people and animals in vivo.

Keywords

e-cigarette · In vivo · In vitro · Aerosol · Nicotine · e-liquid

7.1 Introduction

Electronic cigarettes (e-cigs) became increasingly popular in the early 2000s as people recognized that they were safer alternatives to conventional cigarettes. In 2007, they were granted global patent rights (Regan et al. 2013). That same year, e-cigs were released into the US market, and smokers subsequently came to embrace them (Sapru et al. 2020). According to numerous investigations, e-cigs are the best smokeless tobacco product for helping smokers stop using tobacco (Chen et al. 2023). Drug delivery has recently been performed using e-cigs. For instance, high-power e-cigs were studied by Chaoui et al. (2022) and Pourchez et al. (2017) as potential substitute aerosol-generating devices for terbutaline administration.

The safety and long-term health impacts of e-cigs have become major public health issues due to their rapid popularity. Researchers have employed in vitro and in vivo models, which offer controlled environments to investigate reactions at cellular and organism levels, to comprehend the biological effects of e-cigs.

Through in vitro research, which involves studying cultured cells or tissues outside of a living organism, researchers can examine cellular toxicity, oxidative stress, and inflammatory responses caused by e-cig exposure. Studies have shown that aerosols from e-cigs can cause cell damage, DNA alterations, and inflammation in human lung cells (Gerloff et al. 2017). Other research has found that exposure to e-cig liquid induces oxidative stress and mitochondrial dysfunction in lung epithelial cells (Lerner et al. 2015). These findings indicate a risk of cellular damage, even in the absence of combustion, a process typically associated with conventional cigarettes.

In vivo investigations employ organisms, often animal models such as mice, to investigate systemic effects and interactions that may not be observed in isolated cells. These studies show that aerosols from e-cigs can lead to lung inflammation, compromised immune responses, and cardiovascular issues in rats (Sussan et al. 2015). Some animal studies have also documented changes in lung development, particularly when exposure occurs at critical developmental stages (McGrath-Morrow et al. 2015). Animal models shed information on physiological reactions, but because they might not precisely represent human reactions, care should be used when interpreting the results.

Together, in vitro and in vivo studies help to provide light on the potential health risks connected to e-cigs. The artificial conditions of these models, however, limit their application, underscoring the need for longer-term human studies to fully understand the impacts of e-cigarette use on public health.

A summary of the results from in vitro, in vivo, and clinical research is provided in this chapter. This is imperative as concerns regarding the health hazards to both people and animals have been raised by the increase in e-cigarette usage.

7.2 In Vitro Studies

Experiments conducted in vitro assess the direct effects of e-cig ingredients on tissues and cells by conducting tests outside of a living body, usually with the use of artificial systems or cell cultures, as shown in Table 7.1. Scientists have examined the cytotoxicity, oxidative stress, and inflammatory reactions induced by e-cig usage via in vitro experiments.

7.2.1 Studies on E-liquid Exposure to Cells

Analysis of the physiological consequences of e-liquid has traditionally been performed using cells. These investigations have used a range of methodologies and cell lines to evaluate various aspects of vapor exposure, including cell viability, DNA defects, oxygen radical levels, inflammatory mediator generation, cell shape alterations, cell function modification, gene expression disruption, and tumor-causing activities. Using e-cigs can damage living cells and their genomes. Vapor exposure can also cause oxidative stress, and exposure to vapors is directly associated with an excess of proinflammatory chemicals being produced (Merecz-Sadowska et al. 2020).

Furthermore, there is strong evidence linking vapor exposure to inflammation. Lerner discovered that human fetal lung fibroblasts (HFL1) and epithelial cells of the airway (H292) exposed to e-cig aerosols increased the levels of IL-6 and IL-8, leading to stress and morphological alterations (Lerner et al. 2015). Similar to tobacco smoke, e-cigs negatively affect the metabolome of normal human bronchial epithelial (NHBE) cells. These effects are mediated by oxidative stress (Aug et al. 2015). Vapor from flavorless e-cigs that contain nicotine can disrupt the oxidative stress response, among other negative consequences (Solleti et al. 2017), and the fatty acid triacylglycerol and phospholipid metabolism pathways (Shen et al. 2016); it has also been linked to apoptosis, DNA damage, gene deregulation (Bahmed et al. 2019), and the release of inflammatory proteins (Crotty Alexander et al. 2018; Garcia-Arcos et al. 2016). Inhaling nicotine has a significant impact on mitochondrial function (Kaisar et al. 2018), and the disruption of fluid balance affects pulmonary functioning (Otręba et al. 2018) and mucociliary function (Chung et al. 2019; Rayner et al. 2019). Exposure to e-liquids containing nicotine increases susceptibility to mutations (Lee et al. 2018) and phenotypic alterations (Aufderheide and Emura 2017). Furthermore, the majority of the cellular reaction to nicotine in e-cig vapor comes from its presence. Breathing in nicotine-containing vapors from e-cigs may have negative consequences. This study suggested that aerosols from e-cigs have a minimal mutagenic impact on human or transgenic mouse fibroblasts, despite

Table 7.1 Summary of the negative impacts of e-cigarette vapor exposure on cell lines. Adapted from Merecz-Sadowska et al. 2020, licensed under a Creative Commons Attribution 4.0 International License

Cell type	Nature of e-liquid	Action	Reference
NHBE cells	%PG:%VG-[N] = 50:50–36	– Airway surface liquid hydration decreased. – Viscosity of mucus increased	Chung et al. (2019)
NHBE cells	%PG:%VG-[N] = 55:45–0	– ATP levels inside cells decreased – Glycolysis and mitochondrial respiration are compromised.	Clapp et al. (2019)
NHBE cells THP-1 macrophages	%PG:%VG-[N] = 70:30–18	– Reduced: Efferocytosis, MIP-1α and MIP-1β, TNF-α, IL-6, IP-10 release – Increased: Necrosis and apoptosis	Ween et al. (2017)
Immortalized NHBE cell line—CL-1548	%PG:%VG-[N] = Unknown-0	Reduced: Club, ciliated, and mucus-producing cells [phenotypic alterations]	Aufderheide and Emura (2017)
Human lung carcinoma A549 cells	%PG:%VG-[N] = 50:50–18	Reduction in viability of cell	Otręba et al. (2018)
Human airway epithelium cells	PG:VG with 2.4% of nicotine	Variations in the morphology of secretory function	Carson et al. (2017)
Neutrophils	%PG:%VG-[N] = Unknown-16, 24	Enhanced: Expression of CD11b and CD66b, release of MMP-9 and CXCL8, activity of NE and MMP-9, and activation of p38 MAPK	Higham et al. (2016)

%PG:%VG-[N] = % propylene glycol:% vegetable glycerin–nicotine concentration in mg/L

the majority of in vitro studies suggesting detrimental effects of e-cigs on cells (Tommasi et al. 2017). No inflammatory, genotoxic, or cytotoxic impacts were observed in human lung epithelial cells (A549) subjected to an e-liquid concentration of 0–20 mg/mL over a 24-hour period (Misra et al. 2014). In a similar vein, opinions disagree about the potential harm that e-cig vapor might cause a distinct 3D coculture model of the human respiratory system.

Research using human bronchial epithelial cells (CALU-3) and pulmonary fibroblasts (MRC-5) as multicellular models of human cells has shown that exposure to components in e-liquids such as propylene glycol:vegetable glycerin (PG:VG),

16 mg/mL nicotine, and proprietary flavors has a significant effect on oxidative stress, proinflammatory mediator production, and cell viability (Vasanthi Bathrinarayanan et al. 2018).

7.3 In Vivo Studies

In vivo investigations have examined the physiological and biological effects of e-cig usage on living organisms, such as people and animals. Numerous studies have investigated how e-cig vapor affects the respiratory system, cardiovascular health, and lung function. This section summarizes the results of in vivo research on e-cigs, illuminating their effects on both animal and human health.

7.3.1 E-cig Liquid Exposure in Animal Models

The impacts of e-cig vapor on the respiratory system, including lung function, airway inflammation, and histological alterations, have been the subject of several in vivo investigations. Although some rat models have been utilized, mice have been the primary species used in experimentation. Typically, animals are subjected to an e-liquid that is composed of a base with varied nicotine doses and tastes. Following vapor exposure, the mechanics of the respiratory tract, mitochondrial activity, oxidative stress, genotoxicity, irritation, alterations within an organ or tissue structure and activity, alterations in behavior, abnormalities in genetic expression, modifications in the metabolism of the mother and her offspring, and protumor situations have all been identified.

It has been discovered that e-liquid can increase oxidative stress, cause mitochondrial malfunction, and change the physiology of the lung. Vapor from e-cigs can damage DNA, can cause inflammatory reactions, and is harmful to animal health. Thus, vaping might negatively affect liver and heart function. Mice also exhibited alterations in gene expression and behavior after being exposed to e-liquid vapor. Maternal e-cig usage during pregnancy significantly affects the health of offspring. Furthermore, gathered information suggests that when inhaled, e-liquid can be carcinogenic (Merecz-Sadowska et al. 2020).

E-cig vapor, cigarette smoke, and air were introduced to male C57BL6 mice in vivo. The e-cig vapors included either PG:VG or PG:VG with a nicotine content of 18 mg/mL (PG:VG-N) or PG:VG-N with a flavor (PG:VG-N + F). Following a three- or four-week exposure period, pulmonary mechanics, inflammatory marker, and oxidative stress measurements were taken. Muc5ac production and cellularity in bronchoalveolar lavage fluid (BALF) were shown to be enhanced by PG:VG-N + F, and the levels of lung oxidative stress indicators in BALF and BALF were found to be at least as high as those in the case of cigarette smoke. Although the synthesis of mucin (Muc5ac) is associated with airway hyperreactivity, the composition of BALF is associated with factors related to lung function. These findings unequivocally show that e-cig vapor can cause inflammatory reactions, negatively affect respiratory system processes, and negatively influence lung function (Glynos et al. 2018).

A review of the physiological effects of the nicotine delivery system on the lungs of mice subjected to vapor for an hour over a four-month period revealed that nicotine inhalation increases mucous cell enlargement, airspace expansion, and the discharge of inflammatory intermediaries, which are processes linked to the etiology and progression of COPD (Garcia-Arcos et al. 2016). In another study, mice exposed to aerosols from e-cigs showed elevated levels of proinflammatory cytokines (Lerner et al. 2015), along with enhanced ROS generation and airway inflammation (Sussan et al. 2015). In a rat lung model, e-cig vapor exposure was linked to an increase in the activity of the cytochrome P450 enzyme, excessively high lung free radical levels, decreased antioxidant capacity, and widespread DNA damage (Canistro et al. 2017). The laryngeal mucosa of Wistar albino rats after exposure was shown to be hyperplastic and metaplastic; nonetheless, these results lacked statistical significance (Salturk et al. 2015). Furthermore, e-cig vapor exposure has been connected to altered lung growth (McGrath-Morrow et al. 2015) and epigenetic and cognitive alterations in newborn mice (Nguyen et al. 2018). Exposure to e-cigs appears to modify lung physiology, which in turn affects the organism's general homeostasis. Pregnant individuals are advised against the use of e-cigs. Table 7.2 summarizes the side effects of animals' exposure to e-cig vapor. Crotty Alexander et al. (2018) exposed mice to aerosols from e-cigs and revealed elevated oxidative stress, inflammation, and airway resistance in the lungs. McGrath-Morrow et al. (2015) revealed signs of lung inflammation and compromised lung function when e-cig vapors were introduced to rats.

The cellular and molecular processes underlying the harmful consequences of e-cig exposure have been investigated in vivo. Increased oxidative stress, DNA damage, and changes in gene expression have all been shown in animal models (Lerner et al. 2015). Such molecular alterations may play a role in the emergence of cardiovascular and respiratory disorders, underscoring the need for in vivo research.

7.3.2 Human E-cig Exposure

There is still much to learn about how e-cigs affect human bodies. The consequences of using e-cigs and those caused by conventional smoking are still hard to distinguish (Prieto et al. 2002). Research on this topic shows possible advantages for e-cig aerosols while confirming that they have negative impacts on human health. In a human model, e-cig vapor exposure may affect vascular and pulmonary activities, disrupt lung homeostasis, increase the likelihood of heart attack, and increase inflammation and oxidative stress. Using e-cigs also affects the control of molecular pathways and gene expression.

In a human model, brief exposure to e-cig vapor seems to increase peripheral tracheal resistance, overall resistance to breathing, and respiratory flow resistance while lowering the amount of nitric oxide exhaled. In a test sample of healthy

Table 7.2 Summary of the effects of e-cigarettes on animals. (Adapted from Pisinger and Døssing (2014))

Animal	Sample	Duration of exposure	Action	Reference
Mice	Vapor	5 h for 3 days	Inflammation of the airways, anomalies in the structure of the lungs that point to airway remodeling	Lerner et al. (2015)
Mice	Aerosol	2 h daily for 3 days	Elevated blood pressure and compromised function of endothelium	Wang et al. (2019)
Lactating rats	Aerosol	1 h daily for 16 days	During breastfeeding, the aerosol from e-cigarettes greatly decreased the amount of fat in the Milk. Milk volume and protein content both increased significantly, but neither gain was statistically significant. The aerosol from e-cigarettes had no effect on the lactose level of the milk.	Al-Sawalha et al. (2021)
Mice	Vapor	1.5 h twice per day for 14 days	In comparison to mice exposed to air, animals exposed to the vapor of e-cigarettes had markedly reduced lung bacterial clearance.	Sussan et al. (2015)
Rats	Aerosol	3 h per day for 3 days	Caused oxidative stress in the lungs without causing alterations to the pulmonary inflammatory response or lung structure that could be seen	Wang et al. (2020)

smokers, e-cigs were shown to have a detrimental effect on the respiratory system comparable to smoking tobacco (Vardavas et al. 2012).

Using an untargeted metabolomics method, Assiri et al. (2024) examined the metabolic alterations caused by e-cig ingestion in human lung epithelial cells (A549). After exposing A549 cells to escalating doses of e-cig vapor extract, metabolomic alterations, oxidative stress, and cell viability were evaluated. According to previous research, e-cigs increase oxidative stress and cause cell death in a concentration-dependent manner. Metabolomics studies have shown that e-cigs cause distinct metabolic alterations in important cellular metabolic processes. According to the findings, exposure to e-cigs clearly caused segregation in metabolic responses, which are mostly regulated by a variety of vital metabolites, including pyruvate, amino acids, fatty acids, and glutathione. The metabolomics data revealed an intriguing pattern of unique metabolic alterations generated by e-cigs at every dose, indicating the complexity of e-cig-mediated cytotoxicity. A number of vital physiological processes, including fatty acid biosynthesis; glutathione, pyruvate, nicotinate, and nicotinamide metabolism; and amino acid metabolism, have been associated with disrupted metabolites. These findings underscore the need for more studies to completely understand the consequences of e-cig use over time and its possible negative impacts on cellular metabolism. Overall, in A549 lung epithelial cells, e-cigs not only cause oxidative stress and cell death but also interfere with cellular metabolism (Assiri et al. 2024).

7.4 Clinical Studies

Clinical investigations are essential for comprehending the impact of e-cigs on humans. In these investigations, the health consequences of e-cig use were observed and analyzed. Researchers have gathered information on a range of factors, including alterations in lung function, symptoms related to the respiratory system, cardiovascular health, and overall well-being. Polosa et al. (2016a, b) conducted a clinical investigation to assess the long-term consequences of e-cig usage in asthmatic smokers. According to the study, participants who transitioned from smoking to e-cig use showed improvements in lung function and respiratory symptoms. Additional clinical trials are included in Table 7.3.

Policymakers are informed by clinical research on the possible advantages and disadvantages of using e-cigs. The creation of suitable rules is now a challenge for

Table 7.3 Findings from clinical studies of e-cigs

Clinical study to:	Findings	Reference
Estimate the amount of nicotine exposure in teenagers who said they only used e-cigs	Every examined product had a high nicotine concentration. Users who use some prefilled e-cig systems may inhale a lot of nicotine.	Goniewicz et al. (2019)
Evaluate the traits and users experiences with e-cig and investigate the variations among individuals who used e-cig to partly and totally replace smoking	Of the participants, 81.0% were former smokers who reported completely quitting smoking, whereas current smokers reported cutting back on their daily cigarette intake from 20 to 4.	Farsalinos et al. (2014)
Examine the effects on lung health of smokers with COPD diagnoses who stopped using tobacco products or cut back significantly by using e-cigs on a daily basis.	When using e-cigs, cigarette consumption was shown to have significantly decreased. Within the COPD e-cigs user group, there was a noteworthy decrease in COPD exacerbations.	Polosa et al. (2016a, b)
Ascertain the relative efficacy of e-cigs relative to other alternative smoking cessation techniques.	Higher rates of cessation among e-cig users than among those utilizing nicotine replacement treatment	Hajek et al. (2019)
Investigate the correlation between using e-cigs use and subsequent initiation of smoking	Teenagers using e-cigs had a higher chance of starting to smoke later in life than those who did not, which raises serious concerns about e-cigs' impacts on public health.	Leventhal et al. (2015)
Examine the smoking characteristics of conventional and e-cigs.	While conventional cigarettes showed consistent smoking behavior across brands, the aerosol density of e-cigs decreased with successive use.	Trtchounian, et al. (2010)
Examine the relationship between the usage of e-cigs, the perception of their harmfulness, and the symptoms of anxiety and depression.	The usage of e-cigs was substantially correlated with anxiety/depression, the perception that using them is hazardous, and the relationships between these factors.	Adzrago et al. (2023)

governments and regulatory agencies throughout the globe. The aforementioned research offers insightful information on the many factors to take into account when creating policies pertaining to e-cigs.

7.5 Conclusion

In conclusion, e-cig usage may pose a health risk to the general population based on a rapidly expanding body of information from in vitro, in vivo, and clinical investigations; however, scientific data about the impact of e-cigs on human health are limited. This chapter is based on a review of the body of research on e-cig toxicity. The World Health Organization (WHO) emphasizes the severe health risks associated with e-cigs. Studies examining whether e-cigs are less dangerous than cigarettes are not definitive, even though aerosols from e-cigs may contain fewer toxicants than smoke from cigarettes. Long-term usage and frequent inhalation are safe, although proof is still lacking.

Furthermore, as the health impacts of e-cig use are unlikely to be realized for some time, reliable information should be provided to adequately inform the public about e-cig usage. Finally, a deeper understanding of the long-term impacts and possible health connections of e-cig users might be gained by implementing longitudinal studies to track their health outcomes. Future clinical trials should involve a larger range of individuals from a variety of demographic backgrounds to obtain a deeper understanding of the ways in which e-cig usage affects diverse people.

References

Adzrago D, Fujimoto K, Harrell MB, Jones A, Wilkerson JM (2023) Association between E-cigarette use behaviors and perceived harmfulness of e-cigarettes and anxiety/depression symptoms among black/African American adults. Prev Med Rep 31(February):102080

Al-Sawalha NA, Bdeir R, Sohaib A, Saad M, Inghaimesh T, Khabour OF, Alzoubi KH, Shihadeh A (2021) Effect of E-cigarettes aerosol exposure during lactation in rats: hormonal and biochemical aspects. Environ Toxicol Pharmacol 88(November):103759

Assiri MA, Al Jumayi SR, Alsuhaymi S, Emwas AH, Jaremko M, Alsaleh NB, Al Mutairi MM, Alshamrani AA, As Sobeai H, Alghibiwi H (2024) Electronic cigarette vapor disrupts key metabolic pathways in human lung epithelial cells. Saudi Pharm J 32(1) (January 1)

Aufderheide M, Emura M (2017) Phenotypical changes in a differentiating immortalized bronchial epithelial cell line after exposure to mainstream cigarette smoke and E-cigarette vapor. Exp Toxicol Pathol 69)(6):393–401

Aug A, Altraja S, Kilk K, Porosk R, Soomets U, Altraja A (2015) E-cigarette affects the metabolome of primary Normal human bronchial epithelial cells. PLoS One 10(11):e0142053

Bahmed K, Lin C-R, Simborio H, Karim L, Aksoy M, Kelsen S, Tomar D et al (2019) The role of DJ-1 in human primary alveolar type II cell injury induced by e-cigarette aerosol. Am J Physiol Lung Cell Mol Physiol 317(4):L475–L485

Canistro D, Vivarelli F, Cirillo S, Babot Marquillas C, Buschini A, Lazzaretti M, Marchi L et al (2017) E-cigarettes induce toxicological effects that can raise the cancer risk. Sci Rep 7(1):2028

Carson JL, Zhou L, Brighton L, Mills KH, Zhou H, Jaspers I, Hazucha M (2017) Temporal structure/function variation in cultured differentiated human nasal epithelium associated with acute single exposure to tobacco smoke or E-cigarette vapor. Inhal Toxicol 29(3):137–144

Chaoui M, Chevrel S, Perinel-Ragey S, Prévôt N, Pourchez J (2022) Assessment of high-power electronic nicotine delivery system as an alternative aerosol device for terbutaline delivery. Pharm Res 39(3):587–597

Chen R, Pierce JP, Leas EC, Benmarhnia T, Strong DR, White MM, Stone M, Trinidad DR, McMenamin SB, Messer K (2023) Effectiveness of E-cigarettes as aids for smoking cessation: evidence from the PATH study cohort, 2017–2019. Tob Control 32(e2):e145–e152

Chung S, Baumlin N, Dennis JS, Moore R, Salathe SF, Whitney PL, Sabater J, Abraham WM, Kim MD, Salathe M (2019) Electronic cigarette vapor with nicotine causes airway Mucociliary dysfunction preferentially via TRPA1 receptors. Am J Respir Crit Care Med 200(9):1134–1145

Clapp PW, Lavrich KS, van Heusden CA, Lazarowski ER, Carson JL, Jaspers I (2019) Cinnamaldehyde in flavored E-cigarette liquids temporarily suppresses bronchial epithelial cell ciliary motility by dysregulation of mitochondrial function. Am J Physiol Lung Cell Mol Physiol 316(3):L470–L486

Crotty Alexander LE, Drummond CA, Hepokoski M, Mathew D, Moshensky A, Willeford A, Das S et al (2018) Chronic inhalation of E-cigarette vapor containing nicotine disrupts airway barrier function and induces systemic inflammation and multiorgan fibrosis in mice. American J Physiol Regulatory Integrative Comparative Physiol 314(6):R834–R847

Farsalinos K, Romagna G, Tsiapras D, Kyrzopoulos S, Voudris V (2014) Characteristics, perceived side effects and benefits of electronic cigarette use: a worldwide survey of more than 19,000 consumers. Int J Environ Res Public Health 11(4):4356–4373

Garcia-Arcos I, Geraghty P, Baumlin N, Campos M, Dabo AJ, Jundi B, Cummins N et al (2016) Chronic electronic cigarette exposure in mice induces features of COPD in a nicotine-dependent manner. Thorax 71(12):1119–1129

Gerloff J, Sundar IK, Freter R, Sekera ER, Friedman AE, Robinson R, Pagano T, Rahman I (2017) Inflammatory response and barrier dysfunction by different E-cigarette flavoring chemicals identified by gas chromatography–mass spectrometry in e-liquids and e-Vapors on human lung epithelial cells and fibroblasts. Appl In Vitro Toxicol 3(1):28–40

Glynos C, Bibli S-I, Katsaounou P, Pavlidou A, Magkou C, Karavana V, Topouzis S, Kalomenidis I, Zakynthinos S, Papapetropoulos A (2018) Comparison of the effects of E-cigarette vapor with cigarette smoke on lung function and inflammation in mice. American J Physiology-Lung Cell Mol Physiol 315(5):L662–L672

Goniewicz ML, Boykan R, Messina CR, Eliscu A, Tolentino J (2019) High exposure to nicotine among adolescents who use Juul and other vape pod systems ("pods"). Tob Control 28(6):676–677

Hajek P, Phillips-Waller A, Przulj D, Pesola F, Myers Smith K, Bisal N, Li J et al (2019) A randomized trial of E-cigarettes versus nicotine-replacement therapy. N Engl J Med 380(7):629–637

Higham A, Rattray NJW, Dewhurst JA, Trivedi DK, Fowler SJ, Goodacre R, Singh D (2016) Electronic cigarette exposure triggers neutrophil inflammatory responses. Respir Res 17(1):56

Kaisar MA, Sivandzade F, Bhalerao A, Cucullo L (2018) Conventional and electronic cigarettes dysregulate the expression of iron transporters and detoxifying enzymes at the brain vascular endothelium: in vivo evidence of a gender-specific cellular response to chronic cigarette smoke exposure. Neurosci Lett 682(August):1–9

Lee H-W, Park S-H, Weng M, Wang H-T, Huang WC, Lepor H, Wu X-R, Chen L-C, Tang M (2018) E-cigarette smoke damages DNA and reduces repair activity in mouse lung, heart, and bladder as well as in human lung and bladder cells. Proc Natl Acad Sci 115(7)

Lerner CA, Sundar IK, Yao H, Gerloff J, Ossip DJ, McIntosh S, Robinson R, Rahman I (2015) Vapors produced by electronic cigarettes and E-juices with flavorings induce toxicity, oxidative stress, and inflammatory response in lung epithelial cells and in mouse lung. PLoS One 10(2):e0116732

Leventhal AM, Strong DR, Kirkpatrick MG, Unger JB, Sussman S, Riggs NR, Stone MD, Khoddam R, Samet JM, Audrain-McGovern J (2015) Association of Electronic Cigarette use with Initiation of combustible tobacco product smoking in early adolescence. JAMA 314(7):700

McGrath-Morrow SA, Hayashi M, Aherrera A, Lopez A, Malinina A, Collaco JM, Neptune E et al (2015) The effects of electronic cigarette emissions on systemic cotinine levels, weight and postnatal lung growth in neonatal mice. PLOS One 10(2):e0118344

Merecz-Sadowska A, Sitarek P, Zielinska-Blizniewska H, Malinowska K, Zajdel K, Zakonnik L, Zajdel R (2020) A summary of in vitro and in vivo studies evaluating the impact of E-cigarette exposure on living organisms and the environment. Int J Mol Sci. MDPI AG 21

Misra M, Leverette R, Cooper B, Bennett M, Brown S (2014) Comparative in vitro toxicity profile of electronic and tobacco cigarettes, smokeless tobacco and nicotine replacement therapy products: E-liquids, extracts and collected aerosols. Int J Environ Res Public Health 11(11):11325–11347

Nguyen T, Li GE, Chen H, Cranfield CG, McGrath KC, Gorrie CA (2018) Maternal E-cigarette exposure results in cognitive and epigenetic alterations in offspring in a mouse model. Chem Res Toxicol 31(7):601–611

Otręba M, Kośmider L, Knysak J, Warncke JD, Sobczak A (2018) E-cigarettes: voltage- and concentration-dependent loss in human lung adenocarcinoma viability. J Appl Toxicol 38(8):1135–1143

Pisinger C, Døssing M (2014) "A systematic review of health effects of electronic cigarettes." Prev Med 69(December):248–260. https://doi.org/10.1016/j.ypmed.2014.10.009

Polosa R, Morjaria JB, Caponnetto P, Caruso M, Campagna D, Amaradio MD, Ciampi G, Russo C, Fisichella A (2016a) Persisting long term benefits of smoking abstinence and reduction in asthmatic smokers who have switched to electronic cigarettes. Discov Med 21(114):99–108

Polosa R, Morjaria JB, Caponnetto P, Prosperini U, Russo C, Pennisi A, Bruno CM (2016b) Evidence for harm reduction in COPD smokers who switch to electronic cigarettes. Respir Res 17(1):166

Pourchez J, de Oliveira F, Perinel-Ragey S, Basset T, Vergnon J-M, Prévôt N (2017) Assessment of new-generation high-power electronic nicotine delivery system as thermal aerosol generation device for inhaled bronchodilators. Int J Pharm 518(1–2):264–269

Prieto L, Gutiérrez V, Cervera A, Liñana J (2002) Airway obstruction induced by inhaled acetaldehyde in asthma: repeatability relationship to adenosine 5'-monophosphate responsiveness. J Investig Allergol Clin Immunol 12(2):91–98

Rayner RE, Makena P, Prasad GL, Cormet-Boyaka E (2019) Cigarette and ENDS preparations differentially regulate ion channels and Mucociliary clearance in primary Normal human bronchial 3D cultures. American J Physiol-Lung Cell Mol Physiol 317(2):L295–L302

Regan AK, Promoff G, Dube SR, Arrazola R (2013) Electronic nicotine delivery systems: adult use and awareness of the "e-cigarette" in the USA. Tob Control 22(1):19–23

Salturk Z, Çakır Ç, Sünnetçi G, Atar Y, Kumral TL, Yıldırım G, Berkiten G, Uyar Y (2015) Effects of electronic nicotine delivery system on larynx: experimental study. J Voice 29(5):560–563

Sapru S, Vardhan M, Li Q, Guo Y, Li X, Saxena D (2020) E-cigarettes use in the United States: reasons for use, perceptions, and effects on health. BMC Public Health 20(1):1518

Shen Y, Wolkowicz MJ, Kotova T, Fan L, Timko MP (2016) Transcriptome sequencing reveals E-cigarette vapor and mainstream-smoke from tobacco cigarettes activate different gene expression profiles in human bronchial epithelial cells. Sci Rep 6(1):23984

Solleti SK, Bhattacharya S, Ahmad A, Wang Q, Mereness J, Rangasamy T, Mariani TJ (2017) MicroRNA expression profiling defines the impact of electronic cigarettes on human airway epithelial cells. Sci Rep 7(1):1081

Sussan TE, Gajghate S, Thimmulappa RK, Ma J, Kim J-H, Sudini K, Consolini N et al (2015) Exposure to electronic cigarettes impairs pulmonary anti-bacterial and anti-viral defenses in a mouse model. PLoS One 10(2):e0116861

Tommasi S, Bates SE, Behar RZ, Talbot P, Besaratinia A (2017) Limited mutagenicity of electronic cigarettes in mouse or human cells in vitro. Lung Cancer 112(October):41–46

Trtchounian A, Williams M, Talbot P (2010) Conventional and electronic cigarettes (e-cigarettes) have different smoking characteristics. Nicotine Tob Res 12(9):905–912

Vardavas CI, Anagnostopoulos N, Kougias M, Evangelopoulou V, Connolly GN, Behrakis PK (2012) Short-term pulmonary effects of using an electronic cigarette. Chest 141(6):1400–1406

Vasanthi Bathrinarayanan P, Brown JEP, Marshall LJ, Leslie LJ (2018) An investigation into E-cigarette cytotoxicity in-vitro using a novel 3D differentiated co-culture model of human airways. Toxicol In Vitro 52(October):255–264

Wang Q, Ahmad Khan N, Muthumalage T, Lawyer GR, McDonough SR, Der Chuang T, Gong M, Sundar IK, Rehan VK, Rahman I (2019) Dysregulated repair and inflammatory responses by E-cigarette-derived inhaled nicotine and humectant propylene glycol in a sex-dependent manner in mouse lung. FASEB BioAdvances 1(10):609–623

Wang J, Zhang T, Johnston CJ, Kim S-Y, Gaffrey MJ, Chalupa D, Feng G, Qian W-J, McGraw MD, Ansong C (2020) Protein thiol oxidation in the rat lung following E-cigarette exposure. Redox Biol 37(October):101758

Ween MP, Whittall JJ, Hamon R, Reynolds PN, Hodge SJ (2017) Phagocytosis and inflammation: exploring the effects of the components of E-cigarette vapor on macrophages. Physiol Rep 5(16):e13370

From Analytical Platforms to Biomarkers: Metabolomic Analysis of E-cigarette-Related Samples

8

Vuanghao Lim

Abstract

In recent years, electronic cigarettes (e-cigs) have gained popularity worldwide. Several methods of sample preparation for metabolomics have been reported. However, the analysis of e-cigs by using in vitro/in vivo/clinical samples varies from sample to sample according to the instrument used. Experimental records suggest that exposure to e-liquid may be deleterious to cells in vitro and to animals and humans in vivo. Similarly, different platforms, such as liquid chromatography-mass spectrometry (LCMS), nuclear magnetic resonance (NMR), and capillary electrophoresis (CE), have been employed for analyzing e-cig sample matrices, each of which has specific sample preparation procedures and data analysis techniques. Furthermore, potential biomarkers indicating the early pathogenesis of particular e-cig-associated diseases have been discovered. This research provides an overview of the different technologies used in studying metabolomics in relation to e-cigs. This study highlights how different studies can obtain reliable results by utilizing available software and databases.

Keywords

NMR · LCMS · Ce-MS · Metabolomics · Sampling techniques

V. Lim (✉)
Advanced Medical and Dental Institute, Universiti Sains Malaysia, Bertam, Kepala Batas, Penang, Malaysia
e-mail: vlim@usm.my

Y.-H. Leong, A. S. Mohd Samin (eds.), *E-Cigarettes*,
https://doi.org/10.1007/978-981-96-5087-3_8

8.1 Introduction

Appropriate preparatory measures for investigating e-cigs are essential for practical investigations of metabolites. However, in practice, this aspect is often absent in studies dealing with this technology, even though a sample of high quality and the proper analysis used in subsequent data processing are determined by the sample's background and preparation methods. Hence, every single equilibrium of the metabolites, which are of different quantities and substructures, must be retained to extract unique features that equate to the qualitative execution of e-cig research by means of metabolite analysis. A key part of this process is the correct choice of various sample types and sizes as a wheel as the ensemble of the subsequent analyzed compounds (Zucchet and Schmaltz 2017).

E-cig metabolites are difficult to detect because of the few methods used for identification, detection, and quantification. Metabolomics techniques should be optimized because different metabolites can have different concentrations. This approach is commonly achieved by conducting NMR, gas chromatography-mass spectrometry (GCMS), or LCMS hybrid operations. Although much effort has helped to improve identification, detection, and data treatment, some aspects, such as the time-consuming preparation of samples, are key reasons that new methods have not been developed in the field of analytical chemistry thus far (Titz et al. 2020). The greatest challenge faced during the preparation of samples for metabolomic analysis of e-cigs lies in the absence of any generally accepted references or traditional procedures applicable universally. This gap is mainly attributed to diverse usage, variability in stimulation, and the composition of metabolites; hence, a one-size-fits-all solution is virtually nonexistent. For example, protocols used for disrupting metabolism when extracting in vitro or in vivo e-cig samples may require each product type to be optimized to avoid compromising cellular membrane integrity (Merecz-Sadowska et al. 2020).

Reliable methods must be established to replicate e-cig samples from different laboratories. This chapter seeks to provide complete guidelines for the preparation of metabolomic bio samples of e-cigs prior to analysis via chromatography or electrophoresis, which should be useful to all researchers in this area of interest. To date, much attention has been given to the detection process for techniques such as NMR and MS, which are commonly coupled with GC or CE for metabolomics analysis of e-cigs. However, extensive studies have focused only on such samples. The objective of this chapter is to demonstrate why a broader investigation should be conducted to fill existing gaps. Additionally, a brief description of the various methods used in sample preparations to identify the compounds found in e-cigs is included.

Detailed information from selecting the desired material to the essential manipulations before actual extraction is provided after this initial stage has been completed. The next step involves procedures followed during the extraction itself, eventually leading to analyses conducted with the use of relevant equipment. Subsequent procedures involved equipment analysis.

8.2 NMR Analysis

In vivo: This study explored the effects of both acute and chronic smoke exposure on the lungs of mice infected with five different strains. Three strains were susceptible to chronic obstructive pulmonary disease (COPD), while the other two were resistant. The main goal of this research is to understand how various genetic backgrounds can alter the respiratory health of mice exposed to fumes and to study ways to reduce their impact. A Bruker spectrometer operating at 500.13 MHz was used for tissue analysis. Additionally, the authors investigated the differences in metabolic profiles between susceptible and resistant animals and explored the effects of smoke on these profiles (Izquierdo-García et al. 2010).

The researchers selected six different chemical shifts from the ^{1}H NMR spectrum. Taurine, glutathione, phosphorylcholine, taurine, and glycerophosphocholine are chemicals that are specific to COPD. An enduring predictive model of COPD that uses these chemicals was created. Multivariate analysis revealed that the model could classify diseased mice accurately in 100% of the patients and healthy mice in 78%. This research indicates significant advances in understanding the influence of tobacco smoke on noninvasive samples and could aid in the development of new COPD biomarkers. It can be concluded from the disclosed data that the success of the study lies in obtaining informative information from the approach and reducing metabolic breakdown (Izquierdo-García et al. 2010).

E-cig sample: The application of NMR spectroscopy for detecting multiple different components in e-liquids is efficient and effective. Researchers are interested in discovering the kinds of substances (e.g., solvents, nicotine, and bioactive flavor compounds) that are present in products via NMR spectroscopy. The danger levels of certain elements must be determined. Estimated exposures must be compared against recognized toxic thresholds through probabilistic exposure estimation with a margin of exposure (MOE) approach, which is commonly employed within risk assessments that calculate safety margins based on differences between exposure levels estimated from data and those that would cause harm at some point or other. During the research, two methods of sample preparation were also adopted to obtain complete quantitative information on whether the product was composed of these particular products: (1) In this study, a Bruker Avance 400 Ultrashield spectrometer equipped with Z-gradient coils and a 5 mm SEI probe was used for NMR measurements. (2) The quality of the results was confirmed by carrying out linear regressions over ranges covering concentrations observed during analysis. Thus, this work not only confirms what e-liquid constitutes but also demonstrates potential risk factors (Hahn et al. 2014).

Even though every compound tested presented relatively low risks associated with its use, nicotine stood out as the only one showing high risk with a MOE of <1. This finding underscores the need to regulate e-liquid to guarantee consistent delivery of nicotine. Solvents with better toxicological profiles than ethylene glycol and 1,2-propanediol are generally recommended because they may also fall under the high-risk category if their MOE is <100. These factors can ensure the increased safety and dependability of e-liquid (Hahn et al. 2014).

Tobacco: A new technique has been developed for detecting counterfeit tobacco products and distinguishing them from different genuine international brands. The method uses high-resolution magic angle spinning nuclear magnetic resonance (HRMAS NMR) spectroscopy and chemometrics. In this study, six different types of cigarettes manufactured from various blends sold internationally were investigated. A Bruker Avance spectrometer was used to reveal all the knockoffs associated with each brand. Once identified, these imitations were thoroughly probed using the stated toolkits to establish their respective properties and traits. This approach was able to establish profound insights into the material composition of a tobacco product, including the detection of counterfeits. By applying unsupervised multivariate statistical analysis to the obtained results, counterfeit cigarettes can be distinguished from genuine cigarettes. According to this research, unauthorized products contained relatively high levels of both nicotine and solanesol but relatively low amounts of sugars, which are natural derivatives found within the leaves of plants of the genus *Nicotiana*, including those used to produce cigars or pipe tobacco blends. These findings show how powerful HRMAS NMR can be when used as a means to verify the authenticity of any pack (Shintu et al. 2013).

Humans: Progress has been slow in the study of salivary metabolomics given its high cost, large size, and the need for specialized technical personnel at high-cost NMR facilities. However, low-field NMR spectrometers (60 MHz) have practical advantages for investigating the metabolic profiles of human saliva. Studies have shown that such profiles are able to differentiate between smokers and nonsmokers. According to the research findings, the use of benchtop ^{1}H NMR analysis techniques is a promising method for carrying out bioanalytical and metabolomics investigations. Furthermore, the authors explore potential applications of this non-steady state technique with a special focus on its utility in "point-of-care" testing by using samples from within the mouth. Researchers have proposed that this process be used as part of routine dental check-ups at community pharmacies. Early identification of any disease affecting the teeth or surrounding tissues will facilitate less complicated treatment. In dental surgeries alone, where diagnosis would otherwise be limited given the absence of diagnostic tools other than visual examination, it has the potential to revolutionize both diagnosis and treatment provision systems. Therefore, the adoption of such an approach within studies on salivary metabolomics presents great opportunities given its potential to create less expensive methods that are also more efficient (Percival et al. 2022).

8.3 GC–MS and LC–MS Analyses

Humans: Examining the link between smoking and metabolites is crucial for understanding smoking-related diseases and addiction. However, prior studies were limited in terms of scope, size, and target metabolites. For this issue to be addressed, an untargeted metabolomics analysis was conducted on 892 male and female participants from four studies in Italy, the United States, China, and Finland. Linear regression associated individual metabolites with smoking status and the number of cigarettes smoked per day. Associations with smoking duration, years since quitting, and former vs. current smokers were also evaluated in the US study. Specimens were analyzed using UHPLC and GC with mass spectrometry at Metabolon. Peaks were matched to a database of more than 2500 standards to identify metabolites. These findings provide insights into the biological processes underlying smoking-related illnesses and addiction. LCMS metabolomics has revealed associations between smoking and various metabolites. These findings may aid in understanding smoking-related diseases, but further validation is needed (Gu et al. 2016).

E-cig sample: Interestingly, a precise technique for measuring emerging chemicals of concern (ECCs) in e-liquids is currently used in electronic nicotine delivery systems. It utilizes high performance liquid chromatography mass spectrometry (HPLC-MS), a sophisticated combination approach. This method can detect and quantify various compounds in e-liquids. This highly accurate analysis takes only 10 minutes, and it has a wide dynamic linearity range with detection limits from 0.020 to 10 ng/mL. This reliable tool efficiently quantifies ECCs in e-liquid samples. This approach is ideal for future research in this area, with an average accuracy within ±15% (Wei et al. 2019).

In vitro: An extensive study to examine the effects of exposure to e-liquid on human bronchial epithelial cells (HBECs) has been conducted by a group of researchers. This study also aimed to determine whether antioxidants can protect these cells from potential risks associated with such exposure. Researchers have employed a precipitation technique using mixtures of methanol and water to isolate proteins from cell lysates. The resulting supernatants were then analyzed on a QTRAP 3200 mass spectrometer in both positive and negative enhanced mass scan modes. The samples were analyzed through isocratic flow to obtain complete spectra. This optimization process ensured accurate and reliable analysis. Negative scans were performed by applying the respective negative voltages. This study revealed that e-liquid significantly altered the metabolome of HBECs. The first alteration peaked at the fifth hour and gradually subsided by the seventh hour. A second alteration occurred at the 13th hour. These observations suggest that e-liquids can cause substantial changes in the metabolic profile of HBECs. These changes may have potential health implications. This study highlights the need for further research to understand the long-term effects of e-cig use on respiratory health (Aug et al. 2015).

According to another study, systemic metabolites are altered by both vaping e-cigs and smoking cigarettes. In addition, tricarboxylic acid (TCA) cycle-related metabolites are disrupted by e-cig vaping, whereas sphingolipid metabolites are altered by cigarette smoking. The understanding of the metabolic effects of smoking and vaping was improved by this research. Elevated levels of nicotinic metabolites, which may serve as common biomarkers for the early development of cardiovascular or pulmonary diseases, have been detected in both e-cig users and smokers. Volunteers for the experiment provided comprehensive information about their age, sex, race/ethnicity, smoking history (cigarettes and/or e-cigs), and overall health status. Blood samples were analyzed using an Ultimate 3000 UPLC system interfaced with a Q Exactive HF Orbitrap mass spectrometer to determine the presence of metabolites. Hydrophilic interaction chromatography (HILIC) was employed for sample separation. In general, this work sheds light on the metabolic consequences associated with electronic nicotine delivery systems (ENDS), such as e-cigs, compared with those associated with conventional tobacco products such as cigarettes (Wang et al. 2021).

In vitro: This study examined the metabolic effects of e-cigs on human lung epithelial cells. The nicotine concentration was determined using HPLC to analyze e-cig vapor in water at different intervals via GCMS extraction and analysis. For this study, increasing concentrations of e-cig extracts were used to treat cells, after which metabolites were extracted from them. Small molecules were detected and analyzed with a liquid chromatography–electrospray ionization–tandem mass spectrometry (LC–QTOF–MS/MS) system. Additionally, the MetaboScape software tool was used to further evaluate the metabolites. Chemical compounds present in the human body were identified by database searching methods incorporated into this research. Researchers have found that the level of consumption determines how the body metabolizes vaping products. Compounds such as amino acids, fatty acids, glutathione, and pyruvate play key roles in the observed effects. The amount consumed also has different effects on individuals, making it challenging to predict outcomes (Assiri et al. 2024).

Humans: The health risks associated with cigarette smoking are increasingly concerning, leading many smokers to seek safer alternatives. The use of ENDS appears to be one of the most encouraging substitutes. These systems produce less toxicants in vapor than in cigarette smoke, which makes them seem to be a healthier choice. In a 7-day clinical study aimed at assessing the metabolic profiles of smokers who switched from smoking to ENDS, participants continued to smoke their usual brand for the first 2 days before exclusively using ENDS for the next 5 days. A metabolomic platform based on mass spectrometry was used for gathering and analyzing urine and plasma samples. The primary objective was to compare outcomes between patients who smoked conventional cigarettes and those who used ENDS. The investigators found that individuals who switched to ENDS products had a considerable decrease in oxidative stress-associated metabolites coupled with an increase in their urine and plasma levels of vitamin metabolites plus antioxidants. Additional samples were taken over a 24-h period on the third and fourth days, during which only urine was collected, to gather comprehensive information on metabolic changes. The results showed that if maintained over long periods, the toxicological environment created by the consumption of ENDS products might be less harmful, potentially resulting in health benefits for individuals. In conclusion, this study clearly indicates that electronic cigarettes can serve as safer alternatives than conventional tobacco (Liu et al. 2021).

In vivo: A recent study was conducted to examine the negative effects of e-cigs on metabolic changes and body parts in mice. Researchers have found that exposure to e-cigs causes harm to tissues, including enlargement of intercellular spaces, cell death, and inflammation. This study involved the extraction of serum samples from mice via the administration of an anhydrous methanol solution. Analysis was conducted using an Acquity UPLC HSS T3 column with a ThermoFisher instrument at

35 °C. Researchers have conducted an analysis of metabolites in e-cigs by employing a UPLC system with a gradient mobile phase, followed by mass spectrometry analysis by using ESI in positive and negative modes. Full MS was the scan type, and the collected data ranged from 67 to 1000 m/z. As a result, scientists were able to identify different metabolites, such as L-carnitine and capryloyl glycine, which helped in understanding the changes in these compounds over time according to trend analysis results (Ren et al. 2022).

Pathway enrichment analysis was subsequently conducted. The findings revealed that e-cigs significantly influenced 24 metabolic pathways, predominantly associated with the regulation of amino acid metabolism. This control also affects the tricarboxylic acid cycle. This research provides crucial data on metabolite biomarkers associated with immediate exposure to e-cigs and their potential for disease prevention (Ren et al. 2022).

In vitro: Lung epithelial cells are prone to oxidative stress because of the action of biological and chemical irritants containing reactive oxygen species, which is a constant challenge to the respiratory system. Early detection of biomarkers for reactive oxygen species will aid in managing health conditions and is a sign of seeking treatment before the disease progresses to a respiratory disease. Researchers have sought to identify these biomarkers by conducting experiments aimed at detecting different volatile compounds produced during oxidation. This study involved HBECs that were exposed to hydrogen peroxide (H_2O_2) and cigarette smoke extract (CSE), after which the headspace was analyzed (Yamaguchi et al. 2019).

The scientists applied the headspace sorptive extraction-gas chromatography–mass spectrometry method to investigate the volatile organic compounds (VOCs) linked to oxidative measures. HBECs were treated with H_2O_2 at concentrations of 0.1, 0.5, 10, and 50 mM, and VOCs were collected every 12 h for 48 h during the experiments. Similarly, in the CSE cell peroxidation experiments, the viability of cells treated with different smoke extract concentrations and VOC concentrations every 12 h over a 48 h period were also determined. The experiments were performed to determine the oxidative stress biomarkers that can be used to identify respiratory diseases at an early stage. This identification can aid scientists in understanding the mechanisms of oxidative stress, which in turn will allow them to develop treatments that have greater efficacy and fewer side effects. Consequently, this path can lead to improved respiratory health statistics and decreased respiratory disease load. The experiment describes the extraction of VOCs with Twisters® following their trapping in thermal desorption tubes. For the sample analysis, a secondary internal standard was initially applied, followed by processing using a GCMS system. The data demonstrated by a relevant cell culture model include the results of exhaled breath biomarker exploration and the subsequent development of new medical diagnostic and treatment systems (Yamaguchi et al. 2019).

E-cig samples: A new set of standards has been proposed by the Food and Drug Administration to support the Premarket Tobacco Product Applications requiring the tobacco industry to disclose harmful elements. Examining e-liquids and aerosols can be quite a difficult task because these are products with various tastes, and they are also as complex as the one in the statement. Nevertheless, a novel method to identify ingredients in e-juices and aerosols has been developed. This new technique involves the capture of aerosols via a Borgwaldt LX20 device, which is designed specifically for capturing e-vapor in the absence of contamination (Shah et al. 2021).

E-liquids and aerosols are vigilantly scrutinized for their quality and safe use. A noteworthy stage within this sequence is the sampling of the test sample, which is a kind of GCMS analysis (Fig. 8.1) process. The target extraction weight is the average mass of aerosols obtained from all the repeats of one given product. The latest extraction process of a liquid sample includes the process of obtaining it from the product cartridge or bulk formulation container. The extracted specimen is then homogeneously mixed with the extraction solvent, which is the internal standard. This methodology is vital for the subsequent processes to be consistent and accurate. These actions are essential for determining the safety of an item for safe public usage while respecting industry norms. This research accomplished the paramount goal of creating a highly efficient data processing system, enabling the identification of 46 compounds in e-liquid formulations and 55 compounds in aerosol samples. Notably, more than 50% of the substances were examined using reference standards, which clearly indicates the high accuracy of this technique. An ingenious profiling approach is a large improvement in the analysis of vapors and semivolatile compounds in electronic vapor devices, thus potentially ushering this sector to a new level (Shah et al. 2021).

E-cig samples: Currently, scientists are studying how e-cig aerosols affect our bodies differently than conventional cigarettes. Using LCMS-based nontargeted metabolomics, a new scientific experiment investigated the presence of certain compounds in the urine of people who vape, smoke regular tobacco, or neither. The researchers sought to determine what changes occurred at the molecular level between different groups based on structural identification, chemical and biochemical data, etc. For this purpose, super-advanced LCMS platform was used to analyze the urine samples. A Waters Acquity UPLC HSS T3 analytical column placed in an oven set at 40 °C was employed to inject these samples into the system during analysis while maintaining them under constant temperature conditions throughout until elution. The mobile phase used during analysis was composed of acidified

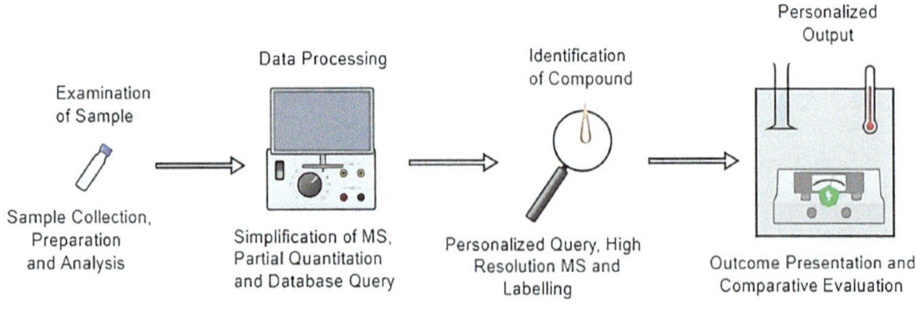

Fig. 8.1 Nontargeted GCMS analysis workflow. (Adapted from Shah et al. (2021) and licensed under the Creative Commons Attribution 4.0 International License)

water and acetonitrile adjusted to a 0.1% formic acid concentration for each component. This LCMS system produced high-resolution results, including detailed information about various structural and/or chemical differences between the different groups. The findings of the study can potentially help future research in the field of biochemical pharmacology, with prospective applications for drug discovery therapy development, among others (Hsiao et al. 2023).

The study's findings reveal unforeseen health risks associated with e-cigs. Although nicotine consumption is lower among people who use "vaping" devices than among those smoking tobacco products (cigarettes), its potential impact on increased lipid metabolism, which can result in inflammation in addition to the absorption of toxic substances such as flavors and plasticizers, remains a concern. Therefore, different safety levels must be considered when evaluating e-cig safety and hazards. Nicotine metabolites are similar in both cigarette smokers and vapers; however, the latter exhibited a dysregulated acylcarnitine pattern coupled with increased levels of cancer-related biomarkers, thus yielding comprehensive urine biochemistry results. Consequently, long-term effects may prove more intricate in addition to being harmful than what had been previously imagined about these products (Hsiao et al. 2023).

8.4 Capillary Electrophoresis-Mass Spectrometry

Clinical: Identifying new biomarkers capable of revealing exposure and early harm is very important for agencies to compare conventional and nonconventional tobacco products. Among the various biomarkers, metabolic biomarkers have proven to be highly useful. In a recent study, researchers analyzed human urine samples from nonsmokers and smokers of cigarettes with two different levels of tar. The authors used capillary electrophoresis-mass spectrometry (CE-MS). This study examined two groups, an ISO tar group, an ISO tar smoking group, and a quality control sample, to confirm system stability and performance (Garcia-Perez et al. 2014).

CE-MS analysis of urine samples from both groups provided the researchers with vital information. The study's findings have the potential to enable authorities in charge of regulation to differentiate effects caused by the use of combustible tobacco from those caused by the use of noncombustible products. A comprehensive study revealed that metabolic biomarkers may be applied to establish severity from exposure to nicotine, among other chemicals found in cigarettes. A fused-silica capillary was employed during separation, which was successively rinsed with ammonium hydroxide solution, deionized water, and running buffer. Fresh capillaries were also conditioned using NaOH followed by deionized water before they were isolated with formic acid, which was similarly prepared. Smoker differentiation models have passed validation tests chemometrically. The primary criterion used for this segregation is the presence of nicotine metabolites in samples collected from individuals' urine. Even after the elimination of nicotine metabolites, smokers can still be distinguished from nonsmokers, suggesting the presence of additional potential biomarkers apart from those already known. Current investigations have shown that Ce-MS can be used to identify new biomarkers in human urinary samples, indicating the possibility for more efficient therapeutic strategies for tobacco addiction treatment. By identifying unique biomarkers, regulatory agencies can tailor safety and control standards for various tobacco products (Garcia-Perez et al. 2014).

8.5 Comparative Studies and Long-Term Health Outcomes

Comparative metabolomic research has yielded important findings on e-cig use and conventional cigarette smoking metabolic pathway perturbations. UPLC, MS, and CE-MS have enabled scientists to pinpoint distinct metabolites influenced by each type of tobacco product. Dysregulation of TCA cycle-related compounds was detected in e-cig vapers, while sphingolipid metabolites affected mainly smokers of regular cigarettes. Both forms of nicotine increase the levels of nicotinic metabolites, which act as general biomarkers for the early development of cardiopulmonary diseases. This study shows how different methods of nicotine delivery can affect a person's metabolism; however, when smokers switch to ENDS, they exhibit reduced levels of oxidative stress-related compounds together with increased levels of vitamin antioxidants, suggesting less toxic physiological conditions. This scenario implies that if such changes continue to occur for some time, then these gadgets would have fewer adverse health effects than conventional cigarettes, making them safer options for smokers.

Additionally, if regulators can detect specific or common metabolic markers among different tobacco products, then this knowledge would be beneficial for enhancing existing regulations to protect public health. Authorities need tools to distinguish individuals who use various kinds of smoke from those who do not. Even if nicotine metabolites are absent after treatment, more efficient strategies for addiction therapy and control policy-making processes must be introduced. Highlighting novel compositions of metabolites in users of electronic cigarettes versus regular cigarettes is crucial because it facilitates the identification of potential advantages for the former and the devising of regulatory frameworks to curtail the effects of the latter.

The use of biomarkers from metabolomics research tailored to e-cigs, including nicotine metabolites and signs of oxidative stress, can potentially correlate to long-term negative effects via extensive longitudinal research and cooperation in clinical trials. Investigating e-cig biomarkers, such as nicotine derivatives and oxidative metabolites, can provide valuable information that researchers can use to track subjects over extended periods and to detect any obstructive changes while undergoing such examinations as lung, heart, and/or other diseases that are longitudinally secured with the conduct of this practice. Longitudinal studies will require baseline data to be collected during follow-ups at regular intervals to ensure the observation of how biomarkers will change over time.

Clinical collaborations and multicenter trials should be incorporated in this analysis to obtain more reliable outcomes with larger sample sizes from different populations to allow the findings to be readily accessible on a universal scale. One should consider the performance of clinical assessments, such as pulmonary function tests or/and cardiovascular monitoring, and the collection of biomarker information to fully understand the impacts of the different pollutants present in the liquid environment, e.g., heavy metals. Moreover, researchers can enhance their work by donating their specimens to the biobank and using the bank for retrospective analysis based on the accumulated historical data. These models will help identify individuals who are most likely to develop certain ailments when they do not intervene in a timely manner. The process is also a pathway for new prevention strategies that can result in heightened regulations on consumption patterns of vaping devices on a general scale.

8.6 Conclusion

Crucial processes of sample preparation in the metabolomics field of e-cigs to yield accurate, aseptic, and fast results are difficult to achieve. Some methods for the detection of products have been given less focus than sample processing steps. Making the correct choices of e-cigs, such as proper storage and preparation, is extremely important for obtaining good quality data. Standardization, advancement, and better comprehension of the techniques can only occur, particularly when untargeted metabolite studies are being performed to unveil the identification of false biomarkers caused by the artifacts that are introduced into the system during quantitative analysis. The problem of validation by NMR/LCMS-based metabolomics cannot be solved except for the preparative steps of the samples, which are the stages that are verified, understood, and controlled in the e-cig technology field. By revealing methods for sample management, one can determine the trustworthiness and exactness of the analysis of metabolites originating from e-cig consumption.

Researchers should begin with developing the analytical pipeline according to the simplest techniques when working on e-cig samples in metabolomics frameworks; this is particularly the case when samples are obtained in vitro or in vivo or when patients are clinically diagnosed. Comprehensive data analysis, especially in the ongoing process of innovation, might be needed. The assumptions behind the chosen models, including a balance between false positives and negatives, must be considered in data interpretation to avoid tightly setting the data. However, the features of the model might also be omitted if extra variables are included without any prior relevant biological knowledge. This process of model interpretation is the only way that causality can be set up by merging statistical or chemical knowledge into chemometric settings, while tools with visual outputs help in the correct understanding of chemical structures in addition to the information extracted from previous analyses.

References

Assiri MA, Al Jumayi SR, Alsuhaymi S, Emwas AH, Jaremko M, Alsaleh NB, Al Mutairi MM, Alshamrani AA, Sobeai HA, Alghibiwi H (2024) Electronic cigarette vapor disrupts key metabolic pathways in human lung epithelial cells. Saudi Pharm J 32(1):101897

Aug A, Altraja S, Kilk K, Porosk R, Soomets U, Altraja A (2015) E-cigarette affects the metabolome of primary normal human bronchial epithelial cells. PLoS One 10(11):e0142053

Garcia-Perez I, Lindon JC, Minet E (2014) Application of CE-MS to a metabonomics study of human urine from cigarette smokers and non-smokers. Bioanalysis 6(20):2733–2749

Gu F, Derkach A, Freedman ND, Landi MT, Albanes D, Weinstein SJ, Mondul AM et al (2016) Cigarette smoking behaviour and blood metabolomics. Int J Epidemiol 45(5):1421–1432

Hahn J, Monakhova YB, Hengen J, Kohl-Himmelseher M, Schössler J, Hahn H, Kuballa T, Lachenmeier DW (2014) Electronic cigarettes: overview of chemical composition and exposure estimation. Tob Induc Dis 12(1):23

Hsiao YC, Matulewicz RS, Sherman SE, Jaspers I, Weitzman ML, Gordon T, Liu CW, Yang Y, Kun L, Bjurlin MA (2023) Untargeted metabolomics to characterize the urinary chemical landscape of E-cigarette users. Chem Res Toxicol 36(4):630–642

Izquierdo-García JL, del Puerto-Nevado L, Peces-Barba G, Pérez-Rial S, Heili S, Villa P, Castejón D, González-Mangado N, Ruiz-Cabello J (2010) A metabonomic approach to evaluate COPD in a model of cigarette smoke exposure in mice. Metabolomics 6(4):564–573

Liu G, Lin CJ, Yates CR, Prasad GL (2021) Metabolomic analysis identified reduced levels of xenobiotics, oxidative stress, and improved vitamin metabolism in smokers switched to Vuse Electronic Nicotine Delivery System. Nicot Tob Res 23(7):1133–1142

Merecz-Sadowska A, Sitarek P, Zielinska-Blizniewska H, Malinowska K, Zajdel K, Zakonnik L, Zajdel R (2020) A summary of in vitro and in vivo studies evaluating the impact of E-cigarette exposure on living organisms and the environment. Int J Mol Sci 21:652

Percival BC, Wann A, Taylor S, Edgar M, Gibson M, Grootveld M (2022) Metabolomics distinction of cigarette smokers from non-smokers using non-stationary benchtop nuclear magnetic resonance (NMR) analysis of human saliva. In: Oral health care-an important issue of the modern society. IntechOpen, London

Ren X, Lin L, Sun Q, Li T, Sun M, Sun Z, Duan J (2022) Metabolomics-based safety evaluation of acute exposure to electronic cigarettes in mice. Sci Total Environ 839:156392

Shah NH, Noe MR, Agnew-Heard KA, Pithawalla YB, Gardner WP, Chakraborty S, McCutcheon N et al (2021) Non-targeted analysis using gas chromatography-mass spectrometry for evaluation of chemical composition of E-vapor products. Front Chem 9:742854

Shintu L, Caldarelli S, Campredon M (2013) HRMAS NMR spectroscopy combined with chemometrics as an alternative analytical tool to control cigarette authenticity. Anal Bioanal Chem 405(28):9093–9100

Titz B, Szostak J, Sewer A, Phillips B, Nury C, Schneider T, Dijon S et al (2020) Multi-omics systems toxicology study of mouse lung assessing the effects of aerosols from two heat-not-burn tobacco products and cigarette smoke. Comput Struct Biotechnol J 18:1056–1073

Wang Q, Ji X, Rahman I (2021) Dysregulated metabolites serve as novel biomarkers for metabolic diseases caused by E-cigarette vaping and cigarette smoking. Metabolites 11(6):345

Wei B, Goniewicz M, O'Connor RJ (2019) Concurrent quantification of emerging chemicals of health concern in E-cigarette liquids by high-performance liquid chromatography-tandem mass spectrometry. ACS Omega 4(13):15364–15372

Yamaguchi MS, McCartney MM, Falcon AK, Linderholm AL, Ebeler SE, Kenyon NJ, Harper RH, Schivo M, Davis CE (2019) Modeling cellular metabolomic effects of oxidative stress impacts from hydrogen peroxide and cigarette smoke on human lung epithelial cells. J Breath Res 13(3):036014

Zucchet A, Schmaltz G (2017) Electronic cigarettes—a review of the physiological health effects. Facets 2(1):575–609

Metabolomics of Toxic Markers in E-cigarettes: Unveiling Molecular Insights

9

Vuanghao Lim, Jer Ping Ooi, Bunleu Sungthong, and Yoke Keong Yong

Abstract

Metabolomic profiling, an advanced tool used to identify different metabolite biomarkers after exposure to different substances, has high sensitivity and reproducibility. This method can detect sensitive changes in biological metabolites under several conditions; thus, it is used to examine the effects of vaping on human health. The main concern with vapes is the potential respiratory issues caused by the detrimental components in their emissions. The new findings not only challenge recognized assumptions but also carry substantial health risks for individuals exposed to them. Metals such as aluminum (Al), antimony (Sb), chromium (Cr), copper (Cu), and lead (Pb) are present in e-liquids according to metabolomics at levels higher than those previously thought in past studies. The highest levels of formaldehyde found in electronic cigarette (e-cig) aerosols due to occupational exposure are far from the safety limits; thus, biological systems should be able to monitor these changes. When nicotine-containing aerosols are inhaled, second-hand exposure occurs. The presence of nicotine-containing aerosols results in second-hand exposure, with flavored liquids, particularly cherry, showing higher levels of benzaldehyde. Furthermore, e-cig emissions contribute to air pollution through the release of particulate matter, but evidence

V. Lim (✉) · J. P. Ooi
Advanced Medical and Dental Institute, Universiti Sains Malaysia, Bertam, Kepala Batas, Penang, Malaysia
e-mail: vlim@usm.my

B. Sungthong
Integrative Pharmaceuticals and Innovative of Pharmaceutical Technology Research Unit, Faculty of Pharmacy, Mahasarakham University, Kantharawichai, Maha Sarakham, Thailand

Y. K. Yong
Department of Human Anatomy, Faculty of Medicine and Health Sciences, Universiti Putra Malaysia, Serdang, Selangor, Malaysia

Y.-H. Leong, A. S. Mohd Samin (eds.), *E-Cigarettes*,
https://doi.org/10.1007/978-981-96-5087-3_9

of their overall impact is conflicting; therefore, additional metabolomics studies aimed at elucidating this issue are needed. The need for continuous research on e-cigs cannot be ignored given our limited knowledge about long-term effects coupled with similarities between toxicity mechanisms exhibited by both conventional cigarettes and electronic ones. Therefore, metabolomic tools should be effectively utilized during such investigations. Regulations for vapes vary from one country to another; hence, strict supervision is needed globally while enhancing community awareness and putting in place preventive measures to minimize the risks associated with their usage.

Keywords

Multiomics · E-cigarette liquid · Metabolites · Markers · Metabolics

9.1 Introduction

Earlier research faced obstacles in detecting reliable biomarkers of exposure to e-cigs (Wang et al. 2021). Metabolomic profiling has recently become popular as a first line of detection for metabolite biomarkers because of its clear visualization, high sensitivity, and reproducibility. Metabolomics can be used to detect subtle biomass fluctuations in living organisms in their respective environments and thus can differentiate varying effects on similar organisms (Dai et al. 2020). Recent studies suggest that the use of e-cigs initiates the nicotine metabolic pathway and, therefore, causes the overproduction of toxic metabolites that disturb the heart and lungs (Wang et al. 2021). Metabolomic profiling is a sensitive and reproducible technique for identifying the smallest indications of changes in metabolites in the human body. Metabolomics also has the potential to differentiate species of the same type of organism (Ren et al. 2022; Gu et al. 2016; Fan et al. 2012; Hsiao et al. 2023). Finally, recent research has suggested that e-cigs might harm organs by generating toxic metabolites via the nicotine metabolic pathway (Titz et al. 2020).

9.2 Multiomics in E-cig Research

Multiomics is one way to integrate and analyze data derived from the genome, epigenome, transcriptome, proteome, and metabolome at the biological level. This approach is an excellent arrangement for those concerned with the study of the intersection of biochemical and toxicological effects, supposing the existence of a "molecular roadmap" would allow scientists to connect the interactions of biological systems together with drugs (Muehlbauer et al. 2023). This approach can reveal standard toxicology endpoints and bridge genotypes and phenotypes. Integrating multiomics theory can lead to a deep understanding of the evidence pathway, allowing a shift from one omic approach to another (Titz et al. 2020). The toxicity of e-cigs has been the main subject of two groundbreaking studies. In the first study, inhalation of e-cig aerosols upregulated the MUC5AC gene, which is associated with blockage of the airways. Another study revealed widespread transcriptome changes in the prefrontal cortex of juvenile and adult offspring born to dams exposed to e-cig aerosols during perinatal stages, indicating the prevalence of learning difficulties and chronic neuropathology (Eshraghian and Al-Delaimy 2021; García-Gómez et al. 2016; Margham et al. 2016).

Titz et al. (2020) recently emphasized the importance of implementing a comprehensive multilayered analysis strategy to measure the damage caused by smoking devices that contain nicotine and are inhaled and penetrate the lungs. Their main point was the possible implications of this model when studying e-cigs. According to researchers, traditional toxicological endpoints may have low sensitivity and be only quasi-informative in terms of toxicological mechanisms (Titz et al. 2020). E-cig devices are characterized by their variability, such as the introduction of different brands, each with its own unique blend of flavors coupled with varying concentrations of nicotine, which hinders the prediction of the kinds or amounts of particular combinations against toxicity. In addition to several other factors, one study revealed that approximately 30,000 different potential combinations could arise alone from changes in these three parameters. Such diversity also extends to study designs and outcomes, which hinders meaningful comparisons between different previous studies across various locations worldwide. Therefore, an all-encompassing multilayered analysis strategy is crucial when deeper insights into potential health risks associated with e-cig products are available (Fig. 9.1) (Titz et al. 2020).

Temporality should be considered when undertaking multiomics analysis for toxicological research. Many different multiomic points of interest in toxicology need to be captured over time. These time intervals are required to obtain a full understanding of what happens as a result of exposure at all levels—from molecules through ecosystems, thus creating a systems toxicology "big picture." The text highlights the importance of comparing molecular changes that occur after acute and chronic exposures to gain a comprehensive understanding of the effects of exposure. Assessing different endpoints at one or multiple timeframes after the final exposure in chronic studies is crucial for obtaining a comprehensive understanding of omics alterations linked to relevant exposures over time (Fan et al. 2012). This method is imperative for conducting accurate toxicological analyses.

9.3 E-cig Components and Metabolic Impact

E-cigs have become more popular because people believe that they are less harmful than conventional cigarettes. This trend has raised concerns about potential health hazards related to vaping e-cigs. Research indicates that short- and long-term exposure to e-cig (Eshraghian and Al-Delaimy 2021; García-Gómez et al. 2016) aerosols can cause inflammation in the lungs and oxidative stress. A significant amount of research has been conducted to establish the link between cigarette smoking and metabolic health disorders. However, few studies have explored the impact of e-cig use on metabolic health. Metabolic diseases are characterized by disturbances in

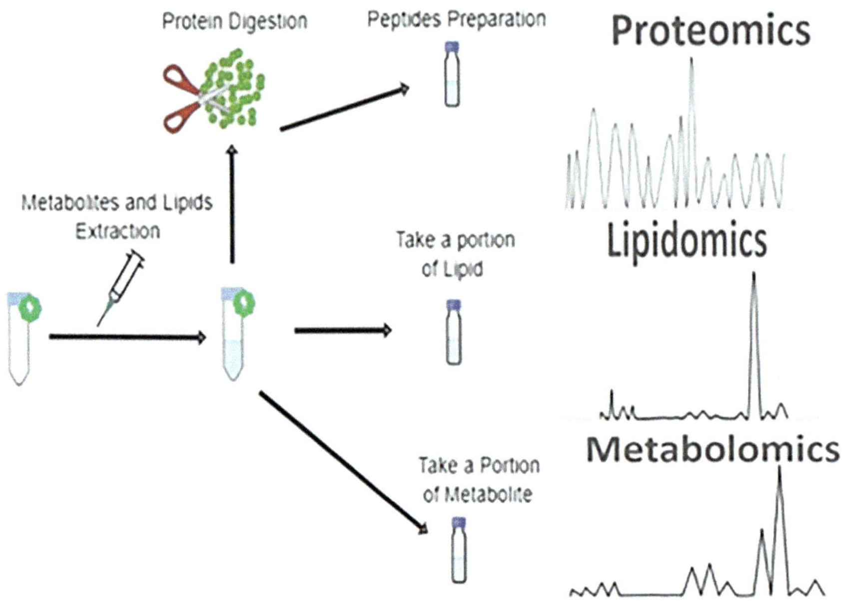

Fig. 9.1 Multiomic sample preparation for mass spectrometry. (Adapted from Muehlbauer et al. 2023)

metabolites, which are molecules produced during metabolic processes. Understanding these disturbances can help identify biomarkers associated with potential health issues. Therefore, researchers have conducted a study to address this gap by identifying dysregulated metabolites in the plasma of both e-cig users and cigarette smokers (Garcia-Perez et al. 2014).

The discovery of the role of metabolites in the degradation of nicotine, sphingolipid metabolism, and the tricarboxylic acid cycle represents the first alteration in the metabolites involved in nicotine degradation, among several others. This type of metabolic pathway is extremely important for the regulation of energy in cells and cellular functions. This research also revealed new metabolites that are affected not only by e-cig aerosols but also by cigarette smoke. The identification of these dysregulated metabolites represents a major step toward a better understanding of the disease mechanisms caused by the use of e-cigs. This study helps to elucidate the close relationships among e-cig vaping, metabolic disorders, and metabolic changes that accompany metabolomics (García-Gómez et al. 2016). The relationship between some metabolites and the respiratory and cardiovascular system is focused on sphingolipid and nicotinic metabolites. Some research shows that sphingolipid metabolites such as sphingosine-1-phosphate (S1P) and ceramide are closely related to the degree of lung inflammation, emphysema, and chronic obstructive pulmonary disease (COPD). In a mouse model of emphysema, scientists detected increased levels of ceramide, and by inhibiting ceramides, airspace enlargement caused by elastase could be reduced (Izquierdo-García et al. 2010).

Moreover, a comparison of e-cig users, cigarette smokers, and healthy individuals revealed that the group that smoked traditional cigarettes had much greater levels of ceramide and sphingosine, which indicates that this method of smoking can have a more severe impact on these sphingolipid metabolites than can e-cig smoking. However, the importance of nicotinic metabolites shows that both e-cig users and cigarette smokers exhibit increased levels of these metabolites. Nicotine, a primary component of both smoking and vaping, is associated with adverse cardiovascular effects, including myocardial remodeling, fibrosis, and an elevated risk of heart-related complications (Liu et al. 2021). According to a recent metabolomics study, the levels of biomarker metabolites differ among cigarette smokers, e-cig users, and healthy individuals. According to the study, cigarette smokers have lower L-methionine levels than do e-cig users and people who are healthy. This decrease in L-methionine has been linked to an increase in metabolic rate, weight loss, and heart attack. These discoveries imply that metabolic outputs can be correlated with a greater risk of heart troubles among individuals who smoke or use e-cigs (Percival et al. 2022).

The effect of e-liquid on human bronchial epithelial cell metabolism was investigated in this study, as was the possible antioxidant-mediated reduction in cellular damage. The changes were likely caused by the e-cig liquid (ECL) in a cyclic manner, which means a dynamic metabolomic response. At the 5th and 13th hours, the study was able to detect biphasic fluctuations because of ECL, which is quite a dynamic metabolomic response. Notably, the number of significantly altered signals was the lowest at the 7th hour, which could indicate a state of equilibrium (Eshraghian and Al-Delaimy 2021). The levels of amino acids, such as glutamate, glutamine, proline, histidine, and arginine, were also found to change after exposure to ECL. These alterations are similar to those observed in one combined study on cigarette smoke extract (CSC) exposure in A549 cells, which confirmed to a certain degree that e-cigs are potentially noxious (Wei et al. 2019; Lorkiewicz et al. 2022; Eshraghian and Al-Delaimy 2021).

This research also revealed an alteration in energy metabolism and a change in energy handling, which is connected to the activation of fatty acid oxidization. Typically, N-acetyl cysteine, which is a free radical, was effective in preventing metabolic struggles induced by ECL or CSC, with the inference that oxidative stress was indeed a pivotal cause of the whole pattern being observed. An e-cig that is intended to be smoked may become a concern to some, as it facilitates nicotine addiction as airway cells cause the nicotine level to rise (Yamaguchi et al. 2019).

9.4 Risk of Metabolites in E-cigs

In the Monte Carlo approach, which is a sophisticated probabilistic method, Table 9.1 describes the necessary risk assessment to be performed, addressing the issue of unknown variability of exposure estimates. In this novel context, researchers have employed a probability method called the Monte Carlo approach to estimate the risk of toxic chemical exposure in the manufacturing of ECL. This study

Table 9.1 Estimated exposure (mg/kg bw/day) of electronic cigarette users using Monte Carlo analysis (10,000 iterations)[a]

Agent	Mean	SD	Median	P5	P95
Nicotine	0.38	0.39	0.25	0.02	1.15
Glycerol	9.0	8.9	6.2	0.6	27.2
1,2-Propanediol	14.5	12.4	11.0	1.3	39.3
Ethylene glycol	2.1	6.7	0.9	0.1	7.1
1,3-Propanediol	0.14	0.51	0	0	1.1
Thujone	1.6E-4	9.7E-4	0	0	Max 0.01
Ethyl vanillin	7.2E-4	2.0E-3	0	0	Max 0.03

[a]Calculated for all agents by using the following formula with the risk functions defined in Hahn et al. (2014):

Exposure = Risk function of concentration * Risk function of e-cig liquid amount per day * Risk function of vaporization percentage/Risk function of bodyweight (kg)

The risk function of the amount of e-cig liquid per day was RiskNormalAlt (25%; 3%; 75%; 5%; RiskTruncate (0;)) based on values from Farsalinos et al. (2013)

The risk function of the vaporization percentage was RiskUniform (6;81) based on values from (Goniewicz et al. 2014)

The risk function of body weight was RiskNormal (73.9–12) according to the average and standard deviation from the EFSA Scientific Committee (EFSA 2012)

leveraged the margin of exposure (MOE) as a tool for risk assessment, adopting a paradigm that generates distributions rather than singular values (Wang et al. 2016). This visualization allows the user to directly compare different exposure scenarios. Scientific researchers took a conservative view by assuming a worst-case scenario, suggesting that 100% of the vaporized liquid is suitable for use via oral or inhalation routes. Given the absence of established inhalation exposure thresholds, toxicological thresholds for oral exposure were utilized (Iskandar et al. 2016).

Furthermore, the possibility of exceeding the acceptable daily intake (ADI) recommended by the European Food Safety Authority (EFSA) for nicotine deposits present in food products should be emphasized. When evaluating the MOE for 1,2-propanediol and ethylene glycol, certain conditions indicate that values may fall below the threshold of 100, which suggests a potential categorization into the "risk" category. The assessment acknowledges that daily conventional cigarette use may lead to nicotine tolerance. This concern raises the possibility that the risk associated with nicotine use in this population may be overestimated. The EFSA selection of heart rate acceleration as an endpoint for nicotine risk assessment has come under scrutiny (EFSA 2023).

However, conclusive epidemiological evidence supporting the direct carcinogenic effect of nicotine is lacking despite increased scrutiny and acknowledgment of uncertainties. This finding draws attention to the ongoing evaluation of nicotine by the World Health Organization's International Agency for Research on Cancer. Currently, the agency has not proven that nicotine is a cancerous component. The analysis performed in the study reflects academic integrity and fidelity in the assessment of the negative effects of nicotine in e-liquids. One of the most important methods to inform people of the possible adverse effects of exposure to nicotine would be to provide a complete survey of the health issues related to this substance that can be faced by the user. This approach enables people to gain a deeper understanding of the subject and implement risk assessment in a better manner, thus delivering the necessary measures for the prevention of a public health crisis (Pisinger and Døssing 2014; Girvalaki et al. 2018).

With respect to clearance, e-cig-deprived individuals should be warned about the use of products containing higher nicotine levels. Moreover, exposure to ethylene glycol surpassed the minimal risk level. Ethylene glycol is considered the only substance not sheltered by the U.S. Food and Drug Administration (FDA), and its risks are rated as much lower than those of nicotine. Thus, the most appropriate measure is to reduce or avoid compounds such as nicotine and use solvents with less toxic contaminants. The detrimental effects of flavor compounds, namely, thujone and ethyl vanillin, are greatly reduced, leading to MOEs greater than 1000; therefore, preventive measures involving these components may not be needed (Kim et al. 2017; Hahn et al. 2014).

The growing use of vaping devices is one of the most worrying problems for the health of the population. For example, histopathology, blood tests, TUNEL assays, and DHE staining are commonly used methods for in vivo assessments that allow researchers to investigate the effects of e-cig vapor on the body. Metabolic dysfunction is a serious condition that can be detected by changes in serum metabolites and, hence, can be used as a diagnostic tool for related disorders caused by e-cig usage. However, these problems are complicated and require rather laborious investigations; therefore, they must be admitted. Research has revealed that e-cigs have adverse toxic effects during acute toxicity tests. This phenomenon can be represented by a shift in the functional parameters of organs, such as the liver and lungs. People who smoke e-cigs often develop health problems in internal organs (for example, the liver) and lipid buildup in organs. Additionally, these blood tests revealed that the levels of total cholesterol, triglycerides, and glucose increased with acute exposure, which indicated that cardiovascular and metabolic outcomes might follow (Hiemstra and Bals 2016; Merecz-Sadowska et al. 2020). According to a previous study, e-cig use is negatively associated with oxidative damage and apoptosis. The TUNEL assay and target DHE staining were used to investigate the increase in reactive oxygen species and the death of cells in the tissue. The study also revealed activation of the mitochondrial cell death pathway, suggesting potential mechanisms of harm (Wang et al. 2022).

The application of metabolomic analysis has enabled the identification of differentially abundant metabolites and enrichment of pathways associated with taurine and hypotaurine metabolism, pulmonary surfactants, and catecholamines. A significant positive correlation among the common metabolites was found, indicating potential regulatory interactions. Trend analysis further revealed changes in metabolites related to nucleic acids and organic acids in terms of exposure time, which suggests potential disruptions in DNA and nucleic acid metabolism due to e-cig exposure (Wang et al. 2021).

9.5 Main Metabolites in E-cigs

Numerous studies have identified the presence of various chemicals in e-liquids and aerosol cigarettes. Acenaphthylene was found to be a prevalent constituent in most of the samples. Acetaldehyde, another significant component, was also detected in these products. Antimony was present in most e-liquid samples but only in some aerosol samples. Furthermore, diacetyl, formaldehyde, glycerol, and propylene glycol were consistently detected in all or most of the samples. Lead and nickel exhibited a near-universal presence and were identified in all or most e-liquid and aerosol samples. The presence of these substances, particularly lead, a substance of regulatory concern, raises concerns about the safety and health implications of using e-cigs and similar products (Turtoi et al. 2023).

The increasing amount of harmful and hazardous matter (HPHCs) in e-liquids is an important health concern, and among these contaminants are the carcinogenic substances acetaldehyde, formaldehyde, lead, and nickel, which have been identified by the FDA. In addition, nicotine, ethyl benzene, crotonaldehyde, chromium, and anabasine are HPHCs that the FDA has reported to be pooled into the list of nascent risks because of the secondary inhalation of these e-cigs. As a result of the presence of these chemicals in e-liquids and their aerosols, many issues have been raised that may affect public health, bringing attention to the need for better control and monitoring of the possible havoc they can cause (Wang et al. 2021). The in-depth study of the different types of complex e-liquid content and their possible health implications has become the key to revealing the main risks that accompany their intake. The present realization that the situation calls for extra care and watch in the use of e-liquids is also supported by a new finding that at the early stages of vaping and e-cig usage, the popularity of e-liquids continues to rise (Moshensky et al. 2021). For this reason, regulatory bodies and public health officers are bound to remain alert about the dangers that might be entailed by these products and to take the necessary corrective steps to protect people's health and well-being.

Various research analyses have established the existence of drugs, pesticides, or caffeine in e-liquids. As microorganisms are present in these formulations, many different kinds of materials are utilized in the production of e-liquids (Lorkiewicz et al. 2022). Many of these analyses focused on the exact substances present in e-liquids, and minor tobacco alkaloids, such as special tobacco chemical compounds (anabasine, anatabine, and myosomine), which are known to be present in tobacco products, were identified. At the same time, e-liquid is an interesting food flavor because diacetyl and vanillin are recognized as two distinct substances that can be added as flavoring agents (Morris et al. 2021). With the discovery of diacetyl in e-cigs, one substance was reported as the main cause of health problems in the users and was also found in the aerosols produced from e-cigs. Many studies have focused on metals in e-cig aerosols. The belief that metals might be one of the products of the evaporation of e-liquids from the metal atomizer was supported by the tests conducted. Nevertheless, the use of small amounts of metals, such as aluminum, antimony, chromium, copper, and lead, has been proposed by other investigations either because of the presence of e-liquids or in contradiction to this hypothesis (Kim et al. 2017).

Studies have shown that e-cig aerosols containing nicotine can lead to second-hand exposure. In addition, the analysis of aerosol samples revealed that one of the aromatic compounds, benzaldehyde, was more abundant in flavored e-cigs than in conventional cigarettes. Benzaldehyde is an aromatic compound that has a high level of emission in cherry-flavored e-cigs. A possible reason is that different flavorings can have different emission amounts (Eshraghian and Al-Delaimy 2021). Recent research studies have indicated the effect of e-cigs on indoor air quality, as particulate matter can be released into the air when used. In addition, the increasing number of fine cigarettes indoors and outdoors has likely caused ambient air pollution. However, some findings are contradictory. A study stressed that indoor vaping does not involve exposure to fine particles. Further investigations are needed to harmonize these contradictory perspectives and to determine the effect of e-cigs on air quality, especially in indoor areas (Lorkiewicz et al. 2022). The potential health effects of e-cig aerosol emissions are a subject of ongoing scientific inquiry, and it is essential to continue investigating this issue to better understand the risks involved.

9.6 Conclusion

E-cigs have become a popular substitute for conventional tobacco cigarettes in the past few years. Despite a significant increase in the e-cig market, the long-term impact of e-cig usage is still mostly unknown, and scientific studies on this topic are lacking, especially in humans. Moreover, the FDA has concerns regarding the harmful effects and carcinogenicity of prefilled pod- or cartridge-based e-cigs flavored in various ways, which has led to increased demand for conducting the next level of experiments on the potential health risks of these materials. Different flavors are one of the things that e-cigs utilize when carcinogens are present. These chemicals can be dangerous for the health of people exposed to them (Morris et al. 2021). The conclusion of this research is important because people can thereby completely comprehend the long-term health effects associated with e-cigs, which allows users to be reassured that these products are safe to consume.

The toxicity of each type of e-cig for safety must be specifically evaluated; however, one cannot underestimate the importance of their comparative assessment. These findings can help in conducting more comprehensive safety assessments and, in turn, facilitate the ability to make informed decisions. In the future, research on lung cells, especially lung cancer cells, makes sense because it is the easiest way to determine whether ongoing prenatal exposure to e-cig aerosols has any effect on the development of lung cancer and respiratory diseases among children (Wang et al. 2022). In previous research, only the influence of e-cigs on the lungs has been taken into account, and it is of utmost importance that the harmful effects on other body parts, that is, on the immune and cardiovascular systems, be equally investigated in depth and that their potential toxicity be elucidated. The variable regulations of e-cig use across different countries, as observed in the United States and the UK, highlight the need for stricter supervision, limited availability, and prevention of misuse among minors and nonsmokers. Efforts to reduce the widespread use of e-cigs must be prioritized to ensure the well-being and health of the public (Woodall et al. 2020).

By utilizing multidimensional analyses, the examination of e-cigs is a solid foundation for obtaining realistic insights; however, transparent limitations or consistency are lacking. The limitation of small or nondiverse sample sizes decreases the power of the study to try and obtain a generalizable understanding of the long-term effects of the use of the e-cig. Moreover, variability in e-cig devices and e-liquids introduces a significant impediment to the standardization and comparison of investigations; in addition, self-reported usage data can contribute to potential biases. Factors such as polysubstance use and other lifestyle variables that are not directly related to biological functioning mask specific biological impacts, while omics data integration is a complex and dynamic process.

Moreover, the field changes constantly, given that each historical dataset is not perfectly covered, and the rapid change in technology products, such as smokeless cigarettes, explains that comparison causes are much more complicated; therefore, the long-term effects are difficult to observe, and indications about its fear of practice might not be convincing. On the one hand, laws that change are stronger; on the other hand, the techniques used by the industry to make products more complicated, hence the limitations of the avidity of longitudinal studies. The evolution of multiomics technology has provided great advantages, such as providing a general overview of the biological effects of e-cigs. Ultimately, these disadvantages and ambiguities arise during the assessment of the risks and the establishment of the required public health regime.

The control of e-cigs is extremely important to avoid the harm of counterfeit products and reduce the health risk that arises from their usage. The development of toxicity evaluation systems and quality criteria is the main driving force behind the technology that is able to do so; hence, sufficient assessments of the toxicity, regulation of the components of e-cigs, and counterfeit prevention measures are needed. In addition, reaching out to consumers can help educate them on the safe use of these products as a method of preventing potential health problems. By implementing these measures, the e-cig industry can provide consumers with a safer alternative to conventional tobacco products (Schwanz et al. 2019).

References

Dai Y, Huo X, Cheng Z, Faas MM, Xu X (2020) Early-life exposure to widespread environmental toxicants and maternal-fetal health risk: a focus on Metabolomic biomarkers. Sci Total Environ 739:139626

EFSA (2023) Statement on the revised targeted risk assessment for certain maximum residue levels for nicotine. EFSA J 21(3):e07883

EFSA Scientific Committee (2012) Guidance on selected default values to be used by the EFSA scientific committee, scientific panels and units in the absence of actual measured data. EFSA J 10(3):2579

Eshraghian EA, Al-Delaimy WK (2021) A review of constituents identified in E-cigarette liquids and aerosols. Tob Prev Cessat 7:1–15

Fan TW-M, Lane AN, Higashi RM (2012) The handbook of metabolomics, vol 291. Humana Press, Totowa

Farsalinos KE, Romagna G, Tsiapras D, Kyrzopoulos S, Voudris V (2013) Evaluating nicotine levels selection and patterns of electronic cigarette use in a group of "vapers" who had achieved complete substitution of smoking. Subst Abuse Res Treat 7:139–146

García-Gómez D, Gaisl T, Barrios-Collado C, Vidal-De-Miguel G, Kohler M, Zenobi R (2016) Real-time chemical analysis of E-cigarette aerosols by means of secondary electrospray ionization mass spectrometry. Chem Eur J 22(7):2452–2457

Garcia-Perez I, Lindon JC, Minet E (2014) Application of CE-MS to a metabonomics study of human urine from cigarette smokers and non-smokers. Bioanalysis 6(20):2733–2749

Girvalaki C, Tzatzarakis M, Kyriakos CN, Vardavas AI, Stivaktakis PD, Kavvalakis M, Tsatsakis A, Vardavas C (2018) Composition and chemical health hazards of the most common electronic cigarette liquids in nine European countries. Inhal Toxicol 30(9–10):361–369

Goniewicz ML, Hajek P, Mcrobbie H (2014) Nicotine content of electronic cigarettes, its release in vapour and its consistency across batches: regulatory implications. Addiction 109(3):500–507

Gu F, Derkach A, Freedman ND, Landi MT, Albanes D, Weinstein SJ, Mondul AM et al (2016) Cigarette smoking behaviour and blood metabolomics. Int J Epidemiol 45(5):1421–1432

Hahn J, Monakhova YB, Hengen J, Kohl-Himmelseher M, Schössler J, Hahn H, Kuballa T, Lachenmeier DW (2014) Electronic cigarettes: overview of chemical composition and exposure estimation. Tob Induc Dis 12(1):23

Hiemstra PS, Bals R (2016) Basic science of electronic cigarettes: assessment in cell culture and in vivo models. Respir Res 17:127

Hsiao YC, Matulewicz RS, Sherman SE, Jaspers I, Weitzman ML, Gordon T, Liu CW, Yang Y, Lu K, Bjurlin MA (2023) Untargeted metabolomics to characterize the urinary chemical landscape of E-cigarette users. Chem Res Toxicol 36(4):630–642

Iskandar AR, Gonzalez-Suarez I, Majeed S, Marescotti D, Sewer A, Xiang Y, Leroy P et al (2016) A framework for in vitro systems toxicology assessment of E-liquids. Toxicol Mech Methods 26(6):389–413

Izquierdo-García JL, del Puerto-Nevado L, Peces-Barba G, Pérez-Rial S, Heili S, Villa P, Castejón D, González-Mangado N, Ruiz-Cabello J (2010) A metabonomic approach to evaluate COPD in a model of cigarette smoke exposure in mice. Metabolomics 6(4):564–573

Kim JJ, Sabatelli N, Tutak W, Giuseppetti A, Frukhtbeyn S, Shaffer I, Wilhide J, Routkevitch D, Ondov JM (2017) Universal electronic-cigarette test: physiochemical characterization of reference e-liquid. Tob Induc Dis 15(1):14

Liu G, Lin CJ, Yates CR, Prasad GL (2021) Metabolomic analysis identified reduced levels of xenobiotics, oxidative stress, and improved vitamin metabolism in smokers switched to Vuse Electronic Nicotine Delivery System. Nicot Tob Res 23(7):1133–1142

Lorkiewicz P, Keith R, Lynch J, Jin L, Theis W, Krivokhizhina T, Riggs D, Bhatnagar A, Srivastava S, Conklin DJ (2022) Electronic cigarette solvents, JUUL E-liquids, and biomarkers of exposure: in vivo evidence for acrolein and glycidol in E-cig-derived aerosols. Chem Res Toxicol 35(2):283–292

Margham J, McAdam K, Forster M, Liu C, Wright C, Mariner D, Proctor C (2016) Chemical composition of aerosol from an E-cigarette: a quantitative comparison with cigarette smoke. Chem Res Toxicol 29(10):1662–1678

Merecz-Sadowska A, Sitarek P, Zielinska-Blizniewska H, Malinowska K, Zajdel K, Zakonnik L, Zajdel R (2020) A summary of in vitro and in vivo studies evaluating the impact of e-cigarette exposure on living organisms and the environment. Int J Mol Sci 21(2):652

Morris AM, Leonard SS, Fowles JR, Boots TE, Mnatsakanova A, Attfield KR (2021) Effects of E-cigarette flavoring chemicals on human macrophages and bronchial epithelial cells. Int J Environ Res Public Health 18(21):11107

Moshensky A, Mulong D, Shin J, Advani I, Gunge D, Mathew D, Alkolla R et al (2021) Vaping-induced metabolomic signatures in the circulation of mice are driven by device type, e-liquid, exposure duration and sex. ERJ Open Res 7(3):00229

Muehlbauer LK, Jen A, Zhu Y, He Y, Shishkova E, Overmyer KA, Coon JJ (2023) Rapid multi-omics sample preparation for mass spectrometry. Anal Chem 95(2):659–667

Percival BC, Wann A, Taylor S, Edgar M, Gibson M, Grootveld M (2022) Metabolomics distinction of cigarette smokers from non-smokers using non-stationary benchtop nuclear magnetic resonance (NMR) analysis of human saliva. In: Oral health care-an important issue of the modern society. IntechOpen, London

Pisinger C, Døssing M (2014) A systematic review of health effects of electronic cigarettes. Prev Med 69:248–260

Ren X, Lin L, Sun Q, Li T, Sun M, Sun Z, Duan J (2022) Metabolomics-based safety evaluation of acute exposure to electronic cigarettes in mice. Sci Total Environ 839:156392

Schwanz TG, Bokowski LVV, Marcelo MCA, Jandrey AC, Dias JC, Maximiano DH, Canova LS, Pontes OFS, Sabin GP, Kaiser S (2019) Analysis of chemosensory markers in cigarette smoke from different tobacco varieties by GC×GC-TOFMS and chemometrics. Talanta 202:74–89

Titz B, Szostak J, Sewer A, Phillips B, Nury C, Schneider T, Dijon S et al (2020) Multi-omics systems toxicology study of mouse lung assessing the effects of aerosols from two heat-not-burn tobacco products and cigarette smoke. Comput Struct Biotechnol J 18:1056–1073

Turtoi E, Jeudy J, Henry S, Dadi I, Valette G, Enjalbal C, Turtoi A (2023) Analysis of polar primary metabolites in biological samples using targeted metabolomics and LC-MS. STAR Protoc 4(3):102400

Wang D-G, Dong Q-Q, Du J, Yang S, Zhang Y-J, Na G-S, Ferguson SG, Wang Z, Zheng T (2016) Using Monte Carlo simulation to assess variability and uncertainty of tobacco consumption in a city by sewage epidemiology. Open 6:10583

Wang Q, Ji X, Rahman I (2021) Dysregulated metabolites serve as novel biomarkers for metabolic diseases caused by E-cigarette vaping and cigarette smoking. Metabolites 11(6):345

Wang L, Wang Y, Chen J, Liu P, Li M (2022) A review of toxicity mechanism studies of electronic cigarettes on respiratory system. Int J Mol Sci 23(9):5030

Wei B, Goniewicz M, O'Connor RJ (2019) Concurrent quantification of emerging chemicals of health concern in E-cigarette liquids by high-performance liquid chromatography-tandem mass spectrometry. ACS Omega 4(13):15364–15372

Woodall M, Jacob J, Kalsi KK, Schroeder V, Davis E, Kenyon B, Khan I, Garnett JP, Tarran R, Baines DL (2020) E-cigarette constituents propylene glycol and vegetable glycerin decrease glucose uptake and its metabolism in airway epithelial cells in vitro. Am J Physiol Lung Cell Mol Physiol 319:957–967

Yamaguchi MS, McCartney MM, Falcon AK, Linderholm AL, Ebeler SE, Kenyon NJ, Harper RH, Schivo M, Davis CE (2019) Modeling cellular metabolomic effects of oxidative stress impacts from hydrogen peroxide and cigarette smoke on human lung epithelial cells. J Breath Res 13(3):036014

Health Outcomes of E-cigarette Use or Vaping

10

Mohd Afiq Mohd Nor, Mohd Hafyzuddin Md Yusuf, and Mohammad Aizuddin Azizah Ariffin

Abstract

Despite being advertised as less harmful than smoking combusted tobacco, when used extensively, e-cigs or vaping likely have several health consequences. Vaping is believed to affect the users' and others' respiratory and cardiovascular systems and may cause nicotine addiction. Vaping irritates the user's lungs; for those who have lung diseases such as asthma and bronchitis, vaping may make their conditions worse. Vaping, even briefly, causes a cough, shortness of breath, and inflammation. EVALI, also known as e-cig or vaping use-associated lung injury, is a severe medical problem caused by a vaping habit. Long-term vaping impact on the heart is currently unknown. Nevertheless, research indicates that it can raise blood pressure and heart rate, thus increasing the risk of heart disease. Nicotine in vapes may also worsen existing cardiovascular illness and is highly addictive. Thus, people are unable to stop using vapes because of dependence. It is important to understand that using both e-cigs and conventional cigarettes can increase overall nicotine addiction rather than reduce health risks. We should also be aware that e-cigs could potentially cause cancer because they contain several harmful chemicals, and nicotine itself in its content can support tumor growth and damage the DNA. Overall, the impact of e-cigs on health is complicated, and more research is needed to determine its long-term effects and policies for public health.

M. A. M. Nor (✉)
Universiti Malaya Medical Centre, Kuala Lumpur, Malaysia
e-mail: afiq@ummc.edu.my

M. H. M. Yusuf · M. A. A. Ariffin
Faculty of Medicine, Universiti Malaya, Kuala Lumpur, Malaysia

Y.-H. Leong, A. S. Mohd Samin (eds.), *E-Cigarettes*,
https://doi.org/10.1007/978-981-96-5087-3_10

Keywords

E-cigarette · EVALI · Cardiovascular effect · Cancer risk · Nicotine dependency

10.1 Introduction

Currently, there is a growing interest in novel methods of utilizing tobacco among smokers, which has garnered significant attention in the media. According to reports, it was deemed significantly less hazardous to human well-being. Additionally, it was available in a diverse range of enticing flavors. However, vaping continues to generate significant media attention through the introduction of novel products such as advanced devices and pod systems. Nonetheless, concerns regarding the possible health risks persist.

In this chapter, we will explore the latest research on the subject, which is always evolving. Our aim is to provide a clear and comprehensible understanding of the topic. The health implications of vaping are intricate and are significantly influenced by the composition of the substance, the method of application, and the individual's physiological response. This science fiction, which can be irritatingly specific at times, depicts the current state of affairs and highlights the need for further development. The task involves carefully considering several issues, such as the potential benefits of harm reduction for smokers and the potential risks of vaping, during a thorough evaluation.

Further research indicates that vaping is dangerous to one's health. For example, it has been revealed that in humans, the consumption of nicotine-containing e-cigs causes acute changes in various hemodynamic indexes, notably an increase in blood pressure and heart rate. Additionally, usage of e-cigs has been tied to other concerns such as worsening bronchitis and asthma, nicotine misuse, sickened immune systems, and an increased risk of serious lung disease. Unfortunately, the long-term usage of e-cigs is very worried about the impact on blood vessels and its outcomes. Moreover, it has been well documented that racial polytoxicomans of conventional and e-cigs exacerbate the risks of use far beyond the sum of hazards faced by each product, thus it is necessary to further medical research investigate the use of e-cigs, in particular, to determine the pathogenesis of the disease using organs, such as the lungs and the circulatory bed.

10.2 Effects of Vaping on Lung

Recent studies have shown that vaping may harm our lungs. While some research suggests that vaping might not have a major impact on lung function in the short term or in controlled settings, other studies indicate that vaping, including second-hand exposure, can cause a variety of respiratory issues. This section explains how inhaling vaporized substances, often containing nicotine and other chemicals, might affect lungs and breathing.

Vaping has been associated with worsening bronchitis and asthma, coughing, difficulty breathing, as well as symptoms like fever, nausea, and vomiting. These could be signs of vaping-related lung illness, which is known as e-cigarette or vaping use-associated lung injury (EVALI), and also can worsen existing lung conditions.

10.2.1 Exposure to Harmful Chemicals

E-cigarette vapor and e-liquids contain thousands of potentially harmful compounds, the vast majority of which are unidentified (Tehrani et al. 2021). The heating process of the standard e-liquid constituents such as propylene glycol (PG), vegetable glycerine (VG), and various types of flavorings in the e-cigarette can release new breakdown compounds with unknown toxicity. Regular exposure to these substances, including fine and ultrafine particles from the vapor, can increase the risk of respiratory disorders and promote pulmonary and systemic inflammation (Winnall et al. 2023). Surprisingly, several compounds observed in e-cigarette vapor are not found in e-liquids or the device material but appear to contribute to the presence of metal and silicate particles in the vapor. All these compounds have been related to respiratory irritation, respiratory function abnormalities, lung edema, airway epithelial injury, EVALI, and carcinogens or potential carcinogens.

One of the common e-liquid constituents is propylene glycol (PG), which commonly makes up 95% of e-cigarette liquid volume. Several pharmaceuticals and food products contain PG, a chemical generally regarded as safe (GRAS) for human ingestion. When heated, PG produces vapor from the device and diacetyl, formaldehyde, and methylglyoxal, which may cause irritation, organ damage, and respiratory difficulties when using at high doses or prolonged use (Klager et al. 2017; Peterson et al. 2021). While chronic inhalation of PG from electronic nicotine delivery devices was discovered to play a function in altering the lipid components of lung surfactant and producing or aggravating pulmonary injury (Hayeck et al. 2021). However, other studies have discovered that low concentrations and short exposures have no significant effect on respiratory function or irritation (Kim et al. 2023; Dalton et al. 2018; Farsalinos et al. 2015).

Another common ingredient in e-liquids is vegetable glycerine (VG), which is derived from vegetable oils. VG is used in many products, including e-cigs, and helps create vapor when heated. It has a thicker consistency than propylene glycol (PG), which leads to bigger vapor clouds and a smoother sensation feeling in the throat. Because VG is so widely used, it is important how it affects respiratory health. Similar to propylene glycol, prolonged inhalation of vegetable glycerine can likewise disrupt the lipid constituents of lung surfactant, which may lead to or worsen pulmonary injury (Hayeck et al. 2021). Researchers have found that exposure to glycerol exhaust gases harms the respiratory system, causing changes in respiratory system mechanics, airway resistance, and lung tissue micromechanics in rats (Serra et al. 2017). In conclusion, inhaling PG and VG, especially from combustion or using them in e-cigs, can have detrimental effects on respiratory health.

10.2.2 Acute Respiratory Illness Associated with E-cig Use

Several acute respiratory illnesses have been described in e-cigarette usage cases, including hypersensitivity pneumonitis, organizing pneumonia, diffuse alveolar hemorrhage, eosinophilic pneumonia, giant cell foreign body reaction, and bronchiolitis obliterans, which is also known as "popcorn lung" (Jonas 2022). Some of these diseases could have been early examples of EVALI, as the pathologic concern is lung damage in the absence of infection (Winnall et al. 2023).

Bronchiolitis obliterans, colloquially known as "popcorn lung," has emerged as a concerning respiratory condition associated with vaping. This condition earned its nickname from an earlier association with inhaling diacetyl, a chemical used to produce buttery flavorings, including those found in popcorn (Harber et al. 2006). Researchers have linked diacetyl inhalation to severe inflammation and scarring of the small airways in the lungs, leading to obliterative bronchiolitis (OB), with varying levels of respiratory tract injury depending on exposure duration and concentration (Landman et al. 2019). A study of sub-acute and sub-chronic diacetyl exposures in mice found that they caused significant epithelial injury and lymphocytic inflammation, which are similar to what happens in humans with OB (Morgan et al. 2008). However, differences in diacetyl-induced airway injury between humans and rats may be due to varying levels of diacetyl penetration into the bronchioles, with humans potentially experiencing higher tissue concentrations and a greater risk of injury (Gloede et al. 2011). One study found that diacetyl levels in the emissions of some e-cigs surpass the recommended acceptable levels of exposure, even though these compounds are present in extremely low amounts in e-cigarette emissions (White et al. 2021). Although diacetyl is no longer very commonly used in e-cigarette liquids, other potentially harmful chemicals are present in some vaping products, raising concerns about the risk of developing popcorn lung. Continued research is essential to completely understanding the relationship between vaping and OB, emphasizing the importance of informed decision-making and regulatory actions to mitigate potential respiratory dangers associated with vaping.

10.2.3 Chronic Respiratory Illness Associated with E-cig Use

Because e-cigs were only introduced in the last decade, there is limited data to determine whether they have long-term impacts on respiratory health (Banks et al. 2023). While e-cigs may pose fewer respiratory health risks compared to tobacco cigarettes, there is evidence of impaired respiratory functioning and cytotoxic effects on lung tissue associated with e-cigarette aerosol inhalation (Ratajczak et al. 2018). We understand that exposure to e-cigarette aerosols can lead to oxidative stress and inflammatory responses in lung cells, which are critical events in the pathogenesis of chronic airway disease (Lerner et al. 2015). Exposure to e-cigarette aerosol during adolescence and early adulthood can cause significant impairments in lung function, similar to tobacco smoke exposure (Larcombe et al. 2017).

Like cigarette smoking, electronic cigarette use alters the profile of innate defense proteins in airway secretions, eliciting distinct and comparable changes (Reidel et al. 2017). Flavored e-liquids, particularly those containing cinnamaldehyde, can impair the function of respiratory innate immune cells, potentially increasing susceptibility to infection and impairing the resolution of inflammation (Clapp et al. 2017). Although many flavorings used in e-cigs are generally recognized as safe when used in food products, concerns have been raised about the potential inhalation toxicity of these chemicals. Cherry-flavored e-cigs expose users to benzaldehyde, a key ingredient in natural fruit flavors, which has been shown to cause irritation of respiratory airways in animal and occupational exposure studies. Given the potential inhalation toxicity of this compound, benzaldehyde in aerosol generated in a laboratory setting was measured from flavored e-cigs purchased online and they have detected benzaldehyde in 108 out of 145 products. Benzene exposure also can lead to numerous non-cancerous health effects, including disruption of vital systems and potential links to chronic diseases like diabetes, breast, and lung cancers (Kosmider et al. 2016; Bahadar et al. 2014). These pathologies have the potential to lead to chronic obstructive pulmonary disease (COPD), chronic bronchitis, emphysema, and especially asthma. The relationship between these diseases and asthma is particularly significant among individuals who do not smoke (Wills et al. 2019; Xie et al. 2020).

According to a longitudinal cohort study, using e-cigs increases the risk of respiratory disease in adults, and using both electronic and conventional products at once is riskier than using one or the other alone (Bhatta and Glantz 2020). While in young people, using e-cigs has been linked to an increased risk of wheezing-related respiratory problems (Xie et al. 2022). Studying the long-term effects of vaping on the respiratory system poses a significant challenge. The diversity of vaping products, including the varying composition of their e-liquids, formulations, and delivery mechanisms, makes it difficult to draw broad conclusions.

10.2.4 E-cig or Vaping Use-Associated Lung Injury (EVALI)

Reports of vaping-associated lung injuries (VALI) or EVALI have raised significant concerns. EVALI first made headlines in 2019 when clusters of severe lung injuries began to emerge across the United States. Initially dismissed as isolated incidents, the severity and frequency of cases soon warranted a closer investigation. From August 2019 to February 2020, over 2800 e-cig users in the United States required hospital admission due to EVALI, of whom 68 died. Patients, predominantly teens and young adults, presented with symptoms such as coughing, chest pain, shortness of breath, and, in severe cases, respiratory failure that required hospitalization (Ellington et al. 2020). Although the United States has reported the majority of EVALI cases, there is evidence of isolated cases in other countries like Canada, Japan, and Germany (Schäfer et al. 2021). The first case in Europe was documented in Barcelona, Spain, involving a patient traveling from the United States (Casanova et al. 2020). One probable case was reported in the UK, and two confirmed cases in Malaysia (Ind 2020).

EVALI is diagnosed by considering the individual's vaping history within the past 90 days, their clinical symptoms, imaging results, and laboratory tests. It is crucial to understand that the diagnosis of EVALI involves ruling out other possible causes. Hence, it is imperative to eliminate any potential factors that could contribute to lung damage, such as pneumonia, congestive heart failure, or other inflammatory etiologies (Karpenko et al. 2022; Jonas and Raj 2020; Smith et al. 2021; Ind 2020). EVALI can be undiagnosed due to a lack of awareness among healthcare providers, especially during the COVID-19 pandemic (Rai et al. 2022).

Research is ongoing to determine the mechanisms causing lung injury in this illness. One of the primary challenges in understanding EVALI lies in the diverse range of substances present in vaping products. As investigations into EVALI unfolded, researchers identified a familiar culprit: Vitamin E acetate, added to vaping products as a diluent, is thought to be one of the main reasons for the EVALI outbreak (Blount et al. 2019). This oily substance, often used as a thickening agent in illicit Tetrahydrocannabinol (THC)-containing vaping products, emerged as a potential cause of lung injury (Lozier et al. 2019). Several cases have also documented the purchase or acquisition of products via informal sources, such as family and friends. Vitamin E acetate can coat the lungs when heated and inhaled, leading to inflammation and impaired respiratory function (Boudi et al. 2019).

On the other hand, substantial evidence links the use of e-cigs, both with and without THC, to EVALI. Researchers found THC in the lung fluid of about 94% of EVALI patients with Broncho-Alveolar Lavage (BAL). In comparison, 51% of EVALI patients reported recent use of both THC and nicotine e-cigs, and 17% reported sole use of nicotine e-cigs (Blount et al. 2019; Marrocco et al. 2022). Taken together, these studies indicate that between 6% and 17% of EVALI patients were not THC-containing e-cig users. This discovery provided crucial insights into the outbreak and underscored the dangers of unregulated vaping products and the need for stricter oversight.

While most patients recover with supportive care, and corticosteroids have been used in severe cases, the long-term health effects and optimal treatment strategies require further study (Smith et al. 2021; Ind 2020; Abbara and Kay 2019; Karpenko et al. 2022). In conclusion, EVALI serves as a sobering reminder of the dangers associated with vaping and the urgent need for greater regulation and awareness.

10.3 Effects of Dual Users of Cigarettes

While e-cigs, or vaping, have been marketed as a potential tool for quitting smoking, many people end up using both combustible cigarettes and e-cigs concurrently—a habit known as dual use. They often do so to reduce their tobacco consumption or quit smoking altogether (Adriaens et al. 2017). However, studies suggest that this behavior may not necessarily lead to a reduction in harm. Dual users may expose themselves to additional health risks rather than reaping the purported benefits of harm reduction (Azagba 2018). Dual use of e-cigs and cigarettes is associated with lower general health scores and increased respiratory health risks (Wang et al. 2018).

One of the primary concerns surrounding dual use is the potential for nicotine dependence. A study has demonstrated that individuals who use both combustible cigarettes and e-cigs see a decrease in the number of combustible cigarettes they smoke. However, their overall consumption of nicotine and level of dependence on nicotine increases (Martínez et al. 2020). High-frequency dual users have a substantially greater likelihood of nicotine dependence compared to individuals who smoke cigarettes (Azagba et al. 2019). This situation signifies a transition or a shift rather than a complete elimination of nicotine addiction (Adriaens et al. 2017).

Furthermore, the inclusion of flavored e-cigs may enhance the appeal and quantity of nicotine consumed, potentially worsening addiction (Snow 2020). Dual use is also linked to the longer time it takes to fall asleep (increased sleep patency), which may indirectly indicate a greater intake of nicotine that affects sleep patterns (Advani et al. 2022). While e-cigs result in lower overall nicotine intake than conventional cigarettes, individuals who use both products have a decrease in withdrawal symptoms and cravings for either product. This indicates a possible adjustment in nicotine addiction (St. Helen et al. 2020). Dual users of cigarettes and e-cigs exhibit unique aspects of nicotine dependency for each product without a shared underlying cause, highlighting the necessity for separate evaluations of dependence (Rest et al. 2021). Overall, using vaping devices and cigarettes simultaneously appears to increase overall nicotine consumption and addiction despite a decrease in conventional cigarette smoking. This behavior may result in the transfer of nicotine dependency from one product to another, rather than effectively reducing total addiction. Engaging in dual-use also exposes individuals to a mixture of noxious substances (Darville and Hahn 2019). Cigarettes contain several chemicals, such as tar and carbon monoxide, that are associated with the development of lung cancer, heart disease, and various other health issues. Although e-cigs may emit fewer dangerous chemicals than conventional cigarettes, they still pose dangers as they release detectable toxic components and volatile organic compounds (VOCs) (Papaefstathiou et al. 2019). As we previously stated, inhaling vaporized chemicals such as propylene glycol and flavorings may still have health risks with uncertain long-term consequences, particularly when combined with the harmful substances in combustible tobacco smoke.

Researchers link the combined use of e-cigs and conventional tobacco to higher levels of nicotine and harmful substances compared to using either product alone. Dual users show similar levels of toxicant biomarkers as individuals who solely smoke cigarettes (Smith et al. 2020). Nevertheless, subsets of individuals who use both cigarettes and alternative products exhibit different degrees of biomarkers signaling exposure and potential harm. This suggests that not all dual users are identical, and the frequency at which they smoke cigarettes is a critical factor in determining their levels of exposure (Smith et al. 2020). In addition to the longer time to fall asleep, dual-use is linked to a greater likelihood of reporting coughing.

This could be because both e-cigs and conventional tobacco contain nicotine, which is a common factor between them (Advani et al. 2022).

Another concern with dual use is its potential to undermine cessation efforts. Some people believe vaping can help them cut down on cigarettes, but research suggests otherwise. A study shows that dual users are more likely to attempt quitting smoking than exclusive smokers, but they tend to relapse more quickly and are not more likely to achieve 30-day smoking cessation (Frost-Pineda et al. 2010). They often struggle to quit smoking altogether, and some may even end up smoking more cigarettes alongside vaping. This could be due to the psychological dependence cigarettes create or the belief that vaping somehow "cancels out" the risks of smoking (Arnold et al. 2021). Vaping can create a false sense of security, reducing the urgency a smoker feels to quit cigarettes completely. They might believe vaping mitigates the harm of smoking, delaying or preventing a total commitment to quitting, indicating that additional cessation resources may be necessary for these individuals (McClave-Regan and Berkowitz 2011). In some cases, dual-use might lead to increased cigarette consumption. The ease of access and perceived "safety" of vapes could lead smokers to feel comfortable indulging in more cigarettes alongside vaping (Öznur Muz et al. 2022).

There is some evidence that dual users who primarily vape might be more likely to quit smoking altogether eventually (Frost-Pineda et al. 2010). However, more research is necessary to confirm this, and it's crucial to remember that even a small amount of cigarette smoking carries significant health risks. In conclusion, the dual use of combustible cigarettes and e-cigs presents a complex and multifaceted challenge in the realm of tobacco control. The current evidence indicates that the dual use of cigarettes and vapes is not a safe alternative to smoking. It exposes users to a double dose of harmful chemicals and may impede successful smoking cessation.

10.4 Carcinogenic Effects of Vaping

This research subchapter focuses on the carcinogenic effects of vaping, which has become a prevalent concern due to the increasing popularity of e-cigs as an alternative to conventional smoking among various age groups, including teenagers and never-smokers. Despite being marketed as a safer option, the safety and long-term health effects of e-cigs are not fully understood, particularly concerning cancer risk.

The potential carcinogenic effects of e-cig use on living organisms are still uncertain. Several different molecular mechanisms have been proposed for the oncogenic impact of vaping fluids—by means of their direct chemical action or derivative products generated by pyrolysis and combustion ranging from epithelial–mesenchymal transition, redox stress and mitochondrial toxicity to DNA breaks and fragmentation (Petrella et al. 2023). A study utilizing a rat lung model to investigate the co-mutagenic and cancer-initiating properties of e-cig vapor demonstrated that exposure to e-cigs could potentially endanger human health, especially among younger individuals who are more vulnerable. Scientists have found that e-cigs significantly increase the effectiveness of enzymes that activate cancer-causing substances in the body, including polycyclic aromatic hydrocarbons (PAHs). They also increase the production of harmful oxygen free radicals and DNA oxidation, specifically 8-hydroxy-2′-deoxyguanosine. Additionally, it was found that e-cigs have a detrimental effect on DNA at both the gene level, as evidenced by point mutations in urine, and at the chromosomal level in peripheral blood, where they induce strand breakage in leucocytes and the production of micronuclei in reticulocytes (Canistro et al. 2017).

The e-liquid mixture contains several chemicals in small quantities that are recognized as carcinogens. Several of these chemicals are synthesized due to the heating process in the e-cig (Marques et al. 2021). Over an extended period, an individual who often uses e-cigs may inhale these compounds multiple times a day, consistently. The extent to which this degree of exposure can lead to the development of cancer remains uncertain. However, there is concern that specific e-cig users may experience the development of cancer due to cellular harm and biomarkers indicating exposure to cancer-causing chemicals in e-cig users. The correlation between lung cancer and tobacco smoking, which required more than 50 years to establish, has yielded valuable insights (Wynder 1988).

Vaping use also has potential carcinogenic effects on the lungs and may cause cancer in individuals with chronic obstructive pulmonary disease (O'Callaghan et al. 2022; Schuller 2019). Vaping devices and e-liquids contain carcinogens such as nicotine derivatives, polycyclic aromatic hydrocarbons, heavy metals, and aldehydes, which may pose cancer risks (Bracken-Clarke et al. 2021). Vaping of vitamin E acetate can produce toxic ketene gas and carcinogen alkenes and benzene, which have adverse long-term medical effects (Wu and O'Shea 2020).

Extensive research has been conducted on nicotine, the addictive substance found in tobacco smoke, to investigate its potential contribution to the development of cancer. Nicotine has the ability to transform into nitrosamine compounds, specifically nitrosamine ketone and nitrosonornicotine. This characteristic of nicotine gives it a strong potential to cause cancer (Tang et al. 2019). The majority of inhaled nicotine is metabolized into cotinine, a compound believed to be non-toxic, and subsequently eliminated through urine. Nevertheless, the body's natural conversion of inhaled nicotine to nitrosamine molecules appears minimal, accounting for approximately 10% (Stepanov et al. 2009; Benowitz et al. 2009).

Studies have shown that nicotine can greatly enhance the advancement and spread of tumors in mice models of lung cancer, using both intraperitoneal injections and transdermal patches. The administration of nicotine resulted in an augmentation in the size of implanted tumors, a greater likelihood of tumor recurrence following surgical removal, and an elevated rate of metastasis to the lungs (Heeschen et al. 2001). Studies have demonstrated that nicotine can stimulate the growth of cells and the formation of new blood vessels, known as angiogenesis. These processes have been shown to contribute to the development and progression of tumors (Dasgupta et al. 2009).

Nicotine has been discovered to stimulate multiple signaling pathways commonly associated with cellular development and viability. Nicotine has the ability to activate the Src and Rb-Raf 1 pathways, which can cause the growth of cells in non-small cell lung malignancies (NSCLCs) (Dasgupta 2006). Additionally, it triggers the phosphorylation of the proapoptotic protein Bad, leading to the inhibition of apoptosis and the promotion of cell survival. This mechanism may have a role in developing lung cancer and resistance to chemotherapy (Jin et al. 2004).

The association between nicotine and cancer extends beyond lung cancer. Nicotine has been associated with various malignancies, such as those affecting the head and neck, stomach, pancreas, gallbladder, liver, colon, breast, cervix, urinary bladder, and kidney (Grando 2014). The tumorigenic effects of nicotine have been linked to its potential to harm the genome, interfere with cellular metabolic processes, and promote the proliferation and spread of altered cells (Davis et al. 2009).

Furthermore, nicotine has been demonstrated to induce angiogenesis, the process of generating new blood vessels, hence facilitating the growth of tumors and atherosclerosis. Furthermore, the impact of nicotine on microRNA expression, particularly the increase in miR-16 and miR-21, has been detected in gastric cancer cells. This proves that nicotine stimulates cell growth through pathways regulated by NF-κB (Shin et al. 2011).

While e-cigs are marketed as a safer alternative because they do not burn tobacco, however studies suggest that the toxic effects of e-cig aerosol exposure are essentially identical to those caused by combustible cigarette smoking. Both increase proinflammatory cytokine expression and affect protein regulation, leading to an increased lung cancer risk (Abelia et al. 2023). However, because vapes contain a higher proportion of nicotine than tobacco cigarettes, they may be more hazardous to lung cancer survival. The A549 cell line was used to examine the effects of various nicotine concentrations. The results indicated that up to 2% nicotine concentration considerably enhanced the vitality of A549 cells, but that more than 5% nicotine concentration caused cell death due to its high toxicity (Choe and Lee 2022).

An experiment was done to find out how much second-hand smoke from regular cigarettes and e-cigs people are exposed to, as well as the lung cancer risk that passive users and vapers are exposed to. They measured exposure levels by looking at the number of particles, PM10 and black carbon concentrations, as well as the distribution of particle sizes, in an indoor setting that was naturally aired while people smoked tobacco and e-cigs. For the different aerosol measures that were being investigated, the particle emission rates of smokers and vapers were checked. A Multiple-Path Particle Dosimetry (MPPD) model was used to estimate the dose that passive smokers and vapers received in an indoor environment that was naturally ventilated. This model could figure out the particle dose that was received in the different parts of the respiratory system. Finally, they conclude that passive vapers at indoor get up to 15 times less than second-hand tobacco aerosol into their lungs and have a five times lower Extra Life Cancer Risk (ELCR) than second-hand smokers (Avino et al. 2018).

Genetic predispositions have a substantial impact on the likelihood of developing cancer, and this risk is further influenced by smoking habits (Suzuki et al. 2007). Individuals who have a genetic predisposition to cancer and also smoke are more likely to acquire cancer compared to non-smokers with a similar genetic makeup. Particular genetic variations can alter the impact of smoking on the likelihood of developing cancer. For instance, several genetic variations greatly enhance the likelihood of developing bladder cancer in individuals who smoke, particularly in those with a high polygenic risk score (PRS). Nevertheless, this potential hazard has not yet been demonstrated in the vaping community. Eliminating smoking might effectively reduce the incidence of bladder cancer in persons with a higher genetic predisposition, hence potentially influencing focused prevention methods (Garcia-Closas et al. 2013).

To summarize, although nicotine does not directly cause cancer, but it can expedite the development of tumors, their dissemination, and the appearance of cancerous cells through various mechanisms. These mechanisms include activating signaling pathways that lead to the growth of new blood vessels, increased cell division, the prevention of cell death, and the initiation of a process called epithelial-mesenchymal transition. Although there is currently a paucity of epidemiological data on the cause of cancer by e-cigs, it does not mean that e-cig consumption cannot lead to cancer. Given lag durations of up to 20 years (extrapolated from tobacco smoke statistics), this could have serious future public health consequences, necessitating additional research and stricter regulation.

10.5 Cardiovascular Effects of Vaping

There is a lack of consensus about the potential impact of e-cig usage on the cardiovascular system. The majority of research conducted on e-cig use is constrained to limited or short-term studies, and there is a lack of long-term studies to definitively determine the association between e-cig use and cardiovascular disease (Stratton et al. 2018).

The Australian Department of Health's Report has determined that there is currently no evidence to demonstrate the impact of e-cig use on the risk of subclinical atherosclerosis-related outcomes, such as carotid intima-media thickness and coronary artery calcification, as well as the risk of clinical cardiovascular disease outcomes, such as myocardial infarction, stroke, or cardiovascular mortality on user (Banks et al. 2023). Even though there is not enough evidence to prove that using e-cigs affects the autonomic control of cardiovascular outcomes in non-smokers, there is some evidence to suggest that e-cig use can temporarily increase heart rate, systolic blood pressure, diastolic blood pressure, and arterial stiffness in smokers (Franzen et al. 2018).

The rapid impact of vaping on cardiovascular parameters prompts inquiries concerning its short-term influence on the circulatory system. Studies suggest that the activation of the sympathetic nervous system by nicotine exposure in e-cigs may exacerbate cardiovascular issues. However, this effect is less pronounced compared to the immediate impact of conventional tobacco smoking (Garcia et al. 2020). Nicotine attaches to α3β4 nicotinic acetylcholine receptors in postganglionic sympathetic nerve terminals and the adrenal medulla. This attachment leads to the discharge of catecholamines, specifically norepinephrine and epinephrine. Catecholamine production stimulates β-adrenergic receptors in the heart, resulting in heightened heart rate, contractility, and workload (Benowitz and Fraiman 2017). When young adult rats are exposed to nicotine for a long time, it hurts their hearts by blocking the process of mitophagy clearance. This could cause heart muscle cells to fail in young adult populations (Jia et al. 2020).

A study conducted in 2018 discovered a correlation between the use of e-cigs and a higher likelihood of experiencing arrhythmia, palpitations, chest pain, and coronary artery disease when compared to individuals who do not use e-cigs. However, it is important to note that the study did not sufficiently consider the extent of past cigarette smoking (Wang et al. 2018).

Daily e-cigarette use was associated with an increased risk of heart attacks, especially among dual users of ecigarettes and combustible cigarettes (Alzahrani et al. 2018). The current umbrella review indicates that the use of e-cigs is linked to a higher likelihood of experiencing adverse cardiac-related clinical events, such as atrial fibrillation and myocardial infarction. However, e-cigs may be a temporary alternative that is less harmful than regular combustible cigarettes for them (Peruzzi et al. 2020).

Although the immediate cardiovascular effects of vaping are becoming increasingly evident, further long-term research is still required. Assessing the enduring effects of vaping on the cardiovascular system, specifically the potential emergence of chronic ailments, is crucial for formulating public health guidelines prior to offering guidance to individuals considering vaping as a substitute (Besaratinia 2021).

10.6 Conclusion

E-cigs can have harmful health consequences, especially for young individuals, children, and non-smokers. These factors may have unanticipated impacts on numerous significant health outcomes. Although e-cigs are not officially recognized as smoking cessation aids, they may offer certain advantages for smokers who use them to rapidly and ultimately quit smoking. There is an urgent need for more rigorous research to properly regulate e-cigs and evaluate their safety and efficacy in helping people quit smoking and vaping.

References

Abbara S, Kay FU (2019) Electronic cigarette or vaping-associated lung injury (EVALI): the tip of the iceberg. Radiol Cardiothorac Imaging 1(4):e190212. https://doi.org/10.1148/ryct.2019190212

Abelia XA, Lesmana R, Goenawan H, Abdulah R, Barliana MI (2023) Comparison impact of cigarettes and e-cigs as lung cancer risk inductor: a narrative review. Eur Rev Med Pharmacol Sci 27(13):6301–6318. https://doi.org/10.26355/eurrev_202307_32990

Adriaens K, Van Gucht D, Baeyens F (2017) Differences between dual users and switchers center around vaping behavior and its experiences rather than beliefs and attitudes. Int J Environ Res Public Health 15(1):12. https://doi.org/10.3390/ijerph15010012

Advani I, Gunge D, Boddu S, Mehta S, Park K, Perera S, Pham J, Nilaad S, Olay J, Ma L, Masso-Silva J, Sun X, Jain S, Malhotra A, Crotty Alexander LE (2022) Dual use of e-cigarettes with conventional tobacco is associated with increased sleep latency in cross-sectional study. Sci Rep 12(1):2536. https://doi.org/10.1038/s41598-022-06445-8

Alzahrani T, Pena I, Temesgen N, Glantz SA (2018) Association between electronic cigarette use and myocardial infarction. Am J Prev Med 55(4):455–461. https://doi.org/10.1016/j.amepre.2018.05.004

Arnold MJ, Nollen NL, Mayo MS, Ahluwalia JS, Leavens EL, Zhang G, Rice M, Pulvers K (2021) Harm reduction associated with dual use of cigarettes and e-cigarettes in Black and Latino smokers: secondary analyses from a randomized controlled e-cigarette switching trial. Nicotine Tob Res 23(11):1972–1976. https://doi.org/10.1093/ntr/ntab069

Avino P, Scungio M, Stabile L, Cortellessa G, Buonanno G, Manigrasso M (2018) Second-hand aerosol from tobacco and electronic cigarettes: evaluation of the smoker emission rates and doses and lung cancer risk of passive smokers and vapers. Sci Total Environ 642:137–147. https://doi.org/10.1016/j.scitotenv.2018.06.059

Azagba S (2018) E-cigarette use, dual use of e-cigarettes and tobacco cigarettes, and frequency of cannabis use among high school students. Addict Behav 79:166–170. https://doi.org/10.1016/j.addbeh.2017.12.028

Azagba S, Shan L, Latham K (2019) Adolescent dual use classification and its association with nicotine dependence and quit intentions. J Adolesc Health 65(2):195–201. https://doi.org/10.1016/j.jadohealth.2019.04.009

Bahadar H, Mostafalou S, Abdollahi M (2014) Current understandings and perspectives on non-cancer health effects of benzene: a global concern. Toxicol Appl Pharmacol 276(2):83–94. https://doi.org/10.1016/j.taap.2014.02.012

Banks E, Yazidjoglou A, Brown S, Nguyen M, Martin M, Beckwith K, Daluwatta A, Campbell S, Joshy G (2023) Electronic cigarettes and health outcomes: umbrella and systematic review of the global evidence. Med J Aust 218(6):267–275. https://doi.org/10.5694/mja2.51890

Benowitz NL, Fraiman JB (2017) Cardiovascular effects of electronic cigarettes. Nat Rev Cardiol 14(8):447–456. https://doi.org/10.1038/nrcardio.2017.36

Benowitz NL, Hukkanen J, Jacob P (2009) Nicotine chemistry, metabolism, kinetics and biomarkers, pp 29–60. https://doi.org/10.1007/978-3-540-69248-5_2

Besaratinia A (2021) From tobacco cigarettes to electronic cigarettes: The two sides of a nicotine coin. Frontiers in Oral Health, 2, 790634. https://doi.org/10.3389/froh.2021.790634

Bhatta DN, Glantz SA (2020) Association of E-cigarette use with respiratory disease among adults: a longitudinal analysis. Am J Prev Med 58(2):182–190. https://doi.org/10.1016/j.amepre.2019.07.028

Blount BC, Karwowski MP, Morel-Espinosa M, Rees J, Sosnoff C, Cowan E, Gardner M, Wang L, Valentin-Blasini L, Silva L, De Jesús VR, Kuklenyik Z, Watson C, Seyler T, Xia B, Chambers D, Briss P, King BA, Delaney L et al (2019) Evaluation of bronchoalveolar lavage fluid from patients in an outbreak of E-cigarette, or vaping, product use–associated lung injury—10 states, August–October 2019. MMWR Morb Mortal Wkly Rep 68(45):1040–1041. https://doi.org/10.15585/mmwr.mm6845e2

Boudi FB, Patel S, Boudi A, Chan C (2019) Vitamin E acetate as a plausible cause of acute vaping-related illness. Cureus 11(12):e6350. https://doi.org/10.7759/cureus.6350

Bracken-Clarke D, Kapoor D, Baird AM, Buchanan PJ, Gately K, Cuffe S, Finn SP (2021) Vaping and lung cancer—a review of current data and recommendations. Lung Cancer 153:11–20. https://doi.org/10.1016/j.lungcan.2020.12.030

Canistro D, Vivarelli F, Cirillo S, Babot Marquillas C, Buschini A, Lazzaretti M, Marchi L, Cardenia V, Rodriguez-Estrada MT, Lodovici M, Cipriani C, Lorenzini A, Croco E, Marchionni S, Franchi P, Lucarini M, Longo V, Della Croce CM, Vornoli A et al (2017) E-cigarettes induce toxicological effects that can raise the cancer risk. Sci Rep 7(1):2028. https://doi.org/10.1038/s41598-017-02317-8

Casanova GS, Amaro R, Soler N, Sánchez M, Badía JR, Barberà JA, Agustí A (2020) An imported case of e-cigarette or vaping associated lung injury in Barcelona. Eur Respir J 55(2):1902076. https://doi.org/10.1183/13993003.02076-2019

Choe J, Lee WR (2022) Comparing the effects of electronic cigarette smoke and conventional cigarette smoke on lung cancer viability. J Emerg Investig. https://doi.org/10.59720/22-043

Clapp PW, Pawlak EA, Lackey JT, Keating JE, Reeber SL, Glish GL, Jaspers I (2017) Flavored e-cigarette liquids and cinnamaldehyde impair respiratory innate immune cell function. Am J Phys Lung Cell Mol Phys 313(2):L278–L292. https://doi.org/10.1152/ajplung.00452.2016

Dalton P, Soreth B, Maute C, Novaleski C, Banton M (2018) Lack of respiratory and ocular effects following acute propylene glycol exposure in healthy humans. Inhal Toxicol 30(3):124–132. https://doi.org/10.1080/08958378.2018.1470207

Darville A, Hahn EJ (2019) E-cigarettes and atherosclerotic cardiovascular disease: what clinicians and researchers need to know. Curr Atheroscler Rep 21(5):15. https://doi.org/10.1007/s11883-019-0777-7

Dasgupta P (2006) Nicotine induces cell proliferation by arrestin-mediated activation of Src and Rb-Raf-1 pathways. J Clin Invest 116(8):2208–2217. https://doi.org/10.1172/JCI28164

Dasgupta P, Rizwani W, Pillai S, Kinkade R, Kovacs M, Rastogi S, Banerjee S, Carless M, Kim E, Coppola D, Haura E, Chellappan S (2009) Nicotine induces cell proliferation, invasion and epithelial-mesenchymal transition in a variety of human cancer cell lines. Int J Cancer 124(1):36–45. https://doi.org/10.1002/ijc.23894

Davis R, Rizwani W, Banerjee S, Kovacs M, Haura E, Coppola D, Chellappan S (2009) Nicotine promotes tumor growth and metastasis in mouse models of lung cancer. PLoS One 4(10):e7524. https://doi.org/10.1371/journal.pone.0007524

Ellington S, Salvatore PP, Ko J, Danielson M, Kim L, Cyrus A, Wallace M, Board A, Krishnasamy V, King BA, Rose D, Jones CM, Pollack LA, Abbas A, Adebayo A, Atti S, Carter E, Chandra G, Eckhaus L et al (2020) Update: product, substance-use, and demographic characteristics of hospitalized patients in a nationwide outbreak of E-cigarette, or vaping, product use–associated lung injury—United States, August 2019–January 2020. MMWR Morb Mortal Wkly Rep 69(2):44–49. https://doi.org/10.15585/mmwr.mm6902e2

Farsalinos KE, Kistler KA, Gillman G, Voudris V (2015) Evaluation of electronic cigarette liquids and aerosol for the presence of selected inhalation toxins. Nicotine Tob Res 17(2):168–174. https://doi.org/10.1093/ntr/ntu176

Franzen KF, Willig J, Cayo Talavera S, Meusel M, Sayk F, Reppel M, Dalhoff K, Mortensen K, Droemann D (2018) E-cigarettes and cigarettes worsen peripheral and central hemodynamics as well as arterial stiffness: a randomized, double-blinded pilot study. Vasc Med 23(5):419–425. https://doi.org/10.1177/1358863X18779694

Frost-Pineda K, Appleton S, Fisher M, Fox K, Gaworski CL (2010) Does dual use jeopardize the potential role of smokeless tobacco in harm reduction? Nicotine Tob Res 12(11):1055–1067. https://doi.org/10.1093/ntr/ntq147

Garcia PD, Gornbein JA, Middlekauff HR (2020) Cardiovascular autonomic effects of electronic cigarette use: a systematic review. Clin Auton Res 30(6):507–519. https://doi.org/10.1007/s10286-020-00683-4

Garcia-Closas M, Rothman N, Figueroa JD, Prokunina-Olsson L, Han SS, Baris D, Jacobs EJ, Malats N, De Vivo I, Albanes D, Purdue MP, Sharma S, Fu Y-P, Kogevinas M, Wang Z, Tang W, Tardón A, Serra C, Carrato A et al (2013) Common genetic polymorphisms modify the effect of smoking on absolute risk of bladder cancer. Cancer Res 73(7):2211–2220. https://doi.org/10.1158/0008-5472.CAN-12-2388

Gloede E, Cichocki JA, Baldino JB, Morris JB (2011) A validated hybrid computational fluid dynamics-physiologically based pharmacokinetic model for respiratory tract vapor absorption in the human and rat and its application to inhalation dosimetry of diacetyl. Toxicol Sci 123(1):231–246. https://doi.org/10.1093/toxsci/kfr165

Grando SA (2014) Connections of nicotine to cancer. Nat Rev Cancer 14(6):419–429. https://doi.org/10.1038/nrc3725

Harber P, Saechao K, Boomus C (2006) Diacetyl-induced lung disease. Toxicol Rev 25(4):261–272. https://doi.org/10.2165/00139709-200625040-00006

Hayeck N, Zoghzoghi C, Karam E, Salman R, Karaoghlanian N, Shihadeh A, Eissenberg T, Zein El Dine S, Saliba NA (2021) Carrier solvents of electronic nicotine delivery systems alter pulmonary surfactant. Chem Res Toxicol 34(6):1572–1577. https://doi.org/10.1021/acs.chemrestox.0c00528

Heeschen C, Jang JJ, Weis M, Pathak A, Kaji S, Hu RS, Tsao PS, Johnson FL, Cooke JP (2001) Nicotine stimulates angiogenesis and promotes tumor growth and atherosclerosis. Nat Med 7(7):833–839. https://doi.org/10.1038/89961

Ind PW (2020) E-cigarette or vaping product use-associated lung injury. Br J Hosp Med 81(4):1–9. https://doi.org/10.12968/hmed.2019.0371

Jia G, Meng Z, Liu C, Ma X, Gao J, Liu J, Guo R, Yan Z, Christopher T, Lopez B, Liu W, Dai H, Lau WB, Jiao X, Zhao J, Wang Z-X, Cao J, Wang Y (2020) Nicotine induces cardiac toxicity through blocking mitophagic clearance in young adult rat. Life Sci 257:118084. https://doi.org/10.1016/j.lfs.2020.118084

Jin Z, Gao F, Flagg T, Deng X (2004) Nicotine induces multi-site phosphorylation of bad in association with suppression of apoptosis. J Biol Chem 279(22):23837–23844. https://doi.org/10.1074/jbc.M402566200

Jonas A (2022) Impact of vaping on respiratory health. BMJ 378:e065997. https://doi.org/10.1136/bmj-2021-065997

Jonas AM, Raj R (2020) Vaping-related acute parenchymal lung injury. Chest 158(4):1555–1565. https://doi.org/10.1016/j.chest.2020.03.085

Karpenko MA, Ovsyannikov DY, Frolov PA, Nikiforova TI, Khandy MV (2022) E-cigarette or vaping use-associated lung injury. Tuber Lung Dis 100(4):52–61. https://doi.org/10.21292/2075-1230-2022-100-4-52-61

Kim MD, Chung S, Baumlin N, Sun L, Silswal N, Dennis JS, Yoshida M, Sabater J, Horrigan FT, Salathe M (2023) E-cigarette aerosols of propylene glycol impair BK channel activity and parameters of mucociliary function. Am J Phys Lung Cell Mol Phys 324(4):L468–L479. https://doi.org/10.1152/ajplung.00157.2022

Klager S, Vallarino J, MacNaughton P, Christiani DC, Lu Q, Allen JG (2017) Flavoring chemicals and aldehydes in E-cigarette emissions. Environ Sci Technol 51(18):10806–10813. https://doi.org/10.1021/acs.est.7b02205

Kosmider L, Sobczak A, Prokopowicz A, Kurek J, Zaciera M, Knysak J, Smith D, Goniewicz ML (2016) Cherry-flavoured electronic cigarettes expose users to the inhalation irritant, benzaldehyde. Thorax 71(4):376–377. https://doi.org/10.1136/thoraxjnl-2015-207895

Landman ST, Dhaliwal I, Mackenzie CA, Martinu T, Steel A, Bosma KJ (2019) Life-threatening bronchiolitis related to electronic cigarette use in a Canadian youth. Can Med Assoc J 191(48):E1321. https://doi.org/10.1503/cmaj.191402

Larcombe AN, Janka MA, Mullins BJ, Berry LJ, Bredin A, Franklin PJ (2017) The effects of electronic cigarette aerosol exposure on inflammation and lung function in mice. Am J Phys Lung Cell Mol Phys 313(1):L67–L79. https://doi.org/10.1152/ajplung.00203.2016

Lerner CA, Sundar IK, Yao H, Gerloff J, Ossip DJ, McIntosh S, Robinson R, Rahman I (2015) Vapors produced by electronic cigarettes and E-juices with flavorings induce toxicity, oxidative stress, and inflammatory response in lung epithelial cells and in mouse lung. PLOS One 10(2):e0116732. https://doi.org/10.1371/journal.pone.0116732

Lozier MJ, Wallace B, Anderson K, Ellington S, Jones CM, Rose D, Baldwin G, King BA, Briss P, Mikosz CA, Austin C, Brown S, Chandra G, Coulliette-Salmond A, Coy K, Currie D, Cyrus A, Danielson M, Fajardo G et al (2019) Update: demographic, product, and substance-use characteristics of hospitalized patients in a nationwide outbreak of E-cigarette, or vaping, product use–associated lung injuries—United States, December 2019. MMWR Morb Mortal Wkly Rep 68(49):1142–1148. https://doi.org/10.15585/mmwr.mm6849e1

Marques P, Piqueras L, Sanz M-J (2021) An updated overview of e-cigarette impact on human health. Respir Res 22(1):151. https://doi.org/10.1186/s12931-021-01737-5

Marrocco A, Singh D, Christiani DC, Demokritou P (2022) E-cigarette vaping associated acute lung injury (EVALI): state of science and future research needs. Crit Rev Toxicol 52(3):188–220. https://doi.org/10.1080/10408444.2022.2082918

Martínez Ú, Martínez-Loredo V, Simmons VN, Meltzer LR, Drobes DJ, Brandon KO, Palmer AM, Eissenberg T, Bullen CR, Harrell PT, Brandon TH (2020) How does smoking and nicotine dependence change after onset of vaping? A retrospective analysis of dual users. Nicotine Tob Res 22(5):764–770. https://doi.org/10.1093/ntr/ntz043

McClave-Regan AK, Berkowitz J (2011) Smokers who are also using smokeless tobacco products in the US: a national assessment of characteristics, behaviours and beliefs of "dual users". Tob Control 20(3):239–242. https://doi.org/10.1136/tc.2010.039115

Morgan DL, Flake GP, Kirby PJ, Palmer SM (2008) Respiratory toxicity of Diacetyl in C57Bl/6 mice. Toxicol Sci 103(1):169–180. https://doi.org/10.1093/toxsci/kfn016

O'Callaghan M, Boyle N, Fabre A, Keane MP, McCarthy C (2022) Vaping-associated lung injury: a review. Medicina 58(3):412. https://doi.org/10.3390/medicina58030412

Öznur Muz F, Metintaş S, Önsüz M, Aydoğan Gedik S, Mutlu A, Arslan Torba T (2022) Evaluation of factors associated with dual use of E-cigarettes in university students. Eur J Pub Health 32(Supplement_3). https://doi.org/10.1093/eurpub/ckac129.036

Papaefstathiou E, Stylianou M, Agapiou A (2019) Main and side stream effects of electronic cigarettes. J Environ Manag 238:10–17. https://doi.org/10.1016/j.jenvman.2019.01.030

Peruzzi M, Biondi-Zoccai G, Carnevale R, Cavarretta E, Frati G, Versaci F (2020) Vaping cardiovascular health risks: an updated umbrella review. Curr Emerg Hosp Med Rep 8(3):103–109. https://doi.org/10.1007/s40138-020-00219-0

Peterson J, Stadlen R, Radke JB (2021) Propylene glycol toxicity from compulsive corn starch ingestion: a case report and review. Am J Emerg Med. https://api.semanticscholar.org/CorpusID:238258733

Petrella F, Rizzo S, Masiero M, Marzorati C, Casiraghi M, Bertolaccini L, Mazzella A, Pravettoni G, Spaggiari L (2023) Clinical impact of vaping on cardiopulmonary function and lung cancer development: an update. Eur J Cancer Prev 32(6):584–589. https://doi.org/10.1097/CEJ.0000000000000797

Rai S, Hormozdyaran S, Burns J, Amodio JB, Quizon AI (2022) Diagnosis of EVALI in adolescents during the COVID-19 pandemic: a case series. Hosp Pediatr 12(5):538–543. https://doi.org/10.1542/hpeds.2021-006110

Ratajczak A, Feleszko W, Smith DM, Goniewicz M (2018) How close are we to definitively identifying the respiratory health effects of e-cigarettes? Expert Rev Respir Med 12(7):549–556. https://doi.org/10.1080/17476348.2018.1483724

Reidel B, Radicioni G, Clapp PW, Ford AA, Abdelwahab S, Rebuli ME, Haridass P, Alexis NE, Jaspers I, Kesimer M (2017) E-cigarette use causes a unique innate immune response in the lung, involving increased neutrophilic activation and altered mucin secretion. Am J Respir Crit Care Med 197(4):492–501. https://doi.org/10.1164/rccm.201708-1590OC

Rest EC, Mermelstein RJ, Hedeker D (2021) Nicotine dependence in dual users of cigarettes and E-cigarettes: common and distinct elements. Nicotine Tob Res 23(4):662–668. https://doi.org/10.1093/ntr/ntaa217

Schäfer M, Steindor M, Stehling F, Dohna-Schwake C (2021) EVALI (E-cigarette or vaping product use associated lung injury): first case report of an adolescent in Europe. Pediatr Pulmonol 56(5):1274–1275. https://doi.org/10.1002/ppul.25265

Schuller HM (2019) The impact of smoking and the influence of other factors on lung cancer. Expert Rev Respir Med 13(8):761–769. https://doi.org/10.1080/17476348.2019.1645010

Serra DS, Evangelista JSAM, Zin WA, Leal-Cardoso JH, Cavalcante FSÁ (2017) Changes in rat respiratory system produced by exposure to exhaust gases of combustion of glycerol. Respir Physiol Neurobiol 242:80–85. https://doi.org/10.1016/j.resp.2017.04.001

Shin VY, Jin H, Ng EKO, Cheng ASL, Chong WWS, Wong CYP, Leung WK, Sung JJY, Chu K-M (2011) NF-κB targets miR-16 and miR-21 in gastric cancer: involvement of prostaglandin E receptors. Carcinogenesis 32(2):240–245. https://doi.org/10.1093/carcin/bgq240

Smith DM, Shahab L, Blount BC, Gawron M, Kosminder L, Sobczak A, Xia B, Sosnoff CS, Goniewicz ML (2020) Differences in exposure to nicotine, tobacco-specific nitrosamines, and volatile organic compounds among electronic cigarette users, tobacco smokers, and dual users from three countries. Toxics 8(4):88. https://doi.org/10.3390/toxics8040088

Smith ML, Gotway MB, Crotty Alexander LE, Hariri LP (2021) Vaping-related lung injury. Virchows Arch 478(1):81–88. https://doi.org/10.1007/s00428-020-02943-0

Snow D (2020) Nicotine addiction and E-cigarettes. J Addict Nurs 31(2):77–78. https://doi.org/10.1097/JAN.0000000000000331

St. Helen G, Nardone N, Addo N, Dempsey D, Havel C, Jacob P, Benowitz NL (2020) Differences in nicotine intake and effects from electronic and combustible cigarettes among dual users. Addiction 115(4):757–767. https://doi.org/10.1111/add.14884

Stepanov I, Carmella SG, Han S, Pinto A, Strasser AA, Lerman C, Hecht SS (2009) Evidence for endogenous formation of N'-nitrosonornicotine in some long-term nicotine patch users. Nicot Tob Res 11(1):99–105. https://doi.org/10.1093/ntr/ntn004

Stratton K, Kwan LY, Eaton DL (eds) (2018) Public health consequences of E-cigarettes. National Academies Press. https://doi.org/10.17226/24952

Suzuki T, Matsuo K, Wakai K, Hiraki A, Hirose K, Sato S, Ueda R, Tajima K (2007) Effect of familial history and smoking on common cancer risks in Japan. Cancer 109(10):2116–2123. https://doi.org/10.1002/cncr.22685

Tang M, Wu X-R, Lee H-W, Xia Y, Deng F-M, Moreira AL, Chen L-C, Huang WC, Lepor H (2019) Electronic-cigarette smoke induces lung adenocarcinoma and bladder urothelial hyperplasia in mice. Proc Natl Acad Sci 116(43):21727–21731. https://doi.org/10.1073/pnas.1911321116

Tehrani MW, Newmeyer MN, Rule AM, Prasse C (2021) Characterizing the chemical landscape in commercial E-cigarette liquids and aerosols by liquid chromatography–high-resolution mass spectrometry. Chem Res Toxicol 34(10):2216–2226. https://doi.org/10.1021/acs.chemrestox.1c00253

Wang JB, Olgin JE, Nah G, Vittinghoff E, Cataldo JK, Pletcher MJ, Marcus GM (2018) Cigarette and e-cigarette dual use and risk of cardiopulmonary symptoms in the Health eHeart Study. PLoS One 13(7):e0198681. https://doi.org/10.1371/journal.pone.0198681

White AV, Wambui DW, Pokhrel LR (2021) Risk assessment of inhaled diacetyl from electronic cigarette use among teens and adults. Sci Total Environ 772:145486. https://doi.org/10.1016/j.scitotenv.2021.145486

Wills TA, Pagano I, Williams RJ, Tam EK (2019) E-cigarette use and respiratory disorder in an adult sample. Drug Alcohol Depend 194:363–370. https://doi.org/10.1016/j.drugalcdep.2018.10.004

Winnall W, Greenhalgh EM, Scollo MM (2023) E-cigarette use and the risk of non-infectious respiratory diseases. Tobacco in Australia | facts & issues

Wu D, O'Shea DF (2020) Potential for release of pulmonary toxic ketene from vaping pyrolysis of vitamin E acetate. Proc Natl Acad Sci 117(12):6349–6355. https://doi.org/10.1073/pnas.1920925117

Wynder EL (1988) Tobacco and health: a review of the history and suggestions for public health policy. Public Health Rep (Washington, D.C.: 1974) 103(1):8–18

Xie Z, Ossip DJ, Rahman I, Li D (2020) Use of electronic cigarettes and self-reported chronic obstructive pulmonary disease diagnosis in adults. Nicotine Tob Res 22(7):1155–1161. https://doi.org/10.1093/ntr/ntz234

Xie W, Tackett AP, Berlowitz JB, Harlow AF, Kathuria H, Galiatsatos P, Fetterman JL, Cho J, Blaha MJ, Hamburg NM, Robertson RM, DeFilippis AP, Hall ME, Bhatnagar A, Benjamin EJ, Stokes AC (2022) Association of electronic cigarette use with respiratory symptom development among U.S. young adults. Am J Respir Crit Care Med 205(11):1320–1329. https://doi.org/10.1164/rccm.202107-1718OC

Toxicological Effects of E-Cigarettes on Cells, Animals, and Humans and Mechanisms of Toxicity

11

Nur Azzalia Kamaruzaman, Norsyifa Harun, and Mazlin Mohideen

Abstract

E-cigarettes (E-cigs), also known as electronic cigarettes or vapes, have risen in popularity as alternatives to traditional tobacco smoking. Marketed as safer choices, they have garnered significant attention for their perceived health advantages. Nonetheless, recent research indicates inherent risks associated with e-cigs. Therefore, understanding the toxicological effects of e-cigs *in vitro* (i.e., cells), *in vivo* (i.e., animals), and in humans is crucial for evaluating their overall safety and developing effective regulatory measures. While the *in vitro* system refers to experiments conducted outside of living organisms such as in culture flasks, petri dishes, or test tubes, the *in vivo* system offers more complex experimentation in living organisms such as in animals or humans. Furthermore, using an e-cig that contains nicotine increases the expression of cytokines, hyperreactivity of the airways, and damage to lung tissue. This chapter also investigates various mechanisms of toxicity from the perspectives of pulmonary, inflammation, oxidative stress, DNA damage, apoptosis, transcriptome alterations, and epithelial-mesenchymal transition (EMT), all of which are directly linked to the effects of e-cigs. As for humans, studies have shown that using e-cigs increases the risk of multiorgan involvement, particularly respiratory, cardiovascular, neurodevelopment, immune system, and heightened cancer risk.

N. A. Kamaruzaman (✉)
Malaysia National Poison Centre, Universiti Sains Malaysia, Penang, Malaysia
e-mail: azzalia@usm.my

N. Harun
Centre for Drug Research, Universiti Sains Malaysia, Penang, Malaysia

M. Mohideen
Faculty of Pharmacy and Health Sciences, Universiti Kuala Lumpur-Royal College of Medicine Perak (UniKL-RCMP), Perak, Malaysia

Y.-H. Leong, A. S. Mohd Samin (eds.), *E-Cigarettes*,
https://doi.org/10.1007/978-981-96-5087-3_11

Keywords

Cardiovascular · DNA damage · Inflammation · *In vitro* · *In vivo* · Rodent model · Oxidative stress · Pulmonary

11.1 Introduction

The use of e-cig is rapidly increasing, and with this growth, so does the urgency to understand its health implications. The adverse health effects of e-cig encompass various systems in the human body. Studies on e-cigs have been utilizing *in vitro* and *in vivo* models of animals as well as clinical. Evidently, all of these models contribute to the discovery and understanding of the effects of e-cigs on human health and how the effects are influenced by various factors. Thus, the need for comprehensive studies on e-cig's short- and long-term health impacts is essential, as current models are just beginning to uncover the depth and breadth of these effects. This chapter highlights different aspects of the toxicological effects of e-cigs as reported in studies of *in vitro*, *in vivo*, and clinical models, as well as the mechanisms of toxicity in human. These insights ultimately aid in the development of public health guidelines and regulatory policies aimed at minimizing the adverse health effects of e-cig use.

11.2 *In Vitro* Models of Toxicity

11.2.1 *In Vitro* Toxicology Assessment Framework

The worrying trend of e-cig use among global consumers and its rapid development call for new assessment methods to ensure that knowledge of toxicity is relevant to the latest technologies. Overall, an *in vitro* toxicology assessment framework approach will improve the ability to extrapolate from *in vitro* toxicity data (cell cultures) to *in vivo* effects (in animals and humans). The *in vitro* system refers to experiments conducted outside of living organisms in a controlled environment, such as studying cells in culture flasks, petri dishes, or test tubes, while the *in vivo* system offers more complex experimentation in whole living organisms such as in animal or human models. High-throughput toxicity screening, which involves the rapid assessment of a wide variety of e-liquids using primary cell culture systems, is one of the three primary levels of assessment. The other two levels are a toxicity-based mechanistic assessment of selected e-liquids—an evaluation focusing on understanding the underlying biological processes and pathways affected by the substances—and a further toxicity-based mechanistic assessment of aerosols (Iskandar et al. 2015).

Here in this chapter, we present a multi-layered framework for the evaluation of e-liquids that complement the above-mentioned battery of classic tests. Following the previously described system toxicological approach to exposure (Sturla et al. 2014), the framework for evaluating e-cigs involves three main levels of assessment, each focusing on different aspects of toxicity and mechanistic understanding of e-liquids and aerosols (Fig. 11.1). The initial level of assessment involves subjecting a wide variety of e-liquids (the solutions used in e-cigs) to high-throughput toxicity screening. High-throughput screening refers to a method of rapidly testing large numbers of substances or samples against biological targets. Then, following the high-throughput screening, a subset of e-liquids showing significant toxicity or potential health risks is selected for a more detailed mechanistic assessment. Mechanistic assessment involves studying the underlying biological mechanisms by which the selected e-liquids exert their toxic effects on cells or organisms. The last phase is the toxicity-based mechanistic assessment of a further selection of aerosols. In this final level of assessment, a narrower selection of e-cig aerosols (the inhalable vapors produced by heating e-liquids) is chosen for detailed mechanistic evaluation and underlying mechanisms of toxicity associated with this specific aerosol.

The framework is not complete without a computational systems toxicology approach, which facilitates a comprehensive examination of the cellular mechanisms related to toxicity in e-liquids and their aerosols. Furthermore, a supplementary analytical chemistry investigation will be performed to collect essential chemical data to support the toxicological evaluation.

The framework in Fig. 11.1 was designed to provide an *in vitro* systematic level of analysis for screening and evaluating mechanistic toxicity of both e-liquids and their aerosols at the cellular level (Iskandar et al. 2016).

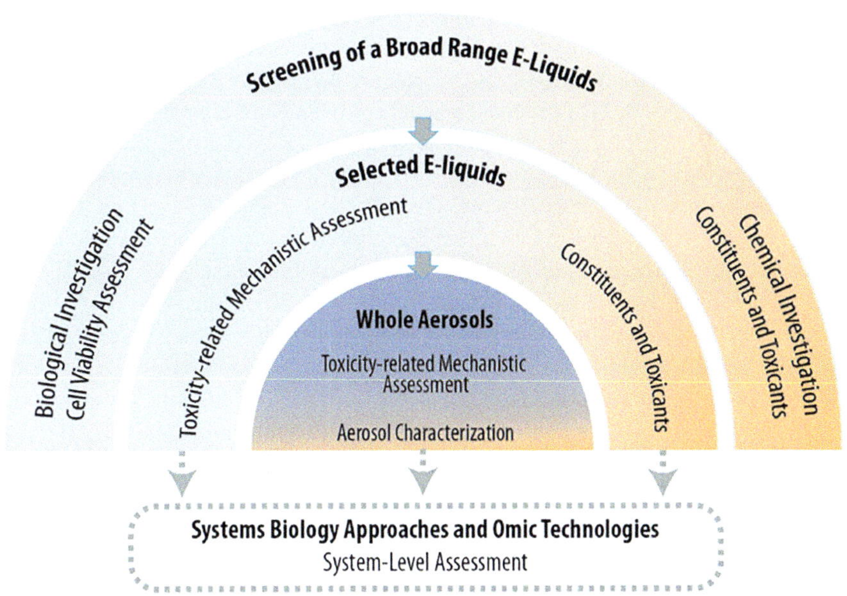

Fig. 11.1 The suggested framework relates to an *in vitro* systems toxicological evaluation of e-liquids, which covers three levels of evaluations that are incorporated into a systems toxicology-based methodology

11.2.1.1 Assessment Level One

The first level of assessment utilizes high-throughput screening of various types of e-liquids in a relevant in vitro biological setting. This level screens through thousands of e-liquids for general toxicity (or cell viability) regardless of the design characteristics of the e-cigs. Such analysis is conducted using relevant *in vitro* cell culture test systems (i.e., primary cell cultures). Primary cell cultures consist of cells directly obtained from living organisms, such as human or animal tissues. This differs from the standard test battery that is often performed with non-human cell lines, such as mouse lymphoma, Chinese hamster lung, or *Salmonella typhimurium* bacteria, which are cell lines that have been cultured and passaged extensively in the laboratory.

Respiratory tract cell cultures are a suitable test system for the proposed toxicological evaluation of e-liquids, as the epithelial cells of the respiratory tract represent the first biological barrier against airborne toxins. Two-dimensional (2D) and three-dimensional (3D) respiratory tract cell culture systems have been employed for toxicological evaluation. 2D cell cultures exist in two dimensions, whereby the cells grow on flat dishes, by adhering and spreading. The method is inexpensive and well established but does not represent a real cell environment. The 3D system is a more advanced cell culture method in which solid tumors naturally grow and cells are able to interact with one another. It generally provides a more relevant cell model, even though the techniques can be expensive, cumbersome, and time-consuming (Jensen and Teng 2020). The use of 2D respiratory tract cell culture systems (i.e., incomplete differentiation of epithelial cells grown in uniform monolayers) is considered suitable for being relatively inexpensive and offering rapid screening. In addition, compared to 3D respiratory tract cell culture systems, the 2D respiratory tract cell culture systems are easier to handle and thus are more suitable for large-scale studies. 2D primary bronchial epithelial cells have been shown to be useful and relevant for the study of pathophysiological toxicity in the airways. In addition, the potential effects of e-liquids on the risk of cardiovascular disease can be investigated using primary 2D vascular endothelial cells. 3D cells on the other hand, are able to differentiate into a pseudostratified airway epithelium which consists of basal, ciliated, and mucus-producing cells (Albino et al. 2004; Catallo et al. 2001; Gonzalez-Suarez et al. 2014, 2016; Kogel et al. 2015; Tan et al. 2012).

The initial stage's assessment is predicated on the utilization of high volumes and high-content screening strategies in the biopharma industry's contemporary drug discovery process. These techniques are currently included in the molecular screening programs of the United States National Institutes of Health (NIH) roadmap (Varma et al. 2011). The goal of using high-throughput and high-content screening is to profile thousands of chemicals quickly and inexpensively. Real-time cellular analysis (RTCA) and high-content screening (HCS) tests can be used to improve multiparametric indicators of general toxicity. Newer methods, such as xCELLigence™ (ACEA Biosciences Ltd., San Diego, CA), allow for real-time detection of cell viability and concurrent observation of the time-dependent exposure response (Iskandar et al. 2015). These advanced techniques enable complex and comprehensive toxicological analysis of e-liquids.

The validity of the biological assay will be substantiated by an analytical chemical assay that is conducted within the initial stratum. This technique is analogous to the traditional battery of tests mandated by regulatory agencies. There are numerous analytical techniques available for detecting ingredients and toxins in e-liquids and cartridges, including high-performance liquid chromatography with diode array detection (HPLC/DAD), liquid chromatography-tandem mass spectrometry (LC-MS/MS), and gas chromatography-mass spectrometry (GC-MS) (Famele et al. 2015). Pharmacopoeia procedures and methods, which are protocols provided by regulatory agencies such as the International Organisation for Standardisation, the US Environmental Protection Agency, and the American Public Health Association, can be valuable for characterizing the chemical components and identifying the presence of toxins as impurities or by-products of compound degradation prior to the aerosolization process (Famele et al. 2015).

11.2.1.2 Assessment Level Two

The second level of the framework is designed to identify the mechanisms associated with toxicity that extend beyond the effects on cell viability that result from exposure to e-liquids. This includes the specific mechanisms associated with cell death and a variety of pathophysiological mechanisms that are associated with e-liquid exposure. The results of the first level of evaluation may be used to identify e-liquids that should be investigated further (those linked with a significant loss in cell viability) (Iskandar et al. 2015).

The second level evaluation can also benefit from the high-content screening (HCS) platforms due to the adaptable properties of the HCS assays. HCS enables the simultaneous detection of various properties including protein phosphorylation, ligand binding to cell surface receptors, molecular uptake and transport, protein expression, cell cycle regulation, enzyme activation, and cell proliferation (Gonzalez-Suarez et al. 2014; Kogel et al. 2015). This permits the simultaneous evaluation of molecular and phenotypic alterations. This technology has been beneficial for assessing the cellular and molecular alterations associated with cigarette smoke and possibly low-risk tobacco product aerosols (Gonzalez-Suarez et al. 2016). Potentially low-risk tobacco product aerosol refers to the aerosol or vapor produced by a tobacco product that is marketed as having a lower risk profile compared to traditional tobacco smoke. This term often arises in discussions related to emerging tobacco and nicotine delivery products, such as e-cigs, which are promoted as alternatives to combustible cigarettes with potentially reduced harm (Anthérieu et al. 2017; Thiriốn-Romero et al. 2019). However, it's important to note that the term "potentially low-risk" implies uncertainty and ongoing debate regarding the actual health impacts of such products, as the long-term effects are still being studied and debated within the scientific and public health communities. Thus, the study on specific cellular pathways, molecular interactions, or physiological responses involved in the toxicity of these e-liquids provides a deeper understanding of toxicity mechanisms and hence provides insights into the potential health impacts of e-cig use and inform regulatory decisions (Thiriốn-Romero et al. 2019).

The HCS technique is more commonly used in 2D cell culture investigations, but it is also being investigated for 3D cultures (Chen et al. 2014; Sirenko et al. 2015). 3D culture systems have certain disadvantages for HCS imaging: they are highly expensive and difficult to culture and maintain, limiting the study's scalability. Their 3D structure also restricts imaging options, affecting measurement sensitivity. Therefore, primary 2D culture systems are deemed a good test system for evaluating the second-layer assessment (Iskandar et al. 2016).

11.2.1.3 Assessment Level Three

The third level of the framework aims to assess the toxicity-related mechanisms of aerosols produced by selected e-liquids. The 3D system poses several advantages over the 2D system due to its ability to mimic the structure and behavior of a cell in vivo while being cultured in vitro. Moreover, the cells in the 3D system are well differentiated, thus providing a more accurate representation of the biological system (Jensen and Teng 2020). 3D airway cell culture systems are suitable and superior to 2D culture systems for evaluating whole aerosols, primarily because they are cultured at the air-liquid interface and allow aerosol exposure at the apical side of the cultures, mimicking the actual "vaping" situation (Constant et al. 2014). The models mimic epithelial tissue in life, with goblet cells that release mucus, columnar ciliated cells, and basal cells (Constant et al. 2014; Iskandar et al. 2015; Mathis et al. 2013; Talikka et al. 2014). By doing so, a more physiologically relevant test system is established. Commercially accessible and ready-to-use human 3D airway epithelial models include the EpiAirway™ from MatTek Corporation in Ashland, MA and the MucilAir™ from Epithelix Sàrl in Geneva, Switzerland. All of these have proven the impact of the 3D respiratory culture system.

3D cell culture systems for the airways make it possible to analyze the complex spatial morphology of the airway epithelium. Histological analysis can be performed with standard staining of tissue sections with hematoxylin and eosin (H&E) and alcian blue. In addition, the proportions of the different cell types can be analyzed using immunohistochemical tests (Talikka et al. 2014). In 3D bronchial or nasal cultures containing functional ciliated cells, the beating frequency of the cilia can be measured (Iskandar et al. 2015) and mucociliary clearance can be assessed (Seagrave et al. 2012). Smoking was reported to reduce ciliary beat frequency and coordination, and the clearance of mucociliary (Leopold et al. 2009). Moreover, endpoints related to toxicity may be assessed in 3D culture systems utilizing well-known conventional cellular assays such as lactate dehydrogenase (LDH) and adenylate kinase (AK) assays (Iskandar et al. 2015; Talikka et al. 2014). Measurement of the quantities of a variety of pro-inflammatory mediators (e.g., cytokines, chemokines, and growth factors) produced into the basolateral medium of culture systems following exposure might provide additional endpoints (Iskandar et al. 2015; Mathis et al. 2013; Talikka et al. 2014). Profiling these mediators can help discover the toxicological pathways linked to exposure. Airway epithelial cells produce numerous mediators in response to stimuli, despite being traditionally thought of as a physical barrier between the environment and the host organism (Proud and Leigh 2011). Furthermore, the induction of xenobiotic-metabolizing enzymes such as cytochrome P450 (CYP) 1A1 and CYP1B1 can be easily measured after exposure to polycyclic aromatic hydrocarbons (PAHs) and nitrosamines (found in CS) by adding luminescence-labeled substrates to the culture medium (Iskandar et al. 2015).

Various aerosol-generating devices are commercially available as depicted in Fig. 11.2. The properties of the aerosol produced (i.e., distribution of particle size and composition) vary based on the device, geometry, and operating conditions. These differences in aerosol properties can lead to different deposition of the aerosol in the test systems, which in turn leads to different toxicity profiles. Therefore, aerosol characterization should be performed as part of the third-layer assessment approach. Such an approach aims to minimize the impact of variability from device to device; therefore, specific aerosol-generating devices can be used. Currently, there is an increasing amount of research on the chemical composition of e-cig aerosols. The primary objective of most investigations is to discover both well-known hazardous substances that are often found in cigarette smoke, such as

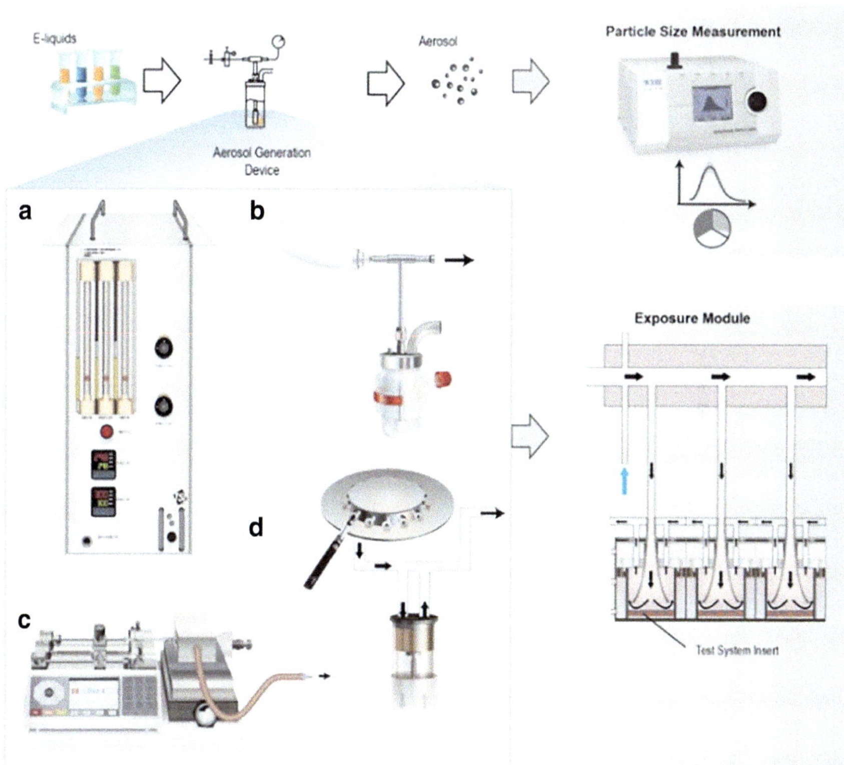

Fig. 11.2 Commercially accessible aerosol-generating equipment. Various commercially available equipment may be used to make aerosols from different e-liquids. The following devices were used in the study: (**a**) a condensation monodisperse aerosol generator, (**b**) a Collison nebulizer, (**c**) a Vitrocell spiking system, and (**d**) smoking machines for vaping e-cig devices. (Iskandar et al. 2016)

tobacco-specific nitrosamines and nicotine-related chemicals, as well as other harmful constituents like carbonyl compounds, volatile organic compounds, metals, and minerals like silicate (Flora et al. 2016). Recent research conducted by Kosmider et al. (2014) examined the impact of battery voltage on the production of potentially hazardous chemicals in aerosols, namely carbonyl compounds. Aside from doing chemical analysis, it is also important to assess the droplet size distribution and the ratio of the GMP (gas-to-particle) phase (Kosmider et al. 2014).

11.3 Animal Models of Toxicity

Studies on the potential *in vivo* toxicity of e-cigs utilized a range of model organisms, e-cig exposure techniques (e-cig aerosols, e-liquids, and ECE), as well as different modes of exposure that range from whole-body to nose-only exposure for durations that are classified as acute, sub-acute, and chronic (Wang et al. 2022). The toxicity of e-cigs affects pulmonary system, cardiovascular system, central nervous system, and other organs, which are further described in the following.

11.3.1 Toxicity Effects of E-cigs on the Pulmonary System

Most of the *in vivo* toxicity research on e-cigs mainly focuses on the lung effects of animal models because e-cigs are first exposed to the respiratory system after inhalation. In a study conducted by Lerner et al. (2015), whole-body inhalation of e-cig aerosols for 5 h per day for 3 days resulted in increased inflammatory responses in mouse lung cells and tissues as well as a decrease in lung glutathione levels, which are essential for maintaining cellular redox balance (Lerner et al. 2015). Another report supports this finding, whereby a study of mice which nose-inhaled e-cig aerosol (1 h/day for 4 weeks) was found to produce increased inflammatory cytokines in bronchoalveolar lavage (BAL) and systemic inflammatory markers (Hwang et al. 2016). Meanwhile, adult rats that were exposed to e-cig aerosols through whole-body inhalation for 1 h each day for 4 weeks showed no significant difference between the exposure and control group in terms of epithelial distribution or inflammation. However, in contrast to the single case of metaplasia in the control group, two occurrences of hyperplasia and four cases of metaplasia were noted in the exposed rats (Salturk et al. 2015).

In order to assess the *in vivo* effects of e-cig vapor on the lung and to compare them to those of cigarette smoke (CS), Glynos et al. (2018) exposed C57BL/6 mice to ambient air, CS, or e-cig vapors for either 3 days or 4 weeks before assessing their levels of oxidative stress, inflammation, and pulmonary mechanics (Glynos et al. 2018). After 3 days, non-nicotine-containing aerosols altered tissue elasticity, static compliance, and airway resistance, whereas after 4 weeks of exposure, CS was the only treatment adversely affecting these parameters. However, e-cigs with nicotine and flavorings increased BAL cellularity, Muc5ac production, BAL, and lung oxidative stress indicators and also resulted in increased airway responsiveness to methacholine to an extent comparable to that of conventional cigarettes. Since flavors are core component of all e-liquids used in vaping devices and nicotine is present in >95% of e-liquids, this finding thus suggests that even with short-term use of e-cigs, the users are at risk of developing airway reactivity, a condition that is mostly associated with chronic obstructive pulmonary disease (COPD) and asthma.

Another study using mice treated with ovalbumin (OVA) to induce asthma-like mouse phenotype with concurrent diluted intratracheal e-cig liquid exposure twice weekly for 10 weeks revealed that e-cig liquid-exposed mice displayed increased airway hyperresponsiveness to methacholine, along with increased serum OVA-specific IgE and increased BAL eosinophils and Th2 cytokines IL-4, IL-5, and IL-13 (Lim and Kim 2014). This finding thus suggested that exposure to e-cigs may enhance the onset of allergic respiratory disease. Furthermore, in another study of chronic exposure, mice exposed to nicotine-containing e-cig liquid for 1 h each day for 4 months exhibited effects typically linked to the onset of COPD, such as cytokine expression, airway hyperreactivity, and lung tissue destruction, which were not present in mice exposed to nicotine-free e-cig (Garcia-Arcos et al. 2016). This finding thus suggested that inhaled nicotine contributes to airway and lung disease.

In an investigation to evaluate the potential risk of e-cig smoke exposure on neonates, one study used neonatal mice for exposure to e-cig smoke, keeping mouse pups with their mothers during aerosol exposure by whole-body inhalation. The pups grew considerably less weight over time compared to control mice after being exposed to both nicotine and non-nicotine e-cig smoke for the first 10 days of life. However, when compared to control mice that were exposed to room air, only pups that had been exposed to nicotine e-cig aerosol showed a significantly reduced lung growth (McGrath-Morrow et al. 2015).

In order to determine whether exposure to e-cig aerosol could impair lung responses to infection, Sussan et al. (2015) conducted a study in which adult mice were exposed to the e-cig aerosol twice daily for two weeks, for 1.5 h each time, before the mice were infected for an hour with *Streptococcus pneumoniae* (Sussan et al. 2015). The study showed a significant impairment of pulmonary bacterial clearance activity in mice exposed to e-cig aerosol compared to air-exposed control mice, supporting the epidemiological association between e-cig exposure and susceptibility to infection.

11.3.2 Toxicity Effects of E-cigs on Cardiovascular, Central Nervous System, and Other Organs

The compounds present in the composition of e-cig liquid pose complex toxicities to the human body. Although the respiratory system is first exposed to e-cig after inhalation, the atomized e-cig smoke is not entirely metabolized in the lung. The substances inhaled are partially absorbed into the bloodstream, where they are subsequently distributed throughout the cardiovascular, central nervous system and immune system, among other systems (Wang et al. 2022).

In vivo toxicity in animal models is classified into three categories based on exposure duration: acute (1–7 days), sub-acute (2–4 weeks), and chronic (2 months or more) (Wang et al. 2022). A sub-acute study by Shi et al. (2019) involving mice exposed to e-cig aerosols for 14 days found no effect on cardiac contractility, as measured by ejection fraction (Shi et al. 2019). However, the study observed an increase in CD31, a biomarker for angiogenesis, in the heart and kidney tissues of the mice. Another study that involved prolonged exposure to e-cig aerosols for up to 8 months (1 h for each exposure, five times per week) found that it increased aortic stiffness and prevented vascular relaxation in mice (Olfert et al. 2018). A chronic e-cig exposure study revealed that rats exposed to e-cigs for 3–6 months exhibited higher systolic blood pressure (SBP), a slower heart rhythm, and signs of cardiac fibrosis (Crotty Alexander et al. 2020). Moreover, Mayyas et al. (2020) found that male Wistar rats exposed to e-cig aerosol, cigarette smoke, and waterpipe smoke for 4 weeks developed cardiac fibrosis compared to a fresh air control group (Mayyas et al. 2020).

Apolipoprotein E knockout ApoE−/− mice, which are susceptible to atherosclerosis, have also been used in a few studies to evaluate the onset of atherosclerosis and alterations in heart function. The research involved subjecting ApoE−/− mice to e-cig aerosols for a duration of 12 weeks. The findings revealed that the aerosols containing nicotine led to a decrease in both the left ventricular fractional shortening and ejection fraction, respectively. Additionally, ApoE−/− mice exposed to nicotine-containing e-cigs also displayed ultrastructural changes suggestive of cardiomyopathy, as well as increased oxidative stress and mitochondrial DNA alterations in cardiac tissue (Espinoza-Derout et al. 2019). In a different study, zebrafish embryos were continuously exposed for 3 days to water-containing electronic cigarette extract (ECE) and cigarette smoke extract (CSE) at equivalent nicotine levels to evaluate cardiac development. Due to the exposure to ECE or CSE, development effects which were dose-dependent were observed, with extreme heart malformation of the heart, pericardial edema, and decreased function of the heart (Palpant et al. 2015).

In addition to cardiovascular toxicity, researchers also discovered that intraperitoneal administration of e-liquid daily for 4 weeks resulted in oxidative stress injuries to both the liver and kidney. The results of the histology analysis revealed that e-liquids seemed to affect the antioxidant defense and cause alterations to kidney and hepatic parameters (El Golli et al. 2016a, b). Using the similar route of administration and exposure duration, e-liquid with or without nicotine also resulted in testis toxicity through disruption of oxidative balance and steroidogenesis (El Golli et al. 2016a, b).

11.3.3 Effects of E-cigs Toxicity to the Reproductive System

Another study that examined the effects of e-cigs on the development of the central nervous system (CNS) found that pregnant mice exposed to e-cig aerosol during gestational and lactation periods induced gene-expression changes and altered CNS development that were linked to chronic neuropathology (Nguyen et al. 2018). This finding was further supported by Zelikoff et al. (2018) when early life exposure to e-cigs induced biological disruptions in the developing CNS (Zelikoff et al. 2018). Additionally, a different study using *Xenopus laevis* embryos revealed that exposure to ECE caused craniofacial abnormalities from cranial cartilage and muscle defects and a reduction in the blood supply to the face, further indicating that e-cig may pose a risk to the developing embryo (Kennedy et al. 2017).

11.4 Mechanisms of Toxicity Caused by E-cigs

E-cig-related toxicity effects are primarily attributed to inflammation response, oxidative stress, and DNA damage, with additional mechanisms that have also been reported. The summary of the major mechanisms is depicted in Fig. 11.3 and is further elaborated in the next sections.

11.4.1 Inflammation Response

E-cigs have the capacity to provoke an inflammatory reaction. Studies show that when exposed to e-cig smoke in a controlled laboratory setting, there is evidence that it can trigger the release of certain substances that cause inflammation in the body. These substances include IL-6, IL-8, CXCL1, and G-CSF. However, the extent of this inflammatory response is found to be lower compared to the effects of conventional cigarette smoke. These findings were observed in experiments using primary normal human bronchial epithelial (NHBE) cells and BEAS-2B cells (Anthérieu et al. 2017; Iskandar et al. 2016). In a similar manner, when human airway epithelial (H292) cells and human fetal lung fibroblasts (HFL1) were exposed to e-cig aerosols in an air-liquid interface, there was an increase in the release of IL-6 and IL-8. This was followed by noticeable changes in morphology. In a study conducted by Scott et al. (2018), it was shown that when human alveolar macrophages from nonsmokers were exposed to e-cig aerosol condensate for 24 hours, there was an increase in cytotoxicity as well as enhanced production of reactive oxygen species (ROS), inflammatory cytokines, and chemokines (Scott et al. 2018). The exposure caused an inflammatory response in the alveolar macrophages in the lung.

11.4.2 Oxidative Stress

It has been documented that e-cigs cause oxidative stress in addition to enhancing the inflammatory response. Unbalanced levels of oxidants and antioxidants cause oxidative stress, and the presence of ROS can weaken or overwhelm the body's natural antioxidant defenses (Barnes 2020). According to Lerner et al. (2015), e-cig aerosols can activate the level of glutathione redox in vivo in the lung to generate

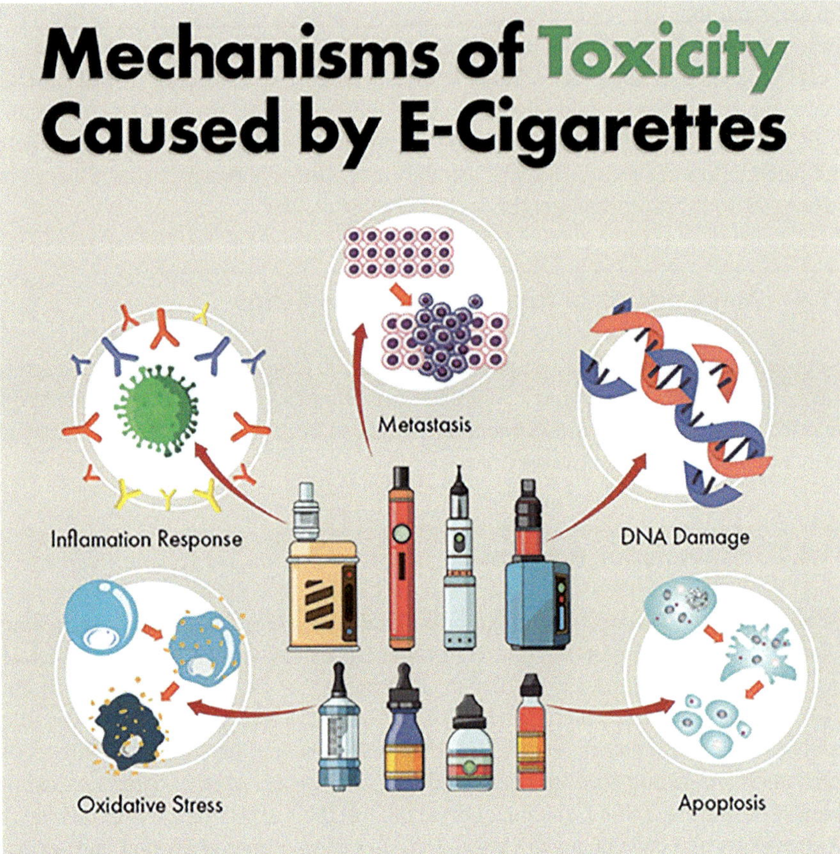

Fig. 11.3 Major mechanisms of toxicity induced by e-cigs include inflammation, oxidative stress, cell apoptosis, DNA damage, and EMT

oxidative stress and can modify oxidative stress indicators in both lung cells and mouse lungs (Lerner et al. 2015).

Moreover, Lerner et al. (2016) discovered that short-term exposure to e-cig aerosol enhanced HFL-1 cells' susceptibility to mitochondrial ROS and decreased the stability of an electron transport chain (ETC) complex IV component, resulting in the induction of mitochondrial stress (Lerner et al. 2016). Compared to cigarette smoke, e-cig aerosols showed relatively modest levels of oxidative stress. However, long-term exposure to e-cigs may result in increased levels of oxidative stress (Dusautoir et al. 2021). Moreover, following a duration of eight weeks of exposure, the lung and plasma of B6C3F1 mice displayed notably elevated concentrations of 8-hydroxy-2′-deoxyguanosine (8-OHdG), a marker of DNA oxidative harm. This suggests that propylene glycol (PG) and vegetable glycerin (VG), the primary solvent carriers employed in e-cigs, have the potential to induce oxidative stress *in vivo*.

Despite various reports on the adverse health effects of e-cigs, one of the factors to its appeal are the variety of enticing flavors in the market, namely Tobacco, Minty Fruit, Fruity, Minty/Cool (Iced), Desserts, Candy, and Drinks/Beverage. Studies have also made comparisons of e-cig-induced oxidative stress in pulmonary cells which is caused by various flavoring chemicals. Different flavors have been shown to generate aerosols with variable levels of reactive oxygen species (ROS) and inflammation, with varying or constant levels of nicotine content (Yogeswaran et al. 2021). Menthol-flavored e-cig exposure to BEAS-2B cells was reported to cause mitochondrial dysfunction, reduce basal and maximal respiration, and consequently affect mitochondrial oxidative stress. It is postulated that some chemical constituents of the menthol-flavored e-cig such as 1,4-dioxane have contributed to the pathology (Lamb et al. 2020). In addition, cinnamon-flavored e-cig aerosols can promote ROS production and greater cytotoxicity than flavorless aerosols (Wavreil and Heggland 2020). This is because cinnamaldehyde has been attributed to be the most potent chemical in cinnamon-flavored e-liquids, which can cause oxidative stress in lung cells (Muthumalage et al. 2018).

11.4.3 DNA Damage

DNA carries genetic information in its stable form, and it must precisely repair damage in order to preserve its integrity. There are two main types of DNA damage: environmental factors like tobacco smoke and outdoor physical radiation, as well as endogenous factors like hydrolysis and oxidative damage (Vijg 2021). Damage to DNA can also result from e-cigs. According to reports, exposure to e-cig aerosol can trigger mitochondrial ROS generation, which in turn can lead to nuclear DNA breakage in HFL-1 cells (Lerner et al. 2016).

In addition, the extract from e-cig aerosol decreased the levels of proteins required for the removal of DNA damage, such as OGG1 and ERCC1, and induced DNA damage in a manner that depended on the dosage, regardless of the quantity of nicotine (Ganapathy et al. 2017; Sun et al. 2021). Furthermore, in a dose-dependent manner, e-cigs produced ROS, damaged DNA, triggered apoptosis, and drastically lowered cell viability (Anderson et al. 2016). E-cig aerosols also suppressed DNA repair and decreased the amount of repair proteins XPC and OGG1/2 in lung tissue in addition to causing carcinogenic γ-OH-PdG and O6-MedG adducts in the lung, bladder, and heart of FVBN mice (Lee et al. 2018). In SD rat lung models, e-cig aerosols may have co-mutagenic and cancer-initiating effects by elevating oxygen-free radical generation and DNA oxidation, which damages DNA (Canistro et al. 2017). Therefore, aerosols from e-cigs may increase the sensitivity to mutations and cause neoplastic transformation.

11.4.4 Apoptosis

A few investigations have also discovered that e-cigs' novel harmful pathways include transcriptome alterations, epithelial-mesenchymal transition (EMT), and cell apoptosis. Cell apoptosis may be induced by e-cig aerosol condensate (Anderson et al. 2016; Scott et al. 2018). According to reports, exposure to the aerosol from e-cigs may cause cell apoptosis, which would lower the rate at which human gingival fibroblasts proliferate—whether or not nicotine is present (Alanazi et al. 2018). Additionally, after repeated exposures, e-cig aerosols raised the proportions of necrotic/apoptotic human gingival epithelial cells and were linked to the caspase-3 pathway. Caspase-3 is a critical executioner protease responsible for cleaving key proteins that lead to cell death during apoptosis (Rouabhia et al. 2017).

11.4.5 Epithelial-Mesenchymal Transition (EMT)

The initial route for the metastasis of cancer cells is called EMT. It was discovered that after 3–8 days of continuous exposure, menthol or tobacco-flavored e-cig liquids or aerosols caused A549 cells to undergo EMT. This resulted in the cells acquiring a fibroblast-like morphology, losing intercellular junctions, internalizing E–cadherin, becoming more motile, and having an increase in EMT markers (Zahedi et al. 2018).

11.4.6 Transcriptome Alterations

Furthermore, in basic human airway epithelial cells, exposure to e-cig aerosols resulted in notable transcriptome alterations linked to ribosomal protein synthesis. Cell growth, proliferation, differentiation, and development depend on ribosome biogenesis and protein translation (Park et al. 2021). Additionally, research has demonstrated that exposure to e-cig aerosol has a significant impact on gene expression, with mRNA, lncRNA, and miRNA expression changing following e-cig treatment (Le et al. 2021).

11.5 Toxicity of E-cigs in Humans

Since e-cigs were first introduced to the global market in 2007, it would take decades to gather epidemiological evidence regarding the long-term health impacts of vaping (Tsai et al. 2020). In the human body, e-cigs are inhaled and first exposed to the respiratory system, but the lungs are not able to fully metabolize atomized e-cig smoke. A portion of the chemicals then enter the bloodstream and are dispersed throughout the immunological, central neurological, and cardiovascular systems, among other systems (Wang et al. 2022).

While animal studies have provided detailed mechanistic insights into e-cig toxicity, human studies offer additional data that contribute to a more comprehensive understanding of e-cig effects from public health and behavioral perspectives. A study by Bowler et al. (2017) found that e-cig use did not result in harm reduction among older adults with a history of cigarette smoking. This study indicated that e-cig use was associated with greater nicotine exposure, rapid decline in lung function, and increased risk of exacerbation, which may be related to the rising use of e-cigs among older adults who smoke cigarettes (Bowler et al. 2017).

Furthermore, using an e-cig that contains nicotine increases the expression of cytokines, hyperreactivity of the airways, and damage of lung tissue—all of which are linked to the onset of COPD (Garcia-Arcos et al. 2016; Reinikovaite et al. 2018). In addition, short-term e-cig exposure affects respiratory symptoms across various types of smokers, including those with COPD, asthma, "healthy" smokers, and healthy non-smokers (Palamidas et al. 2017). Another study found that adolescents using e-cigs experienced higher rates of chronic bronchitic symptoms (McConnell et al. 2017).

Studies have shown that individuals with asthma use significantly more smoke and vape products than non-asthmatics, and using e-cigs is associated with a higher likelihood of self-reported asthma (Bircan et al. 2021; Reid et al. 2018). Another study comparing the immediate respiratory effects of e-cigs to traditional smoking found that both types of smoking caused direct mechanical and inflammatory respiratory impacts in healthy smokers and those with mild asthma, with the effects being more pronounced in individuals with mild asthma (Lappas et al. 2018).

11.6 Conclusion

E-cig liquid typically consists of a mixture of ingredients, including PG, VG, flavorings, and nicotine. When this liquid is heated and atomized by the e-cig device, it produces an aerosol that users inhale. The transformation of e-cig liquid into aerosol alters the toxicity mechanisms, as the heating process leads to chemical reactions and the formation of new compounds, which are harmful to health. Additionally, the inhalation of aerosolized particles has different effects on multi-systems generally, with the respiratory system affected most profoundly. Given the rapid increase in the development and use of e-cigs, new assessment methods are needed to ensure that knowledge of toxicity is relevant to the latest technologies. *In vitro* assessment through a multi-layered framework may enable a systems-level analysis of the cellular mechanistic toxicity of both e-liquids and their aerosols using relevant biological in vitro test systems. In addition, *in vivo* toxicity research on e-cigs mainly focuses on the lung effects in rodents because e-cigs are first exposed to the respiratory system after inhalation, and the relevance of rodents to the human biological system. Recent research has demonstrated that a variety of processes, such as oxidative stress, apoptosis, DNA damage, inflammatory reactions, EMT, and transcriptome alterations, are involved in the toxicity mechanisms of e-cigs. In humans, the use of e-cigs is associated with many health problems and affects multiple systems in the body. Therefore, it is important the toxicological consequences of e-cig use, and more research is needed to fully understand the magnitude of such toxicity.

References

Albino AP, Huang X, Jorgensen E, Yang J, Gietl D, Traganos F, Darzynkiewicz Z (2004) Induction of H2AX phosphorylation in pulmonary cells by tobacco smoke: a new assay for carcinogens. Cell Cycle 3(8):1062–1068

Alanazi H, Park HJ, Chakir J, Semlali A, Rouabhia M (2018) Comparative study of the effects of cigarette smoke and electronic cigarettes on human gingival fibroblast proliferation, migration and apoptosis. Food Chem Toxicol 118: 390–398. https://doi.org/10.1016/j.fct.2018.05.049.

Anderson C, Majeste A, Hanus J, Wang S (2016) E-Cigarette aerosol exposure induces reactive oxygen species, DNA damage, and cell death in vascular endothelial cells. Toxicol Sci 154(2):332–340. https://doi.org/10.1093/toxsci/kfw166

Anthérieu S, Garat A, Beauval N, Soyez M, Allorge D, Garçon G, Lo-Guidice J-M (2017) Comparison of cellular and transcriptomic effects between electronic cigarette vapor and cigarette smoke in human bronchial epithelial cells. Toxicol In Vitro 45:417–425. https://doi.org/10.1016/j.tiv.2016.12.015

Barnes PJ (2020) Oxidative stress-based therapeutics in COPD. Redox Biol 33:101544. https://doi.org/10.1016/j.redox.2020.101544

Bircan E, Bezirhan U, Porter A, Fagan P, Orloff M (2021) Electronic cigarette use and its association with asthma, chronic obstructive pulmonary disease (COPD) and asthma- COPD overlap syndrome among never cigarette smokers. Tob Induc Dis 19(April):1–9. https://doi.org/10.18332/tid/132833

Bowler RP, Hansel NN, Jacobson S, Graham Barr R, Make BJ, Han MK, O'Neal WK, Oelsner EC, Casaburi R, Barjaktarevic I, Cooper C, Foreman M, Wise RA, DeMeo DL, Silverman EK, Bailey W, Harrington KF, Woodruff PG, Drummond MB (2017) Electronic Cigarette use in us adults at risk for or with COPD: analysis from two observational cohorts. J Gen Intern Med 32(12):1315–1322. https://doi.org/10.1007/s11606-017-4150-7

Canistro D, Vivarelli F, Cirillo S, Babot Marquillas C, Buschini A, Lazzaretti M, Marchi L, Cardenia V, Rodriguez-Estrada MT, Lodovici M, Cipriani C, Lorenzini A, Croco E, Marchionni S, Franchi P, Lucarini M, Longo V, Della Croce CM, Vornoli A et al (2017) E-cigarettes induce toxicological effects that can raise the cancer risk. Sci Rep 7(1):2028. https://doi.org/10.1038/s41598-017-02317-8

Catallo WJ, Kennedy CH, Henk W, Barker SA, Grace SC, Penn A (2001) Combustion products of 1,3-butadiene are cytotoxic and genotoxic to human bronchial epithelial cells. Environ Health Perspect 109(9):965–971. https://doi.org/10.1289/ehp.01109965

Chen W, Wong C, Vosburgh E, Levine AJ, Foran DJ, Xu EY (2014) High-throughput image analysis of tumor spheroids: a user-friendly software application to measure the size of spheroids automatically and accurately. J Vis Exp 89. https://doi.org/10.3791/51639

Constant S, Wiszniewski L, Huang S (2014) Methods for conducting connected culture experiments using the Quasi-Vivo R© Chambers. In: Cellular in vitro testing. Jenny Stanford Publishing, pp 17–30. https://doi.org/10.1201/b15630-3

Crotty Alexander LE, Ware LB, Calfee CS, Callahan SJ, Eissenberg T, Farver C, Goniewicz ML, Jaspers I, Kheradmand F, King TE, Meyer NJ, Mikheev VB, Shields PG, Shihadeh A, Strongin R, Tarran R (2020) E-Cigarette or vaping product use–associated lung injury: developing a research agenda. An NIH workshop report. Am J Respir Crit Care Med 202(6):795–802. https://doi.org/10.1164/rccm.201912-2332WS

Dusautoir R, Zarcone G, Verriele M, Garçon G, Fronval I, Beauval N, Allorge D, Riffault V, Locoge N, Lo-Guidice J-M, Anthérieu S (2021) Comparison of the chemical composition of aerosols from heated tobacco products, electronic cigarettes and tobacco cigarettes and their toxic impacts on the human bronchial epithelial BEAS-2B cells. J Hazard Mater 401:123417. https://doi.org/10.1016/j.jhazmat.2020.123417

El Golli N, Jrad-Lamine A, Neffati H, Rahali D, Dallagi Y, Dkhili H, Ba N, El May MV, El Fazaa S (2016a) Impact of e-cigarette refill liquid with or without nicotine on liver function in adult rats. Toxicol Mech Methods 26(6):433–440. https://doi.org/10.3109/15376516.2016.1160963

El Golli N, Rahali D, Jrad-Lamine A, Dallagi Y, Jallouli M, Bdiri Y, Ba N, Lebret M, Rosa JP, El May M, El Fazaa S (2016b) Impact of electronic-cigarette refill liquid on rat testis. Toxicol Mech Methods 26(6):417–424. https://doi.org/10.3109/15376516.2016.1163448

Espinoza-Derout J, Hasan KM, Shao XM, Jordan MC, Sims C, Lee DL, Sinha S, Simmons Z, Mtume N, Liu Y, Roos KP, Sinha-Hikim AP, Friedman TC (2019) Chronic intermittent electronic cigarette exposure induces cardiac dysfunction and atherosclerosis in apolipoprotein-E knockout mice. Am J Phys Heart Circ Phys 317(2):H445–H459. https://doi.org/10.1152/ajpheart.00738.2018

Famele M, Ferranti C, Abenavoli C, Palleschi L, Mancinelli R, Draisci R (2015) The chemical components of electronic cigarette cartridges and refill fluids: review of analytical methods. Nicotine Tob Res 17(3):271–279. https://doi.org/10.1093/ntr/ntu197

Flora JW, Meruva N, Huang CB, Wilkinson CT, Ballentine R, Smith DC, Werley MS, McKinney WJ (2016) Characterization of potential impurities and degradation products in electronic cigarette formulations and aerosols. Regul Toxicol Pharmacol 74:1–11. https://doi.org/10.1016/j.yrtph.2015.11.009

Ganapathy V, Manyanga J, Brame L, McGuire D, Sadhasivam B, Floyd E, Rubenstein DA, Ramachandran I, Wagener T, Queimado L (2017) Electronic cigarette aerosols suppress cellular antioxidant defenses and induce significant oxidative DNA damage. PLoS One 12(5):e0177780. https://doi.org/10.1371/journal.pone.0177780

Garcia-Arcos I, Geraghty P, Baumlin N, Campos M, Dabo AJ, Jundi B, Cummins N, Eden E, Grosche A, Salathe M, Foronjy R (2016) Chronic electronic cigarette exposure in mice induces features of COPD in a nicotine-dependent manner. Thorax 71(12):1119–1129. https://doi.org/10.1136/thoraxjnl-2015-208039

Glynos C, Bibli S-I, Katsaounou P, Pavlidou A, Magkou C, Karavana V, Topouzis S, Kalomenidis I, Zakynthinos S, Papapetropoulos A (2018) Comparison of the effects of e-cigarette vapor with cigarette smoke on lung function and inflammation in mice. Am J Phys Lung Cell Mol Phys 315(5):L662–L672. https://doi.org/10.1152/ajplung.00389.2017

Gonzalez-Suarez I, Sewer A, Walker P, Mathis C, Ellis S, Woodhouse H, Guedj E, Dulize R, Marescotti D, Acali S, Martin F, Ivanov NV, Hoeng J, Peitsch MC (2014) Systems biology approach for evaluating the biological impact of environmental toxicants *in Vitro*. Chem Res Toxicol 27(3):367–376. https://doi.org/10.1021/tx400405s

Gonzalez-Suarez I, Martin F, Marescotti D, Guedj E, Acali S, Johne S, Dulize R, Baumer K, Peric D, Goedertier D, Frentzel S, Ivanov NV, Mathis C, Hoeng J, Peitsch MC (2016) In vitro systems toxicology assessment of a candidate modified risk tobacco product shows reduced toxicity compared to that of a conventional cigarette. Chem Res Toxicol 29(1):3–18. https://doi.org/10.1021/acs.chemrestox.5b00321

Hwang JH, Lyes M, Sladewski K, Enany S, McEachern E, Mathew DP, Das S, Moshensky A, Bapat S, Pride DT, Ongkeko WM, Crotty Alexander LE (2016) Electronic cigarette inhalation alters innate immunity and airway cytokines while increasing the virulence of colonizing bacteria. J Mol Med 94(6):667–679. https://doi.org/10.1007/s00109-016-1378-3

Iskandar AR, Xiang Y, Frentzel S, Talikka M, Leroy P, Kuehn D, Guedj E, Martin F, Mathis C, Ivanov NV, Peitsch MC, Hoeng J (2015) Impact assessment of cigarette smoke exposure on organotypic bronchial epithelial tissue cultures: a comparison of mono-culture and coculture model containing fibroblasts. Toxicol Sci 147(1):207–221. https://doi.org/10.1093/toxsci/kfv122

Iskandar AR, Gonzalez-Suarez I, Majeed S, Marescotti D, Sewer A, Xiang Y, Leroy P, Guedj E, Mathis C, Schaller J-P, Vanscheeuwijck P, Frentzel S, Martin F, Ivanov NV, Peitsch MC, Hoeng

J (2016) A framework for in vitro systems toxicology assessment of e-liquids. Toxicol Mech Methods 26(6):392–416. https://doi.org/10.3109/15376516.2016.1170251

Jensen C, Teng Y (2020) Is it time to start transitioning from 2D to 3D cell culture? Front Mol Biosci 7. https://doi.org/10.3389/fmolb.2020.00033

Kennedy AE, Kandalam S, Olivares-Navarrete R, Dickinson AJG (2017) E-cigarette aerosol exposure can cause craniofacial defects in Xenopus laevis embryos and mammalian neural crest cells. PLoS One 12(9):e0185729. https://doi.org/10.1371/journal.pone.0185729

Kogel U, Gonzalez Suarez I, Xiang Y, Dossin E, Guy PA, Mathis C, Marescotti D, Goedertier D, Martin F, Peitsch MC, Hoeng J (2015) Biological impact of cigarette smoke compared to an aerosol produced from a prototypic modified risk tobacco product on normal human bronchial epithelial cells. Toxicol In Vitro 29(8):2102–2115. https://doi.org/10.1016/j.tiv.2015.08.004

Kosmider L, Sobczak A, Fik M, Knysak J, Zaciera M, Kurek J, Goniewicz ML (2014) Carbonyl compounds in electronic cigarette vapors: effects of nicotine solvent and battery output voltage. Nicotine Tob Res 16(10):1319–1326. https://doi.org/10.1093/ntr/ntu078

Lamb T, Muthumalage T, Rahman I (2020) Pod-based menthol and tobacco flavored e-cigarettes cause mitochondrial dysfunction in lung epithelial cells. Toxicol Lett 333:303–311. https://doi.org/10.1016/j.toxlet.2020.08.003

Lappas AS, Tzortzi AS, Konstantinidi EM, Teloniatis SI, Tzavara CK, Gennimata SA, Koulouris NG, Behrakis PK (2018) Short-term respiratory effects of e-cigarettes in healthy individuals and smokers with asthma. Respirology 23(3):291–297. https://doi.org/10.1111/resp.13180

Le HHT, Liu C, Denaro P, Jousma J, Shao N-Y, Rahman I, Lee WH (2021) Genome-wide differential expression profiling of lncRNAs and mRNAs in human induced pluripotent stem cell-derived endothelial cells exposed to e-cigarette extract. Stem Cell Res Ther 12(1):593. https://doi.org/10.1186/s13287-021-02654-6

Lee H-W, Park S-H, Weng M, Wang H-T, Huang WC, Lepor H, Wu X-R, Chen L-C, Tang M (2018) E-cigarette smoke damages DNA and reduces repair activity in mouse lung, heart, and bladder as well as in human lung and bladder cells. Proc Natl Acad Sci 115(7):E1560. https://doi.org/10.1073/pnas.1718185115

Leopold PL, O'Mahony MJ, Lian XJ, Tilley AE, Harvey B-G, Crystal RG (2009) Smoking is associated with shortened airway Cilia. PLoS One 4(12):e8157. https://doi.org/10.1371/journal.pone.0008157

Lerner CA, Sundar IK, Yao H, Gerloff J, Ossip DJ, McIntosh S, Robinson R, Rahman I (2015) Vapors produced by electronic cigarettes and E-juices with flavorings induce toxicity, oxidative stress, and inflammatory response in lung epithelial cells and in mouse lung. PLoS One 10(2):e0116732. https://doi.org/10.1371/journal.pone.0116732

Lerner CA, Rutagarama P, Ahmad T, Sundar IK, Elder A, Rahman I (2016) Electronic cigarette aerosols and copper nanoparticles induce mitochondrial stress and promote DNA fragmentation in lung fibroblasts. Biochem Biophys Res Commun 477(4):620–625. https://doi.org/10.1016/j.bbrc.2016.06.109

Lim HB, Kim SH (2014) Inhallation of e-cigarette cartridge solution aggravates allergen-induced airway inflammation and hyper-responsiveness in mice. Toxicol Res 30(1):13–18. https://doi.org/10.5487/TR.2014.30.1.013

Mathis C, Poussin C, Weisensee D, Gebel S, Hengstermann A, Sewer A, Belcastro V, Xiang Y, Ansari S, Wagner S, Hoeng J, Peitsch MC (2013) Human bronchial epithelial cells exposed in vitro to cigarette smoke at the air-liquid interface resemble bronchial epithelium from human smokers. Am J Phys Lung Cell Mol Phys 304(7):L489–L503. https://doi.org/10.1152/ajplung.00181.2012

Mayyas F, Aldawod H, Alzoubi KH, Khabour O, Shihadeh A, Eissenberg T (2020) Comparison of the cardiac effects of electronic cigarette aerosol exposure with waterpipe and combustible cigarette smoke exposure in rats. Life Sci 251:117644. https://doi.org/10.1016/j.lfs.2020.117644

McConnell R, Barrington-Trimis JL, Wang K, Urman R, Hong H, Unger J, Samet J, Leventhal A, Berhane K (2017) Electronic cigarette use and respiratory symptoms in adolescents. Am J Respir Crit Care Med 195(8):1043–1049. https://doi.org/10.1164/rccm.201604-0804OC

McGrath-Morrow SA, Hayashi M, Aherrera A, Lopez A, Malinina A, Collaco JM, Neptune E, Klein JD, Winickoff JP, Breysse P, Lazarus P, Chen G (2015) The effects of electronic cigarette emissions on systemic cotinine levels, weight and postnatal lung growth in neonatal mice. PLoS One 10(2):e0118344. https://doi.org/10.1371/journal.pone.0118344

Muthumalage T, Prinz M, Ansah KO, Gerloff J, Sundar IK, Rahman I (2018) Inflammatory and oxidative responses induced by exposure to commonly used e-cigarette flavoring chemicals and flavored e-liquids without nicotine. Front Physiol 8. https://doi.org/10.3389/fphys.2017.01130

Nguyen T, Li GE, Chen H, Cranfield CG, McGrath KC, Gorrie CA (2018) Maternal E-cigarette exposure results in cognitive and epigenetic alterations in offspring in a mouse model. Chem Res Toxicol 31(7):601–611. https://doi.org/10.1021/acs.chemrestox.8b00084

Olfert IM, DeVallance E, Hoskinson H, Branyan KW, Clayton S, Pitzer CR, Sullivan DP, Breit MJ, Wu Z, Klinkhachorn P, Mandler WK, Erdreich BH, Ducatman BS, Bryner RW, Dasgupta P, Chantler PD (2018) Chronic exposure to electronic cigarettes results in impaired cardiovascular function in mice. J Appl Physiol 124(3):573–582. https://doi.org/10.1152/japplphysiol.00713.2017

Palamidas A, Tsikrika S, Katsaounou P, Vakali S, Gennimata S-A, Kaltsakas G, Gratziou C, Koulouris N (2017) Acute effects of short term use of e-cigarettes on airways physiology and respiratory symptoms in smokers with and without airways obstructive diseases and in healthy non smokers. Tob Prev Cessation 3(March):5. https://doi.org/10.18332/tpc/67799

Palpant NJ, Hofsteen P, Pabon L, Reinecke H, Murry CE (2015) Cardiac development in zebrafish and human embryonic stem cells is inhibited by exposure to tobacco cigarettes and E-cigarettes. PLoS One 10(5):e0126259. https://doi.org/10.1371/journal.pone.0126259

Park H-R, Vallarino J, O'Sullivan M, Wirth C, Panganiban RA, Webb G, Shumyatcher M, Himes BE, Park J-A, Christiani DC, Allen J, Lu Q (2021) Electronic cigarette smoke reduces ribosomal protein gene expression to impair protein synthesis in primary human airway epithelial cells. Sci Rep 11(1):17517. https://doi.org/10.1038/s41598-021-97013-z

Proud D, Leigh R (2011) Epithelial cells and airway diseases. Immunol Rev 242(1):186–204. https://doi.org/10.1111/j.1600-065X.2011.01033.x

Reid KM, Forrest JR, Porter L (2018) Tobacco product use among youths with and without lifetime asthma – Florida, 2016. MMWR Morb Mortal Wkly Rep 67(21):599–601. https://doi.org/10.15585/mmwr.mm6721a2

Reinikovaite V, Rodriguez IE, Karoor V, Rau A, Trinh BB, Deleyiannis FW-B, Taraseviciene-Stewart L (2018) The effects of electronic cigarette vapour on the lung: direct comparison to tobacco smoke. Eur Respir J 51(4):1701661. https://doi.org/10.1183/13993003.01661-2017

Rouabhia M, Park HJ, Semlali A, Zakrzewski A, Chmielewski W, Chakir J (2017) E-cigarette vapor induces an apoptotic response in human gingival epithelial cells through the caspase-3 pathway. J Cell Physiol 232(6):1539–1547. https://doi.org/10.1002/jcp.25677

Salturk Z, Çakır Ç, Sünnetçi G, Atar Y, Kumral TL, Yıldırım G, Berkiten G, Uyar Y (2015) Effects of electronic nicotine delivery system on larynx: experimental study. J Voice 29(5):560–563. https://doi.org/10.1016/j.jvoice.2014.10.013

Scott A, Lugg ST, Aldridge K, Lewis KE, Bowden A, Mahida RY, Grudzinska FS, Dosanjh D, Parekh D, Foronjy R, Sapey E, Naidu B, Thickett DR (2018) Pro-inflammatory effects of e-cigarette vapour condensate on human alveolar macrophages. Thorax 73(12):1161–1169. https://doi.org/10.1136/thoraxjnl-2018-211663

Seagrave J, Albrecht HH, Hill DB, Rogers DF, Solomon G (2012) Effects of guaifenesin, N-acetylcysteine, and ambroxol on MUC5AC and mucociliary transport in primary differentiated human tracheal-bronchial cells. Respir Res 13(1):98. https://doi.org/10.1186/1465-9921-13-98

Shi H, Fan X, Horton A, Haller ST, Kennedy DJ, Schiefer IT, Dworkin L, Cooper CJ, Tian J (2019) The effect of electronic-cigarette vaping on cardiac function and angiogenesis in mice. Sci Rep 9(1):4085. https://doi.org/10.1038/s41598-019-40847-5

Sirenko O, Mitlo T, Hesley J, Luke S, Owens W, Cromwell EF (2015) High-content assays for characterizing the viability and morphology of 3D Cancer spheroid cultures. Assay Drug Dev Technol 13(7):402–414. https://doi.org/10.1089/adt.2015.655

Sturla SJ, Boobis AR, FitzGerald RE, Hoeng J, Kavlock RJ, Schirmer K, Whelan M, Wilks MF, Peitsch MC (2014) Systems toxicology: from basic research to risk assessment. Chem Res Toxicol 27(3):314–329. https://doi.org/10.1021/tx400410s

Sun Y-W, Chen K-M, Atkins H, Aliaga C, Gordon T, Guttenplan JB, El-Bayoumy K (2021) Effects of E-cigarette aerosols with varying levels of nicotine on biomarkers of oxidative stress and inflammation in mice. Chem Res Toxicol 34(4):1161–1168. https://doi.org/10.1021/acs.chemrestox.1c00033

Sussan TE, Gajghate S, Thimmulappa RK, Ma J, Kim J-H, Sudini K, Consolini N, Cormier SA, Lomnicki S, Hasan F, Pekosz A, Biswal S (2015) Exposure to electronic cigarettes impairs pulmonary anti-bacterial and anti-viral defenses in a mouse model. PLoS One 10(2):e0116861. https://doi.org/10.1371/journal.pone.0116861

Talikka M, Kostadinova R, Xiang Y, Mathis C, Sewer A, Majeed S, Kuehn D, Frentzel S, Merg C, Geertz M, Martin F, Ivanov NV, Peitsch MC, Hoeng J (2014) The response of human nasal and bronchial organotypic tissue cultures to repeated whole cigarette smoke exposure. Int J Toxicol 33(6):506–517. https://doi.org/10.1177/1091581814551647

Tan X-L, Marquardt G, Massimi AB, Shi M, Han W, Spivack SD (2012) High-throughput library screening identifies two novel *NQO1* inducers in human lung cells. Am J Respir Cell Mol Biol 46(3):365–371. https://doi.org/10.1165/rcmb.2011-0301OC

Thirión-Romero I, Pérez-Padilla R, Zabert G, Barrientos-Gutiérrez I (2019) Respiratory impact of electronic cigarettes and low-risk tobacco. Rev Investig Clin 71(1):17. https://doi.org/10.24875/RIC.18002616

Tsai M, Byun MK, Shin J, Crotty Alexander LE (2020) Effects of e-cigarettes and vaping devices on cardiac and pulmonary physiology. J Physiol 598(22):5039–5062. https://doi.org/10.1113/JP279754

Varma H, Lo DC, Stockwell BR (2011) High-throughput and high-content screening for Huntington's disease therapeutics. In: Neurobiology of Huntington's disease: applications to drug discovery, p 5

Vijg J (2021) From DNA damage to mutations: all roads lead to aging. Ageing Res Rev 68:101316. https://doi.org/10.1016/j.arr.2021.101316

Wang L, Wang Y, Chen J, Liu P, Li M (2022) A review of toxicity mechanism studies of electronic cigarettes on respiratory system. Int J Mol Sci 23(9):5030. https://doi.org/10.3390/ijms23095030

Wavreil FDM, Heggland SJ (2020) Cinnamon-flavored electronic cigarette liquids and aerosols induce oxidative stress in human osteoblast-like MG-63 cells. Toxicol Rep 7:23–29. https://doi.org/10.1016/j.toxrep.2019.11.019

Yogeswaran S, Muthumalage T, Rahman I (2021) Comparative reactive oxygen species (ROS) content among various flavored disposable vape bars, including cool (ICED) flavored bars. Toxics 9(10):235. https://doi.org/10.3390/toxics9100235

Zahedi A, Phandthong R, Chaili A, Remark G, Talbot P (2018) Epithelial-to-mesenchymal transition of A549 lung cancer cells exposed to electronic cigarettes. Lung Cancer 122:224–233. https://doi.org/10.1016/j.lungcan.2018.06.010

Zelikoff JT, Parmalee NL, Corbett K, Gordon T, Klein CB, Aschner M (2018) microglia activation and gene expression alteration of neurotrophins in the hippocampus following early-life exposure to E-cigarette aerosols in a murine model. Toxicol Sci 162(1):276–286. https://doi.org/10.1093/toxsci/kfx257

Chemicals in Electronic Nicotine Delivery Systems (ENDS) 12

Kee Xuan Yeo and Nuan Ping Cheah

Abstract

This chapter provides an overview of the chemical contents, emissions, and factors influencing the chemicals found in both the e-liquid and the aerosol of electronic nicotine delivery systems (ENDS). A section of the chapter examines the chemical composition of ENDS, detailing the various additives, flavorings, and other substances commonly found in e-liquids, as well as exploring the diversity of ingredients used. A discussion on the emissions produced during the use of ENDS, focusing on the formation of toxicants and other harmful compounds in the aerosol, is also presented. It covers the factors affecting the emissions generated, including the effects of temperature, device type, and e-liquid composition. By understanding these processes, readers will gain insights into the complexities and dangers associated with ENDS use.

Keywords

ENDS · Nicotine · Glycerol · Propylene glycol · Flavors

12.1 Introduction

Electronic nicotine delivery systems (ENDS), commonly called electronic cigarettes (e-cigs) or vaporizers, are typically battery-operated or puff-activated devices that are designed and used in the same manner as a conventional cigarette with a

K. X. Yeo · N. P. Cheah (✉)
Cigarette Testing Laboratory, Pharmaceutical Division, Applied Sciences Group, Health Sciences Authority, Singapore, Singapore
e-mail: cheah_nuan_ping@hsa.gov.sg

Y.-H. Leong, A. S. Mohd Samin (eds.), *E-Cigarettes*,
https://doi.org/10.1007/978-981-96-5087-3_12

mechanism to rapidly deliver nicotine to its users. Alternative versions include, but not limited to, electronic- or puff-activated cigars and shisha pipes.

There are various types of e-cigarettes, which are the most prevalent forms of electronic nicotine delivery systems (ENDS) and electronic non-nicotine delivery systems (ENNDS). These devices heat a liquid to produce aerosols that users inhale. These e-liquids, which do not contain tobacco, may or may not contain nicotine.

ENDS deliver nicotine to users by heating a solution, commonly known as e-liquid, to create an aerosol. The e-liquid solution consists primarily of a mixture of propylene glycol and glycerol. In some products, it contains entirely either propylene glycol or glycerol. Propylene glycol and glycerol serve as carrier solutions that form an aerosol when heated, delivering the addictive chemical nicotine, along with flavors, additives, and numerous other known and unknown chemicals generated during the heating process, to the user.

Although generally considered a single product class, ENDS constitute a diverse group with potentially significant differences in the production of toxicants and delivery of nicotine. There are several coexisting types of devices on the market: first-generation or so-called cigalike, second-generation tank systems, and even larger varieties of third-generation personal vaporizers. The latest generation includes puff-activated (also known as inhale- activated nicotine vapes), which are designed as such that it is activated upon puffing or inhalation without the need for buttons or settings adjustments.

The earlier generation of ENDS devices is categorized into closed and open systems, depending mainly on the degree of control that users have over the e-liquid used, the voltage and resistance applied to heating the e-liquid, and ventilation features. The e-liquid cartridges are available in various flavors such as tobacco, menthol, strawberry, apple, chocolate, vanilla, and many others. Nicotine refill cartridges are available in different concentrations ranging from zero nicotine, low and mid-range doses (e.g., 6–8 mg/mL and 10–14 mg/mL respectively), to high and extra high doses (e.g., 16–36 mg/mL and beyond) (Cheah et al. 2014). The concentration labels are often printed on the bottle or cartridge.

ENDS devices consist of a small heating element, or coil, that vaporizes e-liquid and a wicking material that draws liquid onto the coil, upon puffing. When the user inhales, the in-built flow sensor activates the heating element which will then vaporize the liquid. It comes as either puff-activated or manual activation via a push-button.

The number of puffs an ENDS can deliver depends on several factors, including the capacity of the e-liquid reservoir, battery life, and the user's puffing profile (such as puff duration, interval, and inhalation volume). On average, a standard ENDS with a 2 mL e-liquid capacity can deliver between 400 and 600 puffs when fully used, according to our laboratory research. However, the exact number of puffs can vary depending on the device's design and individual user habits. For example, ENDS devices with larger tank capacities, such as 16–20 mL, can deliver thousands of puffs, influenced by factors like the device's power, voltage, and the user's puffing behavior (including frequency, inhalation volume, and intervals between puffs).

Many ENDS devices do not resemble conventional cigarettes. Both ENDS and conventional cigarettes are similar in that they deliver nicotine to users.

12.2 Chemical Composition of E-cig Liquids

12.2.1 Overview of Common Ingredients in E-liquids

The e-liquid typically contains nicotine, propylene glycol, and/or glycerol as solvent, flavorings chemicals, and additives. Nicotine in e-liquids comes in various forms, including tobacco-derived nicotine, nicotine salts, and synthetic nicotine. Chemical additives are included in e-liquid formulations to provide users with a consistent and smooth nicotine delivery, enhancing user satisfaction and sustaining addiction (Cipriano et al. 1975; Talih et al. 2020a, b).

One effective method to achieve smoother nicotine delivery in e-liquids is by converting free nicotine, also known as freebase nicotine, into nicotine salts. This conversion is accomplished by adding acids to the e-liquids, which protonate the nicotine, resulting in a less harsh experience for users when inhaling the aerosols. According to a patent, nicotine salts formed using acids such as benzoic acid, salicylic acid, and levulinic acid provide a consistent and efficient delivery of nicotine. This process leads to a rapid increase in nicotine plasma levels, enhancing the user's experience (Bowen and Xing 2015).

12.2.2 Nicotine and Nicotine Salts

Nicotine, the key ingredient in nicotine-containing ENDS products, is primarily derived from tobacco plants. As a naturally occurring alkaloid, nicotine is a tertiary amine and exhibits chirality, meaning it exists in two non-superimposable mirror image forms, known as enantiomers: (*R*)-nicotine and (*S*)-nicotine. Historically, the chirality of nicotine made its chemical synthesis expensive, so extraction from tobacco plants remained the most cost-effective source. However, recent advances in chemical synthesis have changed this landscape (Dorotheo et al. 2024). Synthetic nicotine used in laboratories mimics the effects of naturally occurring nicotine. It comes in three different compositions: a mix of both (*S*)- and (*R*)-nicotine forms (or isomers), pure (*R*)-nicotine, or pure (*S*)-nicotine. The increased commercial interest in synthetic nicotine has driven up its demand. Manufacturers aim to bypass existing tobacco control laws by using chemically synthesized nicotine, which may be perceived as safer.

The industry continues to refine techniques, achieving success in developing efficient and economical processes for synthetic nicotine production. Several published patents detail commercially viable methods for synthesizing nicotine on an industrial scale, further supporting the growing market for synthetic nicotine.

A new form of nicotine, known as (S)-6-methyl nicotine (6-MN), has been identified in ENNDS products. This nicotine analog, synthetically derived (Zou et al. 2022), is a structural variant of nicotine, featuring a methyl group substitution at position 6 of the pyridine ring. It has a molecular formula of $C_{11}H_{16}N_2$ and a molecular weight of 176 (Jordt et al. 2023) (Fig. 12.1).

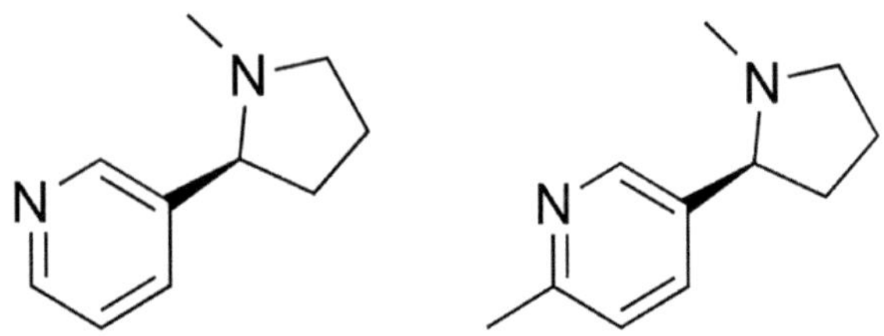

Fig. 12.1 (S)-nicotine (left) and (S)-6-methyl nicotine (right)

Due to its novelty, there is currently limited information available on (1) the chemical synthesis route and corresponding chemical impurities, (2) the pharmacological effects, (3) its addictiveness, and (4) the toxicity profile (Metatine 2023). An in vitro study reported that compared to nicotine, 6-MN displayed more sensitive cytotoxicity on human bronchial epithelial cells (BEAS-2B cells) (Qi et al. 2023).

Published industry patents reported that the most suitable acids for nicotine salt formation are from salicylic acid, formic acid, sorbic acid, acetic acid, benzoic acid, pyruvic acid, lauric acid, and levulinic acid. The melting points of nicotine salts formed from these acids are below 160 °C, which facilitates high nicotine uptake during the initial inhalation phase, particularly within the first few puffs. This lower melting point ensures that the nicotine salts vaporize quickly and efficiently when the e-liquid is heated, allowing users to experience a rapid onset of nicotine delivery. As a result, the user can achieve a satisfying nicotine hit almost immediately, enhancing the overall effectiveness and appeal of the ENDS. This rapid nicotine uptake is particularly significant for smokers transitioning to ENDS, as it mimics the immediate nicotine satisfaction provided by conventional cigarettes.

Nicotine salt such as nicotine salicylate is reported to have little to no flavor and is often used in nicotine containing e-liquid (Pennings et al. 2023). Nicotine salts formulations can accommodate high nicotine delivery to the user, without causing discomfort to the user. Nicotine salts are easily absorbed by the body, allowing faster delivery of nicotine to the bloodstream compared to freebase nicotine. Benzoic acid and levulinic acid are the most commonly used acid to form nicotine salt. This is because nicotine benzoate had a greater ability to increase dopamine release (Li et al. 2023), while nicotine levulinate desensitizes the upper respiratory tract, increasing the potential for the aerosol to be inhaled deeper into the lungs (Levulinic Acid 1991).

12.2.3 Flavors

Most of the flavoring chemicals added to these e-liquids are "generally recognized as safe" (GRAS) when consumed orally (CFR—Code of Federal Regulations Title 21, 2023). However, the breakdown of these compounds, when heated and generated in e-cig aerosols has yet to be assessed, particularly in a toxicological context. The Flavor Extracts Manufacturers Association (FEMA) has emphasized that they do not evaluate flavor ingredients for use in e-cigs or any other exposures other than ingestion (Safety and Regulatory Authority to Use Flavors – Focus on Vaping

Products 2021). The list of flavoring chemicals added to e-liquid was comprehensively reviewed by a study published by Krüsemann et al. (2021).

Besides the power, design features, and e-liquid capacity of the e-cig device, the high variability among the thousands of e-liquid products available arises from differences in flavors.

A study found that there are over 20,000 unique products on the market, with even similarly named flavors differing significantly in their actual flavoring molecules. The most common flavors marketed are fruit-based followed by tobacco and dessert-flavored (Galimov et al. 2021; Hallingberg 2022; Hammond et al. 2022; Havermans et al. 2021).

Numerous flavoring agents have been reported in e-liquid samples. One study identified a total of 126 flavoring agents in 16 different e-liquid samples. For example, diacetyl, which imparts a buttery aroma and is commonly found in food products such as popcorn, is used in e-liquids. Additionally, ethyl esters, which provide a citrus flavor, are also commonly added.

While many of these flavoring agents are generally regarded as safe (GRAS) for consumption, prolonged and repeated inhalation can pose health risks. Specifically, inhalation of diacetyl can cause respiratory problems (Harber et al. 2006).

With some jurisdictions restricting the use of flavors in e-liquids, the industry has turned to alternatives. To circumvent laws prohibiting flavoring in ENDS, manufacturers have begun adding chemicals that impart flavor. These additives mimicing the sensory experience of conventional flavorings is evident as some e-liquid products have been found to contain synthetic coolants, such as 2-isopropyl-N,2,3-trimethylbutyramide (WS-23), N-ethyl-p- menthane-3-carboxamide (WS-3) (Wong et al. 2023), and menthone 1,2-glycerol ketal (MGK) (Klupinski et al. 2023). A published study reported that flavors concentration in ENDS ranged from 1 mg/mL to as high as 34.3 mg/mL (Omaiye et al. 2022). The concentration of WS-3 and WS-23 ranged from 1.5 to 15.5 and 0.9 to 35.9 mg/mL, with combined concentration of both coolants to be as high as 55.8 mg/mL (Omaiye et al. 2022). The high concentration of flavors and coolants exposes users to unknown health risks. The margin of exposure (MOE), a risk indicator to evaluate carcinogenic risk or chemical safety based on predicted or estimated exposure levels, calculated for 1 mL of e-liquid/daily, for WS-3 and WS-23 are both below 100, thereby presenting a health risk to consumers (Margin of Exposure 2023; Omaiye et al. 2022). Another study found the presence of narcotic drugs in the e-cigs seized (159 out of 188, or 84%), with the most common drug being delta-9-tetrahydrocannabinol (THC), which was present in 98% of the positive samples (Almazrouei et al. 2022).

Numerous studies have identified the inclusion of Vitamin E Acetate, also referred to as α- tocopherol acetate, in ENDS or ENNDS containing cannabis. Vitamin E Acetate, serving as an oil-based solvent, has been reported to be present in these products (Muthumalage et al. 2020). The inclusion of Vitamin E Acetate has been linked to the formation of ketene, a highly toxic and unstable chemical that forms under vaping conditions (Council 2014). Furthermore, the presence of ketene, formed in the aerosol during vaping, from commonly used flavoring chemicals containing acetate esters such as benzyl acetate, ethyl acetate, and isoamyl acetate may be more prevalent than presently understood.

12.2.4 Aerosolizing Chemicals

Glycerol and propylene glycol are commonly used in e-liquid formulations, also known as aerosolizing chemicals or aerosolizing agents. These chemicals possess distinct properties, facilitating their rapid aerosolization and efficient transfer of nicotine to the user. The cost disparity between pharmaceutical-grade and industrial-grade glycerol and propylene glycol can be significant, primarily due to the additional purification steps required to achieve higher levels of purity in pharmaceutical-grade chemicals. Industrial-grade chemicals, with their higher impurity levels, are typically less expensive but may not meet the strict quality standards necessary for consumer products. These lower-grade chemicals may contain higher impurities due to less stringent purification processes, making them unsuitable for consumption but still viable for industrial purposes. Among the potential contaminants in propylene glycol and glycerol are diethylene glycol and ethylene glycol (Altamimy et al. 2024). Diethylene glycol has been found to be present in some e-liquid, as a substitute to glycerol and propylene glycol, primarily due to its cheaper cost. This is particularly prevalent in mass market and illicit sources (Varlet et al. 2015).

12.2.5 Other Chemicals

Nicotine derived from the tobacco plant contains several known impurities related to nicotine, including anabasine, anatabine, cotinine, nicotine-N-oxide, β-nicotyrine, nornicotine, and myosmine (Crooks and Byrd 1999; Davis and Curvali 1999; Jacob and Byrd 1999; Langone et al. 1999; Tsai and Gorrod 1999). Metals can also be present in both synthetic nicotine, due to the use of metal-based reagents during synthesis, and in tobacco-derived nicotine, originating from the soil in which the tobacco is grown (Cheetham et al. 2022). Additionally, volatile chemicals such as benzene, toluene, ethylbenzene, and xylenes are commonly reported in e-liquids (Palmisani et al. 2020). These volatile organic compounds (VOCs) are likely introduced during the formulation of e-liquids due to the varying grades of solvents used in the production process.

12.3 Emissions from ENDS

12.3.1 Overview

When the e-liquid in an ENDS is heated either through battery power or puff activation, it undergoes aerosolization, forming what is commonly referred to as an emission. This aerosol encompasses a diverse array of substances, including nicotine, flavoring agents, propylene glycol, glycerol, and carbonyls such as formaldehyde, acetaldehyde, acrolein, glyoxal, and methylglyoxal. Furthermore, it contains volatile organic compounds (VOCs) like benzene and glycidol, metal-containing particles, particulate matter, polycyclic aromatic hydrocarbons, and tobacco-specific nitrosamines (TSNAs) such as 4-(methylnitrosamino)-1-(3-pyridyl)-1-butanone (NNK), *N*′-nitrosonornicotine (NNN), *N*′-nitrosoanabasine (NAB), and *N*′-nitrosoanatabine (NAT). Free radicals and reactive oxygen species, including ketene, are also present (Bekki et al. 2014; Ebersole et al. 2020; El Hajj Moussa et al. 2024; EL-Hellani et al. 2018; Haddad et al. 2019; Hasan et al. 2020; Jin et al. 2022; Kwak et al. 2020; Pappas et al. 2021; Uchiyama et al. 2016, 2020).

Key factors influencing the amount of nicotine generated in the aerosols include, but not limited to, battery voltage, coil resistance, puff topography (puff duration and frequency), nicotine concentration, and the ratio of propylene glycol to glycerol in the e-liquid. The ratio of propylene glycol to glycerol plays a crucial role in nicotine delivery. E-liquids with a higher proportion of propylene glycol tend to deliver more nicotine, resulting in users taking shorter and fewer puffs. This is because propylene glycol facilitates more efficient vaporization and nicotine absorption compared to glycerol. Consequently, the balance between these two components can significantly affect the overall vaping experience and nicotine intake (Spindle et al. 2018). ENDS with high-powered battery capacity when paired with low nicotine concentration liquids are capable of maintaining cigarette-like nicotine exposure (Hiler et al. 2017; Wagener et al. 2017).

Another factor influencing emissions is the number of puffs taken, as it directly affects the amount of e-liquid vaporized. It was reported that, for several e-cig brands tested, the amount of e-liquid vaporized decreased with an increasing number of puffs. Although this trend was not consistent across all e-cig samples, it remains a significant factor that can influence emissions (Talih et al. 2023). Additionally, the practice of removing and reinserting an e-liquid pod during vaping also impacts emissions. A study noted an increase in particulate matter collected when the pod was removed and reinserted multiple times (Soule et al. 2022). This suggests that user behavior, such as how often a pod is handled during use, can significantly affect the quantity and quality of emissions. These observations highlight the complexity of factors contributing to overall exposure to harmful substances in e-cig aerosols. However, the design features of e-cigs are not within the scope of discussion for this chapter and will not be discussed.

12.3.2 Carbonyls

Aerosols generated from ENDS contain toxic and carcinogenic carbonyl compounds. Both the type of solvent used and the battery output voltage significantly affect the levels of carbonyl compounds in the aerosol vapor. Notably, the levels of carbonyls such as formaldehyde and acetaldehyde rapidly increase with higher battery output voltages (Kosmider et al. 2014). Higher voltage settings in ENDS result in increased levels of formaldehyde, acetaldehyde, and acrolein in the generated aerosols (Bitzer et al. 2019; Zhou et al. 2021). Additionally, higher temperatures were correlated with decreasing coil resistance in pods, which led to the production of larger amounts of aerosols and the formation of more carbonyl compounds in these aerosols (Tran et al. 2023).

The study also noted that higher power settings contribute to increased concentrations of carbonyls in the emissions (EL-Hellani et al. 2018; Li et al. 2021). Furthermore, the ratio of propylene glycol to glycerol in the e-liquid significantly affects the emissions profile. Variations in this ratio can influence the formation of toxic compounds, potentially resulting in higher levels of harmful substances, including those not yet identified.

12.3.3 Metals

In ENDS, the heating elements are often made from materials such as stainless steel (comprising chromium, iron, nickel, and manganese), nichrome (a combination of nickel and chromium), brass (composed of copper and zinc), and kanthal (which includes aluminum, chromium, and iron) (Pappas et al. 2021). These heating elements are a critical component connected to the battery. During heating, metals can leach from these components into the e-liquids, with this transfer potentially increasing under conditions of higher airflow, higher nicotine concentration, and greater glycerol content (Zervas et al. 2020). Additionally, metals may be present in the nicotine itself. This is due to the use of metal-based reagents in the synthesis of synthetic nicotine and the presence of metals in the soil where tobacco is cultivated (Cheetham et al. 2022). High levels of lead (up to 130 μg/g), chromium (up to 0.70 μg/g), nickel (up to 5.1 μg/g), and copper (up to 100 μg/g) were reported in a study, on e-liquid, with levels far exceeding the United States Pharmacopoeia limit set on inhaled drug products (lead = 0.2 μg/g, chromium = 0.3 μg/g, nickel = 0.5 μg/g and copper = 3 μg/g) (Neu et al. 2020).

Silicon was the most abundant element in the aerosols of all e-cig brands, accounting for over 50% of the total elemental weight in most samples, with concentrations ranging from 0.09 to 6.8 μg per 10 puffs. The likely sources of aerosolized silicon were the sheath and wick, composed of delicate, finely woven silicate glass threads (fiberglass) that can break easily during handling or heating and cooling cycles (Williams et al. 2017). Other metals detected in ENDS aerosol include boron, iron, zinc, magnesium, arsenic, antimony, indium, strontium, zirconium, lanthanum, barium, aluminum, potassium, selenium, calcium, germanium, tin, rubidium, and tungsten (Williams et al. 2017).

12.3.4 Tobacco-Specific Nitrosamines

Tobacco-specific nitrosamines (TSNAs) are a class of carcinogenic compounds found in e-cig aerosols and tobacco smoke (IARC 2024). These include

4-(methylnitrosamino)-1-(3-pyridyl)-1-butanone (NNK), N′-nitrosonornicotine (NNN), N′-nitrosoanabasine (NAB), and N′-nitrosoanatabine (NAT) (Brunnemann et al. 1977; Farsalinos et al. 2015). TSNAs are formed through the nitrosation of nicotine and related alkaloids during the e-liquid heating process (Jin et al. 2022). Generally, the presence of TSNAs in ENDS is at trace level (Konstantinou et al. 2018).

12.3.5 Reactive Oxygen Species (ROS)

Reactive oxygen species (ROS) encompass a group of highly reactive and typically short-lived radicals, including hydroxyl radicals, superoxide anions, singlet oxygen, alkoxyl, and alkylperoxy radicals (Zhao and Hopke 2012). Among the most common ROS are the hydroxyl radical, superoxide anion, and hydrogen peroxide.

The generation of ROS in aerosols emitted from ENDS is likely attributed to the pyrolysis of nicotine, propylene glycol, glycerol, flavoring agents, organic acids, and the chemical activity of metals present in the heating elements (Guan et al. 2003; Lerner et al. 2015; Zhao et al. 2018).

Studies indicate that a higher proportion of glycerol relative to propylene glycol results in lower aerosol emissions and smaller particle mass but also leads to increased ROS levels generated (El Hajj Moussa et al. 2024; Haddad et al. 2019; Jabba et al. 2020; Li et al. 2021).

ENDS emissions studied in high-powered battery-operated devices with a large coil surface area were found to produce significant levels of ROS species, comparable to those produced by conventional cigarettes (Haddad et al. 2019). It is evident that various factors, including design features, battery capacity, heating coil characteristics, and chemical formulation, influence the generation of ROS in ENDS aerosols.

12.3.6 Free Radicals

A free radical is a chemical species that has an unpaired electron, making it highly unstable and reactive. Free radicals can act as oxidants by accepting an electron or as reductants by donating an electron. They are produced by various normal biological processes, including aerobic metabolism and pathogenic defense mechanisms. External factors like radiation, pollutants, cigarette smoke, and aerosols from ENDS or ENNDS can also contribute to the formation of free radicals.

The production of reactive free radicals in e-cig aerosols depends on both the chemicals used in the formulation of the e-liquid and the temperature, as demonstrated in a study by Bitzer et al. (2018a, b). The study found that free radical production increases with higher temperatures, with disproportionately high levels of free radicals observed at temperatures $\geq$100 °C, even when aerosol production is limited. Additionally, the presence of propylene glycol in the e-liquid contributes to the increase in free radical production. Besides temperature and propylene glycol, flavoring chemicals such as γ-decalactone, citral, ethyl maltol, piperonal, d-limonene, and linalool added to e-liquids have been found to contribute to increased free radical presence in aerosols emitted from vape devices (Bitzer et al. 2018a, b; Woo et al. 2024).

In another study, free radicals were detected in vape devices containing benzoic acid, tartaric acid, and citric acid, as quantified by spin-trapping and electron paramagnetic resonance (EPR) spectroscopy. The study found that nicotine salts contribute to the generation of reactive free radical species, with nicotine benzoate salts producing the highest ROS yield in aerosols generated from a fourth-generation vape (pod-mod type) device (Tran et al. 2024).

These findings suggest that both the chemical composition of e-liquids and the conditions under which they are vaporized significantly influence the generation of harmful free radicals.

12.3.7 Others

Benzene can be formed via a few pathways: dehydration of glycerol, decarboxylation of benzoic acid, and decarboxylation of carboxylic acids formed when aldehydes present are oxidized. Benzene levels were found to increase with higher-powered devices (Pankow et al. 2017). Glycidol, formed through the thermal decomposition of glycerol, was detected exclusively in e-liquids formulated with only glycerol (without propylene glycol) (Lorkiewicz et al. 2022; Uchiyama et al. 2020). Propylene oxide, acetol, and 2-propen-1-ol were produced in the aerosols from e-liquids with propylene glycol (Sleiman et al. 2016). Glyoxal and acrolein were identified in e-liquids with only glycerol, whereas formaldehyde and methyl glyoxal were present in e-liquids containing both glycerol and propylene glycol (Uchiyama et al. 2020).

12.4 Other Factors Influencing the Presence and Levels of Toxic Chemicals in ENDS

The design of the vaping device, encompassing the type of coil, temperature control, and power settings, influences the amount and type of toxic chemicals produced. Higher temperatures can increase the formation of harmful substances such as formaldehyde, acetaldehyde, and acrolein. The coil heating temperature is a design feature and operational parameter that impacts both the quantity and composition of the aerosol emitted from e-cigs, as well as the sensory experience delivered to the users (Geiss et al. 2016; Talih et al. 2017). As the puffing progresses, the liquid composition in the heated portion of the wick tends to become enriched with the less volatile substances during a puff. This causes the boiling temperature to gradually increase toward that of the least volatile component present in the liquid. Consequently, within a single puff, the vaporization process can alter the aerosol composition from being enriched with the most volatile components to being enriched with the least volatile ones compared to the original liquid (Talih et al. 2017).

In addition to temperature, the power output of the device significantly influences the composition of chemicals produced in ENDS emissions. Higher device power levels can lead to increased heating of the e-liquid, which in turn affects the vaporization and decomposition of its components. This dynamic process can alter the quantities and types of chemicals emitted into the aerosol, potentially influencing both the flavor and the chemical profiles inhaled by the user (Gillman et al. 2016). The power level of the device plays a critical role in determining the chemical composition of ENDS emissions, in addition to temperature. Higher power settings result in greater heating of the e-liquid, influencing the vaporization and breakdown of its constituents. This process can lead to variations in the types and amounts of chemicals released into the aerosol, affecting both the sensory experience and potential health implications for users (Kosmider et al. 2014; Zelinkova and Wenzl 2020). Coils with lower ohm ratings produce more vapor and flavor but require more e-liquid and demand greater battery power compared to their higher ohm counterparts. In contrast, higher ohm ENDS coils produce less vapor, use less e-liquid, and are more battery-efficient (Hiler et al. 2020). This was demonstrated in a study where users vaped high-wattage e-liquids with low nicotine concentrations, consuming large quantities of such e-liquids unknowingly increasing their exposure to toxic chemicals (Smets et al. 2019).

How the user operates the device, including puff duration, frequency, and intensity, can affect the amount of toxic chemical exposure. More intense or frequent use can lead to higher temperatures and increased production of harmful compounds. ENDS devices that are capable of adjusting the coil heating temperature can result in greater exposure to non-nicotine toxicants in particular and excessive amounts of aldehydes, especially formaldehyde, as reported in numerous studies (EL-Hellani et al. 2018; Talih et al. 2016; Talih et al. 2020a, b). The level of aldehydes generated in the aerosols was found to correlate with the power of the device, with higher-powered devices producing higher total levels of aldehydes (EL-Hellani et al. 2018; Geiss et al. 2016).

The way e-liquids are stored, including exposure to light, heat, and air, can lead to chemical changes and degradation, resulting in the formation of toxic compounds over time. Nicotine stability in e-liquids was found to be influenced by the nicotine form (pH) of formulation and its chemical composition. A study has shown that the presence of organic acids, which lower the pH, enhances the stability of nicotine by facilitating the formation of more stable nicotine salt products at lower pH levels (Kosarac et al. 2023). Additionally, stress tests on nicotine revealed that the compound degrades more rapidly under alkaline conditions, with 88% degradation observed after 5 days. In contrast, under acidic conditions, 97% of nicotine remained intact even after 10 days. These findings highlight the impact of pH on the stability of nicotine (Bansal et al. 2018).

12.5 Conclusion

Advancements in science have facilitated the industry's rapid synthesis of nicotine, nicotine analogs, and the introduction of new chemicals into e-liquid formulations, often outpacing regulatory oversight. However, synthetically produced nicotine and chemicals carry their own set of impurities, depending on the specific synthetic route employed. This presents challenges for regulatory authorities striving to navigate the ever-changing landscape of e-cig products. Keeping up with these developments demands a substantial allocation of resources.

Ongoing independent research through collaboration is crucial for understanding the unknown health risks associated with the use of e-cigs. Independent research provides unbiased scientific evidence on the content and emissions of e-cigs, offering invaluable insights into the specific chemicals present and their toxicological effects. This knowledge is essential for identifying new potential health risks and understanding how various factors, such as temperature, device type, and e-liquid composition, influence the formation of toxicants.

Furthermore, the unbiased scientific information gathered can serve as a vital resource for policymakers, users, and non-users, informing them about the dangers of using ENDS (and/or ENNDS) and its associated health risks. Continuous communication with the public about the harm posed by these products is essential to safeguard public health.

References

Almazrouei ES, Bintamim AA, Khalil SEA, Alremeithi R, Gewily S (2022) The identification of drugs of abuse in E-cigarette samples seized in Dubai between 2016 and 2020. Forensic Sci Int 333:111233. https://doi.org/10.1016/j.forsciint.2022.111233

Altamimy MA, Alshehri YM, Aldawsari FS, Altalyan NH, AlShmmari SK, Alzaid SF, Al-Qahtani NA, Alsagoor TM, Al Shahrani MS (2024) A selective gas chromatography–Tandem mass spectrometry method for quantitation of ethylene and diethylene glycol in paediatric syrups. Heliyon 10(7). https://doi.org/10.1016/j.heliyon.2024.e27559

Bansal M, Sharma M, Bullen C, Svirskis D (2018) A stability indicating HPLC method to determine actual content and stability of nicotine within electronic cigarette liquids. Int J Environ Res Public Health 15(8). https://doi.org/10.3390/ijerph15081737

Bekki K, Uchiyama S, Ohta K, Inaba Y, Nakagome H, Kunugita N (2014) Carbonyl compounds generated from electronic cigarettes. Int J Environ Res Public Health 11(11):11192–11200. https://doi.org/10.3390/ijerph111111192

Bitzer ZT, Goel R, Reilly SM, Elias RJ, Silakov A, Foulds J, Muscat J, Richie JP (2018a) Effect of flavoring chemicals on free radical formation in electronic cigarette aerosols. Free Radic Biol Med 120:72–79. https://doi.org/10.1016/j.freeradbiomed.2018.03.020

Bitzer ZT, Goel R, Reilly SM, Foulds J, Muscat J, Elias RJ, Richie JP Jr (2018b) Effects of solvent and temperature on free radical formation in electronic cigarette aerosols. Chem Res Toxicol 31(1):4–12. https://doi.org/10.1021/acs.chemrestox.7b00116

Bitzer ZT, Goel R, Reilly SM, Bhangu G, Trushin N, Foulds J, Muscat J, Richie JP Jr (2019) Emissions of free radicals, carbonyls, and nicotine from the NIDA standardized research electronic cigarette and comparison to similar commercial devices. Chem Res Toxicol 32(1):130–138. https://doi.org/10.1021/acs.chemrestox.8b00235

Bowen A, Xing C (2015) Nicotine salt formulations for aerosol devices and methods thereof (Patent US9215895B2). United States Patent

Brunnemann KD, Yu L, Hoffmann D (1977) Assessment of carcinogenic volatile N- nitrosamines in tobacco and in mainstream and sidestream smoke from cigarettes. Cancer Res 37(9):3218–3222

Cheah NP, Chong NWL, Tan J, Morsed FA, Yee SK (2014) Electronic nicotine delivery systems: regulatory and safety challenges: Singapore perspective. Tob Control 23(2):119. https://doi.org/10.1136/tobaccocontrol-2012-050483

Cheetham AG, Plunkett S, Campbell P, Hilldrup J, Coffa BG, Gilliland S III, Eckard S (2022) Analysis and differentiation of tobacco-derived and synthetic nicotine products: addressing an urgent regulatory issue. PLoS One 17(4):e0267049. https://doi.org/10.1371/journal.pone.0267049

Cipriano JJ, Kounnas CN, Spielberg HL (1975) Manipulation of nicotine delivery by addition of acids to filler. https://www.industrydocuments.ucsf.edu/tobacco/docs/#id=zpgp0180

Council NR (2014) Acute exposure guideline levels for selected airborne chemicals, vol 16. The National Academies Press. https://doi.org/10.17226/18707

Crooks PA, Byrd GD (1999) Chapter 7 – use of high-performance liquid chromatographic–mass spectrometric (LC–MS) techniques for the determination of nicotine and its metabolites. In: Gorrod JW, Jacob P (eds) Analytical determination of nicotine and related compounds and their metabolites. Elsevier Science, pp 225–264. https://doi.org/10.1016/B978-044450095-3/50008-5

Davis RA, Curvali M (1999) Chapter 14 – determination of nicotine and its metabolites in biological fluids: in vivo studies. In: Gorrod JW, Jacob P (eds) Analytical determination of nicotine and related compounds and their metabolites. Elsevier Science, pp 583–643. https://doi.org/10.1016/B978-044450095-3/50015-2

Dorotheo EU, Arora M, Banerjee A, Bianco E, Cheah NP, Dalmau R, Eissenberg T, Hasegawa K, Naidoo P, Nazir NT, Newby LK, Obeidat N, Skipalskyi A, Stępińska J, Willett J, Wang Y (2024) Nicotine and cardiovascular health: when poison is addictive – a WHF policy brief. Glob Heart 19:14. https://doi.org/10.5334/gh.1292

Ebersole J, Samburova V, Son Y, Cappelli D, Demopoulos C, Capurro A, Pinto A, Chrzan B, Kingsley K, Howard K, Clark N, Khlystov A (2020) Harmful chemicals emitted from electronic cigarettes and potential deleterious effects in the oral cavity. Tob Induc Dis 18(May):41. https://doi.org/10.18332/tid/116988

El Hajj Moussa F, Hayeck N, Hajir S, El Hage R, Salman R, Karaoghlanian N, Saliba NA (2024) Enhancement of benzene emissions in special combinations of electronic nicotine delivery system liquid mixtures. Chem Res Toxicol 37(2):227–233. https://doi.org/10.1021/acs.chemrestox.3c00251

EL-Hellani A, Salman R, El-Hage R, Talih S, Malek N, Baalbaki R, Karaoghlanian N, Nakkash R, Shihadeh A, Saliba NA (2018) Nicotine and carbonyl emissions from popular electronic cigarette products: correlation to liquid composition and design characteristics. Nicotine Tob Res 20(2):215–223. https://doi.org/10.1093/ntr/ntw280

Farsalinos KE, Gillman G, Poulas K, Voudris V (2015) Tobacco-specific nitrosamines in electronic cigarettes: comparison between liquid and aerosol levels. Int J Environ Res Public Health 12(8):9046–9053. https://doi.org/10.3390/ijerph120809046

Galimov A, Leventhal A, Meza L, Unger JB, Huh J, Baezconde-Garbanati L, Sussman SY (2021) Prevalence of disposable pod use and consumer preference for e- cigarette product characteristics among vape shop customers in Southern California: a cross-sectional study. BMJ Open 11(10):e049604. https://doi.org/10.1136/bmjopen-2021-049604

Geiss O, Bianchi I, Barrero-Moreno J (2016) Correlation of volatile carbonyl yields emitted by e-cigarettes with the temperature of the heating coil and the perceived sensorial quality of the generated vapours. Int J Hyg Environ Health 219(3):268–277. https://doi.org/10.1016/j.ijheh.2016.01.004

Gillman IG, Kistler KA, Stewart EW, Paolantonio AR (2016) Effect of variable power levels on the yield of total aerosol mass and formation of aldehydes in e-cigarette aerosols. Regul Toxicol Pharmacol 75:58–65. https://doi.org/10.1016/j.yrtph.2015.12.019

Guan Z-Z, Yu W-F, Nordberg A (2003) Dual effects of nicotine on oxidative stress and neuroprotection in PC12 cells. Neurochem Int 43(3):243–249. https://doi.org/10.1016/S0197-0186(03)00009-3

Haddad C, Salman R, El-Hellani A, Talih S, Shihadeh A, Saliba NA (2019) Reactive oxygen species emissions from supra- and sub-ohm electronic cigarettes. J Anal Toxicol 43(1):45–50. https://doi.org/10.1093/jat/bky065

Hallingberg B (2022) E-cigarette flavors, devices, and brand preferences among youths in Canada, England, and the United States: the value and challenges of comparing international survey data. Am J Public Health 112(7):1011–1013. https://doi.org/10.2105/AJPH.2022.306903

Hammond D, Reid JL, Burkhalter R, Bansal Travers M, Gravely S, Hyland A, Kasza K, McNeill A (2022) E-cigarette flavors, devices, and brands used by youths before and after partial flavor restrictions in the United States: Canada, England, and the United States, 2017–2020. Am J Public Health 112(7):1014–1024. https://doi.org/10.2105/AJPH.2022.306780

Harber P, Saechao K, Boomus C (2006) Diacetyl-induced lung disease. Toxicol Rev 25(4):261–272. https://doi.org/10.2165/00139709-200625040-00006

Hasan F, Khachatryan L, Lomnicki S (2020) Comparative studies of environmentally persistent free radicals on total particulate matter collected from electronic and tobacco cigarettes. Environ Sci Technol 54(9):5710–5718. https://doi.org/10.1021/acs.est.0c00351

Havermans A, Krüsemann EJZ, Pennings J, de Graaf K, Boesveldt S, Talhout R (2021) Nearly 20 000 e-liquids and 250 unique flavor descriptions: an overview of the Dutch market based on information from manufacturers. Tob Control 30(1):57. https://doi.org/10.1136/tobaccocontrol-2019-055303

Hiler M, Breland A, Spindle T, Maloney S, Lipato T, Karaoghlanian N, Shihadeh A, Lopez A, Ramôa C, Eissenberg T (2017) Electronic cigarette user plasma nicotine concentration, puff topography, heart rate, and subjective effects: influence of liquid nicotine concentration and user experience. Exp Clin Psychopharmacol 25(5):380–392. https://doi.org/10.1037/pha0000140

Hiler M, Karaoghlanian N, Talih S, Maloney S, Breland A, Shihadeh A, Eissenberg T (2020) Effects of electronic cigarette heating coil resistance and liquid nicotine concentration on user nicotine delivery, heart rate, subjective effects, puff topography, and liquid consumption. Exp Clin Psychopharmacol 28(5):527–539

IARC Monographs on the Identification of Carcinogenic Hazards to Humans (2024). https://monographs.iarc.who.int/agents-classified-by-the-iarc/

Jabba SV, Diaz AN, Erythropel HC, Zimmerman JB, Jordt S-E (2020) Chemical adducts of reactive flavor aldehydes formed in E-cigarette liquids are cytotoxic and inhibit mitochondrial function in respiratory epithelial cells. Nicotine Tob Res 22(Supplement_1):S25–S34. https://doi.org/10.1093/ntr/ntaa185

Jacob P, Byrd GD (1999) Chapter 6 – use of gas chromatographic and mass spectrometric techniques for the determination of nicotine and its metabolites. In: Gorrod JW, Jacob P (eds) Analytical determination of nicotine and related compounds and their metabolites. Elsevier Science, pp 191–224. https://doi.org/10.1016/B978-044450095-3/50007-3

Jin XC, Wagner KA, Melvin MS, Smith DC, Pithawalla YB, Gardner WP, Avery KC, Karles GD (2022) Influence of nitrite on formation of tobacco- specific nitrosamines in electronic cigarette liquids and aerosols. Chem Res Toxicol 35(5):782–791. https://doi.org/10.1021/acs.chemrestox.1c00417

Jordt SE, Jabba SV, Silinski P, Berman ML (2023) An electronic cigarette pod system delivering 6-methyl nicotine, a synthetic nicotine analog, marketed in the United States as "PMTA exempt." MedRxiv. https://doi.org/10.1101/2023.11.21.23298778

Klupinski TP, Strozier ED, Koebel DD, Lucas EA (2023) Detection of the synthetic coolant menthone 1,2-glycerol ketal in an e-liquid and in electronic waterpipe aerosols therefrom. Chem Res Toxicol 36(8):1355–1360. https://doi.org/10.1021/acs.chemrestox.3c00089

Konstantinou E, Fotopoulou F, Drosos A, Dimakopoulou N, Zagoriti Z, Niarchos A, Makrynioti D, Kouretas D, Farsalinos K, Lagoumintzis G, Poulas K (2018) Tobacco-specific nitrosamines: a literature review. Food Chem Toxicol 118:198–203. https://doi.org/10.1016/j.fct.2018.05.008

Kosarac I, Katuri GP, Kubwabo C, Siddique S, Mischki TK (2023) Quantitation and stability of nicotine in canadian vaping liquids. Toxics 11(4). https://doi.org/10.3390/toxics11040378

Kosmider L, Sobczak A, Fik M, Knysak J, Zaciera M, Kurek J, Goniewicz ML (2014) Carbonyl compounds in electronic cigarette vapors: effects of nicotine solvent and battery output voltage. Nicotine Tob Res 16(10):1319–1326. https://doi.org/10.1093/ntr/ntu078

Krüsemann EJZ, Havermans A, Pennings JLA, de Graaf K, Boesveldt S, Talhout R (2021) Comprehensive overview of common e-liquid ingredients and how they can be used to predict an e-liquid's flavor category. Tob Control 30(2):185. https://doi.org/10.1136/tobaccocontrol-2019-055447

Kwak S, Choi Y, Na H, Bae C, Song S-Y, Kim Y-D (2020) Glyoxal and methylglyoxal as E-cigarette vapor ingredients-induced pro-inflammatory cytokine and mucins expression in human nasal epithelial cells. Am J Rhinol Allergy 35:194589242094696. https://doi.org/10.1177/1945892420946968

Langone JJ, Gjika HB, Van Vunakis H (1999) Chapter 8 – use of immunoassay techniques for the determination of nicotine and its metabolites. In: Gorrod JW, Jacob P (eds) Analytical determination of nicotine and related compounds and their metabolites. Elsevier Science, pp 265–283. https://doi.org/10.1016/B978-044450095-3/50009-7

Lerner CA, Sundar IK, Yao H, Gerloff J, Ossip DJ, McIntosh S, Robinson R, Rahman I (2015) Vapors produced by electronic cigarettes and E-juices with flavorings induce toxicity, oxidative stress, and inflammatory response in lung epithelial cells and in mouse lung. PLoS One 10(2):e0116732. https://doi.org/10.1371/journal.pone.0116732

Levulinic Acid (1991) https://www.industrydocuments.ucsf.edu/docs/rymk0089

Li Y, Burns AE, Tran LN, Abellar KA, Poindexter M, Li X, Madl AK, Pinkerton KE, Nguyen TB (2021) Impact of e-liquid composition, coil temperature, and puff topography on the aerosol chemistry of electronic cigarettes. Chem Res Toxicol 34(6):1640–1654

Li X, Lu L, He Y, Zhang H, Zhang Y, Sheng H, Chen M, Ru J, Gao Y (2023) Pharmacological effects of nicotine salts on dopamine release in the nucleus accumbens. Front Pharmacol 14

Lorkiewicz P, Keith R, Lynch J, Jin L, Theis W, Krivokhizhina T, Riggs D, Bhatnagar A, Srivastava S, Conklin DJ (2022) Electronic cigarette solvents, JUUL E-liquids, and biomarkers of exposure: in vivo evidence for acrolein and glycidol in E-Cig-derived aerosols. Chem Res Toxicol 35(2):283–292. https://doi.org/10.1021/acs.chemrestox.1c00328

Margin of Exposure (2023) European food safety authority. https://www.efsa.europa.eu/en/topics/topic/margin-exposure

Metatine (2023). https://metatine.com/

Muthumalage T, Friedman MR, McGraw MD, Ginsberg G, Friedman AE, Rahman I (2020) Chemical constituents involved in E-cigarette, or vaping product use-associated lung injury (EVALI). Toxics 8(2). https://doi.org/10.3390/toxics8020025

Neu HM, Lee A, Brandis JEP, Patel V, Schneider A, Kane MA, Dalby RN, Michel SLJ (2020) Cigalike electronic nicotine delivery systems e-liquids contain variable levels of metals. Sci Rep 10(1):11907. https://doi.org/10.1038/s41598-020-67789-7

Omaiye EE, Luo W, McWhirter KJ, Pankow JF, Talbot P (2022) Disposable puff bar electronic cigarettes: chemical composition and toxicity of E-liquids and a synthetic coolant. Chem Res Toxicol 35(8):1344–1358. https://doi.org/10.1021/acs.chemrestox.1c00423

Palmisani J, Abenavoli C, Famele M, Di Gilio A, Palmieri L, de Gennaro G, Draisci R (2020) Chemical characterization of electronic cigarette (e-cigs) refill liquids prior to EU tobacco product directive adoption: evaluation of BTEX contamination by HS-SPME-GC-MS and identification of flavoring additives by GC- MS-O. Atmos 11(4). https://doi.org/10.3390/atmos11040374

Pankow JF, Kim K, McWhirter KJ, Luo W, Escobedo JO, Strongin RM, Duell AK, Peyton DH (2017) Benzene formation in electronic cigarettes. PLoS One 12(3):e0173055. https://doi.org/10.1371/journal.pone.0173055

Pappas RS, Gray N, Halstead M, Valentin-Blasini L, Watson C (2021) Toxic metal-containing particles in aerosols from pod-type electronic cigarettes. J Anal Toxicol 45(4):337–347. https://doi.org/10.1093/jat/bkaa088

Pennings JLA, Havermans A, Pauwels CGGM, Krüsemann EJZ, Visser WF, Talhout R (2023) Comprehensive Dutch market data analysis shows that e-liquids with nicotine salts have both higher nicotine and flavor concentrations than those with free-base nicotine. Tob Control 32(e1):e78. https://doi.org/10.1136/tobaccocontrol-2021-056952

Qi H, Chang X, Wang K, Xu Q, Liu M, Han B (2023) Comparative analyses of transcriptome sequencing and carcinogenic exposure toxicity of nicotine and 6-methyl nicotine in human bronchial epithelial cells. Toxicol In Vitro 93:105661. https://doi.org/10.1016/j.tiv.2023.105661

Safety and Regulatory Authority to Use Flavors – Focus on Vaping Products (2021). https://www.femaflavor.org/safety-regulatory-authority-use-flavors-focus-vaping-products

Sleiman M, Logue JM, Montesinos VN, Russell ML, Litter MI, Gundel LA, Destaillats H (2016) Emissions from electronic cigarettes: key parameters affecting the release of harmful chemicals. Environ Sci Technol 50(17):9644–9651. https://doi.org/10.1021/acs.est.6b01741

Smets J, Baeyens F, Chaumont M, Adriaens K, Van Gucht D (2019) When less is more: vaping low-nicotine vs high-nicotine E-liquid is compensated by increased wattage and higher liquid consumption. Int J Environ Res Public Health 16(5). https://doi.org/10.3390/ijerph16050723

Soule EK, Sousan S, Streuber D, Fresquez SE, Mooring R, Salman R, Talih S, Pender J (2022) Increased JUUL emissions from initial puffs after removing and reinserting pod. Chem Res Toxicol 35(3):383–386. https://doi.org/10.1021/acs.chemrestox.2c00017

Spindle TR, Talih S, Hiler MM, Karaoghlanian N, Halquist MS, Breland AB, Shihadeh A, Eissenberg T (2018) Effects of electronic cigarette liquid solvents propylene glycol and vegetable glycerin on user nicotine delivery, heart rate, subjective effects, and puff topography. Drug Alcohol Depend 188:193–199. https://doi.org/10.1016/j.drugalcdep.2018.03.042

Talih S, Balhas Z, Salman R, Karaoghlanian N, Shihadeh A (2016) "Direct dripping": a high-temperature, high-formaldehyde emission electronic cigarette use method. Nicotine Tob Res 18(4):453–459. https://doi.org/10.1093/ntr/ntv080

Talih S, Balhas Z, Salman R, El-Hage R, Karaoghlanian N, El-Hellani A, Baassiri M, Jaroudi E, Eissenberg T, Saliba N, Shihadeh A (2017) Transport phenomena governing nicotine emissions from electronic cigarettes: model formulation and experimental investigation. Aerosol Sci Technol 51(1):1–11. https://doi.org/10.1080/02786826.2016.1257853

Talih S, Salman R, El-Hage R, Karam E, Salam S, Karaoghlanian N, El-Hellani A, Saliba N, Shihadeh A (2020a) A comparison of the electrical characteristics, liquid composition, and toxicant emissions of JUUL USA and JUUL UK e-cigarettes. Sci Rep 10(1):7322. https://doi.org/10.1038/s41598-020-64414-5

Talih S, Salman R, El-Hage R, Karaoghlanian N, El-Hellani A, Saliba N, Shihadeh A (2020b) Effect of free-base and protonated nicotine on nicotine yield from electronic cigarettes with varying power and liquid vehicle. Sci Rep 10(1):16263. https://doi.org/10.1038/s41598-020-73385-6

Talih S, Karaoghlanian N, Salman R, Hilal E, Patev A, Bell A, Fallah S, El-Hage R, Saliba NA, Cobb C, Barnes A, Shihadeh A (2023) Effects of aftermarket electronic cigarette pods on device power output and nicotine, carbonyl, and ROS emissions. Chem Res Toxicol 36(12):1930–1937. https://doi.org/10.1021/acs.chemrestox.3c00213

Tran LN, Chiu EY, Hunsaker HC, Wu K, Poulin BA, Madl AK, Pinkerton KE, Nguyen TB (2023) Carbonyls and aerosol mass generation from vaping nicotine salt solutions using fourth- and third-generation E-cigarette devices: effects of coil resistance, coil age, and coil metal material. Chem Res Toxicol 36(10):1599–1610. https://doi.org/10.1021/acs.chemrestox.3c00172

Tran LN, Rao G, Robertson NE, Hunsaker HC, Chiu EY, Poulin BA, Madl AK, Pinkerton KE, Britt RD, Nguyen TB (2024) Quantification of free radicals from vaping electronic cigarettes containing nicotine salt solutions with different organic acid types and concentrations. Chem Res Toxicol 37(6):991–999. https://doi.org/10.1021/acs.chemrestox.4c00065

Tsai MC, Gorrod JW (1999) Chapter 15 – determination of nicotine and its metabolites in biological fluids: in vitro studies. In: Gorrod JW, Jacob P (eds) Analytical determination of nicotine and related compounds and their metabolites. Elsevier Science, pp 645–668. https://doi.org/10.1016/B978-044450095-3/50016-4

Uchiyama S, Senoo Y, Hayashida H, Inaba Y, Nakagome H, Kunugita N (2016) Determination of chemical compounds generated from second-generation E-cigarettes using a sorbent cartridge followed by a two-step elution method. Anal Sci 32(5):549–555. https://doi.org/10.2116/analsci.32.549

Uchiyama S, Noguchi M, Sato A, Ishitsuka M, Inaba Y, Kunugita N (2020) Determination of thermal decomposition products generated from E-cigarettes. Chem Res Toxicol 33(2):576–583. https://doi.org/10.1021/acs.chemrestox.9b00410

Varlet V, Farsalinos K, Augsburger M, Thomas A, Etter J-F (2015) Toxicity assessment of refill liquids for electronic cigarettes. Int J Environ Res Public Health 12(5):4796–4815. https://doi.org/10.3390/ijerph120504796

Wagener TL, Floyd EL, Stepanov I, Driskill LM, Frank SG, Meier E, Leavens EL, Tackett AP, Molina N, Queimado L (2017) Have combustible cigarettes met their match? The nicotine delivery profiles and harmful constituent exposures of second-generation and third-generation electronic cigarette users. Tob Control 26(e1):e23. https://doi.org/10.1136/tobaccocontrol-2016-053041

Williams M, Bozhilov K, Ghai S, Talbot P (2017) Elements including metals in the atomizer and aerosol of disposable electronic cigarettes and electronic hookahs. PLoS One 12(4):e0175430. https://doi.org/10.1371/journal.pone.0175430

Wong M, Martinez T, Tran M, Zuvia C, Gadkari A, Omaiye EE, Luo W, McWhirter KJ, Sha J, Kassem A, Wohlschlegel J, Talbot P (2023) A synthetic coolant (WS-23) in disposable electronic cigarettes impairs cytoskeletal function in EpiAirway microtissues exposed at the air liquid interface. Sci Rep 13(1):16906. https://doi.org/10.1038/s41598-023-43948-4

Woo W, Tian L, Lum M, Canchola A, Chen K, Lin YH (2024) Ozonolysis of terpene flavor additives in vaping emissions: elevated production of reactive oxygen species and oxidative stress. Chem Res Toxicol 37(6):981–990. https://doi.org/10.1021/acs.chemrestox.4c00051

Zelinkova Z, Wenzl T (2020) Influence of battery power setting on carbonyl emissions from electronic cigarettes. Tob Induc Dis 18(September):1–5. https://doi.org/10.18332/tid/126406

Zervas E, Matsouki N, Kyriakopoulos G, Poulopoulos S, Ioannides T, Katsaounou P (2020) Transfer of metals in the liquids of electronic cigarettes. Inhal Toxicol 32(6):240–248. https://doi.org/10.1080/08958378.2020.1776801

Zhao J, Hopke PK (2012) Concentration of reactive oxygen species (ROS) in mainstream and sidestream cigarette smoke. Aerosol Sci Technol 46:191–197

Zhao J, Zhang Y, Sisler JD, Shaffer J, Leonard SS, Morris AM, Qian Y, Bello D, Demokritou P (2018) Assessment of reactive oxygen species generated by electronic cigarettes using acellular and cellular approaches. J Hazard Mater 344:549–557. https://doi.org/10.1016/j.jhazmat.2017.10.057

Zhou Y, Irshad H, Dye WW, Wu G, Tellez CS, Belinsky SA (2021) Voltage and e-liquid composition affect nicotine deposition within the oral cavity and carbonyl formation. Tob Control 30(5):485. https://doi.org/10.1136/tobaccocontrol-2020-055619

Zou Y, Zou J, Liu M, Luo W, Liu R, Lü H (2022) Synthetic method of 6-methyl nicotine (Patent CN114437031A). China National Intellectual Property Administration

Recent Developments in Qualitative and Quantitative Analysis of E-cigarette Liquids and Aerosols

13

Doreen N. B. Chandra Siri and Yong Foo Wong

Abstract

The chapter initiates with a short introduction to electronic cigarettes (e-cigs) with their developmental history being outlined. The reported harmful and potentially harmful chemical constituents associated with vaping are briefly introduced. Next, recently reported chromatographic and spectroscopic approaches for the detection and quantitation of chemical entities in e-cigs and their resulting aerosols are discussed. The applicability of direct analysis in real-time mass spectrometry without the need for chromatographic separation was highlighted. The findings of these studies on the basis of identification and/or quantitation of chemical entities and their concentrations in e-liquids and aerosols are summarized. The chapter concludes by providing insights into the importance of applying analytical approaches for a more comprehensive surveillance monitoring of e-cigs to deduce their potential toxicity or carcinogenicity, which could present potential health risks for users.

Keywords

E-liquids · Vaping · Volatile organic compounds · Heavy metals · Polycyclic aromatic hydrocarbons

D. N. B. Chandra Siri
Center for Research on Multidimensional Separation Science, School of Chemical Sciences, Universiti Sains Malaysia, Penang, Malaysia

Toxicology Section, Forensic Division, Department of Chemistry Malaysia, Penang, Malaysia

Y. F. Wong (✉)
Center for Research on Multidimensional Separation Science, School of Chemical Sciences, Universiti Sains Malaysia, Penang, Malaysia
e-mail: wongyongfoo@usm.my

Y.-H. Leong, A. S. Mohd Samin (eds.), *E-Cigarettes*,
https://doi.org/10.1007/978-981-96-5087-3_13

13.1 Introduction

In the past decade, electronic cigarettes, commonly referred to as e-cigarettes (e-cigs), have gained popularity among smokers due to the recognition of the numerous health risks and addictive nature of tobacco smoking. The concept of a smokeless, non-tobacco cigarette was introduced by Herbert A. Gilbert in 1963, which aims to provide a safe and harmless means of smoking (Gilbert 1965). However, the contemporary e-cig model was devised by the Chinese pharmacist Hon Lik in the early 2000s and introduced to the global market as an electronic atomizing cigarette (Hon 2003). Since its inception, the reshaping of public perceptions, policy landscapes, and individual behaviors has accelerated the evolution of e-cigs, as depicted in Fig. 13.1.

The modern e-cig devices are available in different shapes and sizes, resembling cigarettes, pens, USB flash drives, and other items. These devices are alternatively referred to as "vapes," "mods," "e-hookahs," and "vape pens" (Breland et al. 2017). The initial e-cig devices have undergone substantial innovation and evolution in design, progressing from simple devices resembling conventional cigarettes to second-generation "clearomizers," third-generation "mods," and fourth-generation "pods" (Strongin 2019; Almeida-da-silva and Matshik 2021; Omaiye et al. 2021). These customizable devices now have larger storage tanks and increased voltage or power capabilities (Strongin 2019). The modifications to these designs have led to

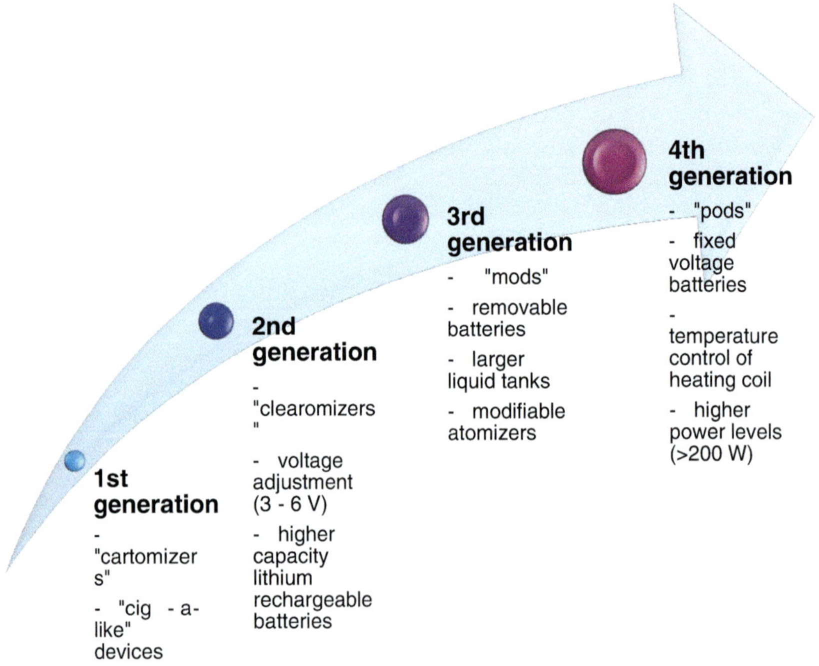

Fig. 13.1 Evolution of the e-cig design

an increase in the production of aerosols and enhanced nicotine delivery to users (Walley and Wilson 2021). The main components of an e-cig device include a battery, a thermal element containing a vaporizing chamber, a liquid reservoir, and a mouthpiece. The e-liquid formulation may or may not contain nicotine, propylene glycol (PG), vegetable glycerine (VG), flavors, and additives (Orellana-Barrios et al. 2015). Additives are frequently added to the liquid to augment the flavor and ease the harshness of inhalation (Huerta et al. 2017). E-liquid formulations have likewise undergone substantial modifications in terms of solvent ratios, concentration and variety of flavors, the addition of nicotine acids and salts, and the introduction of synthetic cooling agents (e.g., WS-3 or WS-23) and synthetic nicotine (Omaiye et al. 2022; Jabba et al. 2022). Customizable devices enable users to create e-liquid formulations according to their preferences by incorporating commercially available additives and flavors. In particular, the lack of comprehensive legislation on the manufacturing and retailing of these numerous flavored products (Walley and Wilson 2021) is a cause of concern. Medical practitioners have expressed concerns over the attractiveness of flavored e-cigs to young people, the potential risks associated with inhaling flavor compounds, and the formation of harmful substances when vaping at high temperatures (Tierney et al. 2016).

E-cigs are often touted as an alternative nicotine delivery or electronic nicotine delivery systems (ENDS) that function without the combustion of tobacco, circumventing the many health issues related to tobacco smoking (Wagner et al. 2018; Halstead et al. 2019). Hence, ENDS are widely promoted as smoking cessation aids. These devices deliver nicotine to the human body through the inhalation of aerosols that are produced from the heating of the e-liquid, imitating the act of smoking (Clapp and Jaspers 2017). E-cig usage, also known as vaping, is gaining popularity among youth and young adults (Du et al. 2020) and can be attributed to its perceived trendiness or modernity. Despite the rapid growth of e-cigs, there are still major concerns about the potential health risks, their effectiveness as smoking cessation aids, regulatory and policy oversight, and the increased propagation of smoking in youth. Alarmingly, there is an increased tendency among youth who have experimented with e-cigs to progress to combustible tobacco product usage (Kong et al. 2022).

While e-cig aerosols are acknowledged to contain far fewer toxic substances and carcinogens than tobacco smoke, it is important to note that the aerosols have not been conclusively shown to be entirely harmless. Its toxicity may be influenced by factors such as device settings, liquid composition, coil material, vaping behavior, and the presence of metals and particulates in the aerosol (Sleiman et al. 2016; Stephens 2018; Lechasseur et al. 2017). E-cigs have been linked to detrimental effects on the respiratory, cardiovascular, nervous, and immune systems, among others (Cao et al. 2021). The harmful and potentially harmful constituents (HPHCs) that have been associated with e-cig use are carbonyl compounds, heavy metals, volatile organic compounds (VOC), and polycyclic aromatic hydrocarbons (PAH) (Jo and Kim 2016; Zhao et al. 2019; Cheng 2014). Tobacco-specific nitrosamines (TSNA), which are known carcinogens, can also be metabolized from the nicotine present in e-liquids upon heating (Harris 2018).

13.2 Analytical Methods for the Analysis of E-liquids and Aerosol

One of the primary concerns regarding e-liquids is the considerable variation in their formulations and additives. The production of HPHCs is dependent on the concentration of flavors and additives. Hence, the growing popularity of e-cig usage is of concern (Khlystov and Samburova 2016). Considering this, accurate qualitative and quantitative determination of selected chemical entities in e-liquids at trace levels is crucial for subsequent investigations concerning legal confirmation judgments and exposure assessment. Many analytical methodologies have been reported over the years for the qualitative and/or quantitative analysis of e-liquids and aerosol. Non-targeted approaches in assessing unregulated e-liquids and aerosol could yield the most informative data on the presence of potential HPHCs and adulterants. Briefly, the methods can be separated into chromatographic and spectroscopic techniques, as highlighted in Fig. 13.2.

13.2.1 Chromatographic Methods

Analytical separation via chromatographic techniques, especially gas chromatography (GC) and high-performance liquid chromatography (HPLC), is necessary to resolve the wide variety of chemical constituents in e-liquids and aerosols for effective quantitative or qualitative analysis. Of the various separation techniques, GC is

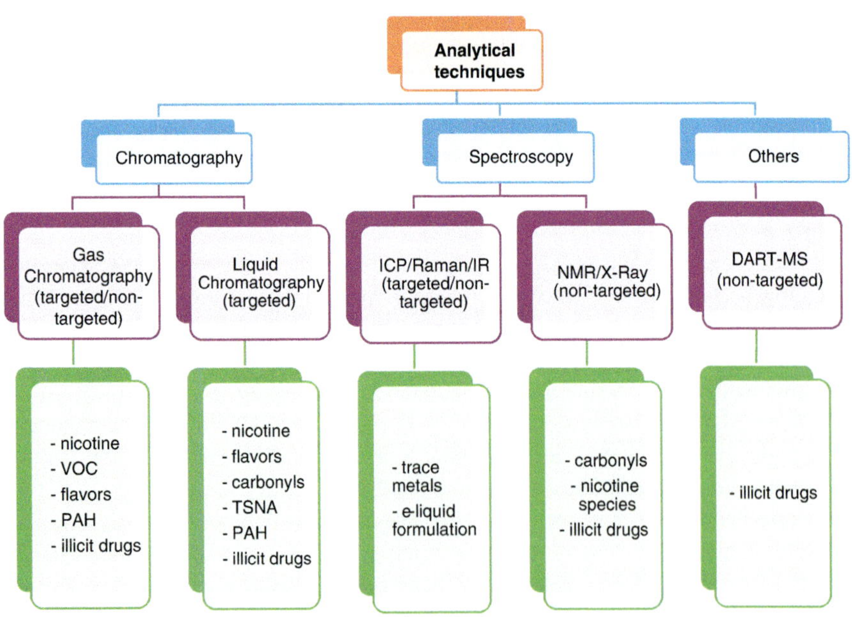

Fig. 13.2 Analytical methods for analysis of e-liquids and aerosols

appreciated as the most widely used chromatographic technique due to its compatibility in analyzing volatile and semi-volatile compounds constituted in e-liquids and aerosols. However, the strong adsorption of highly polar compounds to the GC stationary phase can result in peak broadening, which affects the separation performance. Liquid chromatography techniques with hyphenation to mass spectrometry have also been routinely applied to the analysis of e-liquids, which is helpful for compounds exhibiting low thermal stability (Choi et al. 2022; Fan et al. 2021).

In a recent study reported by Holt et al. (2021), 350 chemical constituents were found in the analysis of 241 regulated and unregulated products utilizing gas chromatography-mass spectrometry (GC–MS) and direct analysis in real-time (DART) time of flight mass spectrometry (TOFMS), followed by quantitation via GC–MS, headspace gas chromatography flame ionization detector (GC–FID) or liquid chromatography–tandem mass spectrometry (LC–MS/MS). Apart from the listed active ingredients (nicotine, Δ9-THC, and CBD) on product labels, the authors also detected other psychoactive substances including cannabinoids (Δ8-THC, CBG, CBDA, CBN, CBC, THCV, and THCA), caffeine, kratom alkaloids (mitragynine and 7-hydroxy-mitragynine), synthetic cannabinoids (5F-ADB, MDMB-FUBINACA, FUB- AMB, EMB-FUBINACA, MFUBINAC, MMB-FUBICA, 4F-MDMB-BINACA, 5F-EDMB-PINACA), and 4-fluoroisocathinone. Ethanol, used in flavor manufacturing or as a solvent to dilute e-liquids, was found in the assessed products at concentrations ranging from absence to 217.2 mg/mL. It is important to note that the inhalation of ethanol along with nicotine may lead to some impairment of the psychomotor skills of e-cig users (Valentine et al. 2016).

An investigation into adulterants or contaminants in illicit e-liquids seized from patients with lung injury was conducted through non-targeted analyses employing GC–MS and liquid chromatography time of flight mass spectrometry (LC–TOFMS). Key findings included low cannabinoid content, the presence of pesticides, and the identification of vitamin E acetate (VEA; Duffy et al. 2020). Interestingly, vaping-associated pulmonary illness has been intriguingly linked to the presence of VEA, a diluent in illicit cannabis e-liquids. This aligns with the findings of this study, as VEA was detected in 64% of the cannabinoid-containing liquids. Another study investigated the presence of VEA and its primary thermal breakdown products in aerosol using a GC–high-resolution mass spectrometry (GC–HRMS) approach (Kovach et al. 2022). During the aerosolization process, VEA underwent notable thermal degradation, yielding duroquinone (DQ) and 2,6,10,14-tetramethyl-1-pentadecene (TMPD), which were confirmed through retention time matching and mass-to-charge fragmentation analysis.

The primary components of e-liquids and aerosols, including terpenes, terpenoids, flavor compounds, cannabinoids, and other additives, were investigated using solid phase microextraction (SPME) coupled with the GC–MS approach (Guo et al. 2021). More than 100 terpenes and natural extracts, along with 19 cannabinoids, were identified. The vaporization of e-liquids yielded additional terpenes and minor cannabinoids from the breakdown of the constituents, while Δ9-THC and potentially toxic additives were detected at lower levels in the vapor and aerosol compared to the e-liquid. An LC–HRMS approach was used for the untargeted and quantitative analysis of chemical constituents in tobacco-flavored e-liquids and aerosols (Tehrani et al. 2021). The study observed an increase in the number of compounds from e-liquids to aerosols, with the rise in the proportion of condensed hydrocarbon-like compounds suggesting that combustion-like processes may have occurred despite lower temperatures. Tributylphosphine oxide (an industrial chemical), caffeine, and lipid-like compounds associated with severe respiratory effects were detected in aerosols and e-liquids.

Palmisani et al. (2020) demonstrated the applicability of headspace solid phase microextraction (HS–SPME) and GC–MS approaches for the extraction and determination of benzene, toluene, ethylbenzene, and xylenes (BTEX) in 97 refill e-liquids. The hybrid analytical-sensory technique is equipped with an olfactory detection port to facilitate the identification of odor-active compounds, particularly flavoring additives. Total concentration levels of BTEX exhibited significant variability from 3.2 to 32,151 μg/L across brands, leading to calculated exposure levels that were deemed unacceptable in certain brands of e-liquids. Additionally, the study explored sensory responses by correlating GC–MS findings with odor-active compounds characterized by descriptors like sweet, bitter, citrus, fresh, and others. Ethoxyethane contributed to a characteristic sweet and pleasant flavor, while diacetyl, linked to respiratory illness (bronchiolitis obliterans), contributed to a caramel-like flavor.

Budzyńska et al. (2020) developed a GC–ion mobility spectrometry (GC–IMS) methodology for the analysis of forbidden substances and other chemical components in e-cig liquids. By analyzing the reference standards with both the GC–IMS and GC–MS, retention dependencies for both methods were evaluated on the basis of their retention indices, which allows the correlation of the substances and facilitates the identification of unknowns within the GC–IMS chromatograms. The utilization of the retention index for GC–IMS in flavor identification was further explored by Augustini et al. (2021). The established indices were validated to illustrate their applicability in correlating with GC–MS data, which significantly enhanced the tentative identification capabilities. In a subsequent study, the same group developed a rapid sampling technique by integrating a condensation aid into GC–IMS to collect and quantify aerosol condensates from a single puff (Augustini et al. 2023). Using this approach, they successfully identified low concentration substances, and small molecule fragments or degradation products.

More recently, researchers have been exploring the applicability of a comprehensive two-dimensional gas chromatography (GC × GC) approach for the high-resolution separation of chemical constituents in e-liquids and aerosols. Patel et al. (2021) employed a combination of thermal desorption and GC × GC–TOFMS approaches to analyze the volatile organic compounds in e-cig fluid, demonstrating superior resolving power and improved detection capabilities for the separated chemical entities. Using a combination of first-dimension mid-polar and second-dimension polar column combination, a total of 1064 volatile compounds were tentatively identified, of which 164 were toxic compounds. Interestingly, acetone, methacrolein, benzene, *n*-butanol, formaldehyde, glycerol, propylene glycol, and toluene were identified as the HPHCs detected in the analyzed samples. Additionally, notable diversity was observed in the number and composition of HPHCs between different flavors. Therefore, the utilization of GC × GC significantly expanded the peak capacity, allowing more compounds to be separated and identified, which allowed a more comprehensive understanding of the aerosol constituents compared to other techniques.

13.2.2 Spectroscopic Methods

Despite being less prevalent in e-liquid and aerosol analysis, spectroscopic methods offer rapid analysis and valuable molecular structure and composition information. Additionally, spectroscopic methods offer quantitative capabilities that complement those of chromatographic techniques. An approach utilizing ^{1}H nuclear magnetic resonance (NMR) for the detection and quantitation of nicotine, glycerol, 1,2-propylene glycol, and water was successfully developed (Crenshaw et al. 2017). 1,2,4,5-tetrachloro-3-nitrobenzene was used as an internal standard for the ^{1}H NMR-based measurements, providing quantitative results that are in good agreement with those obtained from GC–MS analysis. In another study, Wu et al. (2021) reported the applicability of ^{1}H NMR methodology for the detection of matrix signals, synthetic cannabinoids (SCs), and flavoring compounds in e-liquids. The study investigated the potential of low-field ^{19}F NMR, a benchtop cryogen-free instrument for quantitative analysis of SCs, yielding findings that are in good agreement with established methodologies. Kerber et al. (2022) utilized ^{1}H NMR spectroscopy to analyze aerosols, assessing the formation of HPHCs from flavors, nicotine, and the PG-VG solvents, present individually or in combination. They determined that PG-VG degradation could be enhanced in certain flavors, while nicotine could inhibit or enhance HPHCs formation in the presence of different flavors and concentrations.

The degree of nicotine delivery is influenced by the fractionation of nicotine, leading to the formation of volatile free base or non-volatile protonated nicotine. In the presence of acids, the protonation of nicotine is elevated, reducing the harshness of inhalation and promoting a higher intake of nicotine (Harvanko et al. 2020). A simple and direct ^{1}H NMR spectroscopy was employed to determine α_{fb} (free base form), utilizing protonation-dependent chemical shifts for nicotine, therefore overcoming matrix effects and dilution factors of previous studies (Duell et al. 2018). The α_{fb} values of e-liquids were $0.03 \leq \alpha_{fb} \leq 0.84$, correlating with the values obtained from aerosols analysis. Yassine et al. (2023) proposed a simplified method to determine the free base fraction in e-liquid, by determining the pH and dielectric constant of the e-liquid to form a mathematical relationship. This method eliminates the need for dilution or complex NMR procedures, thereby facilitating the determination of other physicochemical properties such as the real pH of solution, pKa, and others.

A more recent study detailed the application of X-ray spectroscopy to investigate the protonation of nicotine on the surface and core of aerosol particles (Berkeley and Ahmed 2023). The study determined that the chemical speciation of nicotine is influenced by pH levels, specific additives, and PG/VG presence. A novel qualitative approach utilizing runtime cavity ringdown spectroscopy (rtCRDS), a laser-based method, was used for the rapid detection of oxidative products in aerosol (Rajapaksha et al. 2021). Accurate identification of additives and products was performed by matching the molecular fingerprint spectra obtained to the established database, which is based on IR spectra of molecules in the 6.8–11 μm mid-IR range. The formation of oxidative products was elevated with the addition of benzoic acids, ethyl maltol, nicotine, and spiked PG/VG, as evidenced by the increase in spectral intensities of carbonyl compounds.

13.2.3 Direct Analysis in Real-Time Mass Spectrometry

Analysis utilizing DART-MS offers the advantages of direct analysis without any sample preparation or solvent extraction and rapid high-throughput analysis with an analysis time of typically less than 3 min (Dou et al. 2023). Recently, Forbes and Krauss (2021) reported a DART-MS approach for aerosol and spiked drugs analysis that eliminates the need for sample preparation. Rapid chemical analysis of aerosol samples was performed in <30 s, which was achieved by direct aerosolization into a devised confined DART-MS configuration. The DART-MS approach was also explored in a multi-analyte (nicotine, CBD, Δ^9-THC, MCTs, PG, and VG) rapid screening of e-liquid samples, yielding data similar to LC–MS analysis (Falconer and Morales-Garcia 2023). Although this approach offers a notable advantage in terms of analysis time and sample throughput, the approach is unable to distinguish isobaric compounds, which include isomeric species of nicotine, THC, and CBD, therefore necessitating additional chromatographic separation or spectroscopic detection.

13.3 Analysis of Commonly Detected Compounds in E-liquids and Aerosols

E-liquid formulations are often modified by users according to individual preferences with the addition of flavors, nicotine amount, and other additives. Unregulated e-liquid components channeled into a high-powered device can produce harmful compounds through thermal decomposition (Johne et al. 2023). Hence, chemical analysis of unregulated liquids and aerosols offers valuable insights into their composition and the chemical constituents produced during vaping. This information can help mitigate health risks associated with HPHCs and enables informed decision-making regarding their usage. Legislation and policies can be tailored to address the use of harmful flavors and adulterated e-cig liquids, allowing targeted public health interventions to be implemented effectively.

The primary components in e-liquid formulations, namely PG, VG, and most flavorings, have been classified as generally recognized as safe (GRAS) by the FDA for food consumption. However, it is important to note that this classification does not extend to inhalation, and permissible limits on these constituents and the vapor emissions from vaping have not been clearly defined (Helen and Eaton 2018). A total of 91 components have been detected in e-liquids and aerosols, including HPHCs that have the potential to induce lung and pulmonary diseases, malignancies, cardiovascular issues, respiratory problems, and gastrointestinal disorders (Eshraghian and Al-Delaimy 2021). Table 13.1 summarizes the latest developments in analytical techniques for analyzing various forms of chemical entities in e-liquids and aerosols. Various sample preparation techniques are utilized to analyze compounds in e-liquids and aerosols, reflecting the diversity of analytes and the complexity of encountered matrices. Compounds of interest, such as nicotine, flavoring compounds, VOC, TSNA, PAH, heavy metals, and illicit drugs, will be discussed in detail in the following sections.

13.3.1 Nicotine

Nicotine remains the primary psychoactive substance in e-liquids, detected even in some liquids marketed as nicotine-free (Lee et al. 2020; Chien et al. 2023). Nicotine in e-liquid formulations is derived from tobacco, resulting in the inevitable co-extraction of small amounts of nicotine-related alkaloids (NRAs), which may contribute to the presence of these minor alkaloids or may originate from the oxidative degradation of nicotine. Concentrations of nicotine and the NRAs (nornicotine, cotinine, anabasine, anatabine, and myosmine) were assessed using a simple dilution technique (ratio of 1:10 in methanol) (Palazzolo et al. 2019). The developed method offered an optimized trace level detection utilizing PDA, with detection limits ranging from 18 ng/mL for nornicotine to 122 ng/mL for nicotine, highlighting salient features to quantify low-level NRAs in aerosol and smoke. LC–MS and GC–MS methodologies were employed in a study to determine the acids of nicotine salts in 23 e-liquid samples sourced from online retailers. Nicotine salts formed from the addition of weak acids provide better consumer satisfaction than free-base nicotine formulations, with varying levels of satisfaction offered by different salt formulations (Harvanko et al. 2020). Results of the study indicated that the most frequent acids used in the nicotinic salt formulations were lactic, benzoic, and levulinic acids, followed by salicylic, malic and tartaric acids.

GC coupled to a nitrogen-phosphorus detector (GC–NPD) has been known to offer superior accuracy, precision, specificity, and sensitivity for nitrogen- or phosphorus-containing compounds, providing commendable quantitation capabilities. A study utilizing GC–NPD to quantitate nicotine in e-liquids sold in several countries was reported, highlighting the limit of quantitation (LOQ) value of 1.0 mg/mL (Fix et al. 2021). No measurable nicotine was detected in e-liquids sold in Canada and Australia, whereas in the US and England, the mean nicotine concentrations were 16.2 and 11.9 mg/mL, respectively, in compliance with the relevant legislation. Kubica (2023) proposed a simple dilute-and-shoot approach utilizing LC-MS/MS for the quantitation of nicotine in e-liquids, with a LOD of 0.032 μg/mL and satisfactory recovery (96–105%) and precision ($<$ 4.2%). The nicotine quantity in the

Table 13.1 Sample preparation techniques and analytical methods for analysis of e-liquids and aerosols

Sample	Analyte	Sample preparation	Analytical method	References
Chromatography techniques				
Liquid	Cannabinoids; CBD, CBDA, Δ9-THC, Δ9-THCA	Dilution	LC–DAD LC–QTOFMS	Barhdadi et al. (2023)
Liquid	Cannabinoids; CBD, CBDA, CBN, THC, THCA	LLE	GC–MS LC–PDA	Grafinger et al. (2020)
Liquid	SC; MDMB-CHMICA, 5F-CUMYL-PINACA, 5F-APINACA, FUB-APINACA, AMB-FUBINACA, MMB-CHMICA	Dilution	LC–DAD LC-QTOFMS NMR GC–MS	Apirakkan et al. (2020)
Liquid	SC; ADB-BRINACA	Dilution	GC–MS LC–QTOFMS NMR	Choi et al. (2022)
Liquid	SC; MDMB-4en-PINACA, ADB-BUTINACA, and ADB-4en-PINACA	Thermal-assisted carbon fiber	LC–MS/MS	Dou et al. (2023)
Liquid	VOC	Headspace	GC × GC–TOFMS	Patel et al. (2021)
Liquid	VOC	Headspace	GC–MS GC–IMS	Budzyńska et al. (2020), Augustini et al. (2021)
Liquid	VOC	Headspace	GC–MS	Omaiye et al. (2019), Haworth-duff et al. (2023)
Liquid	VOC	HS-SPME	GC–MS	Palmisani et al. (2020)
Liquid	Nicotine and TSNA; NNN, NNK, NAT, NAB	Dilution	GC–NPD LC–MS/MS	Fix et al. (2021)
Liquid	Flavor	Headspace	GC–IMS	Augustini et al. (2022)
Liquid	Flavor	Dilution	GC–MS	Krüsemann et al. (2020), Omaiye et al. (2019), Barhdadi et al. (2021)

(continued)

Table 13.1 (continued)

Sample	Analyte	Sample preparation	Analytical method	References
Liquid	Flavor	Dilution	GC–TOFMS	Page and Goniewicz (2021)
Liquid	Nicotine salts	LLE	GC–MS LC–MS	Harvanko et al. (2020)
Liquid	Nicotine	Dilution	LC–MS/MS	Pawel Kubica (2023)
Liquid	VEA, cannabinoids, pesticides, mycotoxins	Dilution	GC–MS LC–TOFMS LC–MS/MS	Duffy et al. (2020)
Liquid	Nicotine, vitamin E, CBD, Δ9-THC	Dilution Headspace	GC–MS DART-TOFMS GC–FID LC–MS/MS	Holt et al. (2021)
Liquid	Nicotine, PG, VG, ethylmaltol, diacetyl, and acetylpropionyl	Derivatization using (i) BSTFA and TMCS (ii) o-phenylenediamine	GC–MS	Dagla et al. (2023)
Liquid	TSNA; NNN, NNK, NAT, NAB	Ultrasound-assisted Solvent extraction of porous membrane packed sample	LC–MS/MS	Paweł Kubica (2019)
Aerosol	VOC	Headspace	GC–MS GC–IMS	Augustini et al. (2023)
Aerosol	VOC	DNPH cartridge	LC–UV	Son et al. (2020)
Aerosol	VOC	SPE	GC–MSGC-TCD LC–PDA	Uchiyama et al. (2020)
Aerosol	PAH	SPME-NTD	GC–FID/MS	Doosti et al. (2023)
Aerosol	Vitamin E	Dilution	GC–MS	Kovach et al. (2022)
Liquid/aerosol	Nicotine and NRA; nornicotine, cotinine, anabasine, anatabine, and myosmine	Dilution	LC–PDA	Palazzolo et al. (2019)
Liquid/aerosol	PAH	SPE	LC–FLD	Kim et al. (2021)
Liquid/aerosol	VOC	Derivatization and SPME	GC–MS	Chen et al. (2021)
Liquid/aerosol	VOC	LLE	LC–MS	Melvin et al. (2020)

(continued)

Table 13.1 (continued)

Sample	Analyte	Sample preparation	Analytical method	References
Liquid/ aerosol	TSNA; NNN, NNK, NAT, NAB	Dilution Aerosol capture filter	LC–MS/MS	Lee et al. (2018)
Liquid/ aerosol	Terpenes, VEA, cannabinoids, PG, triglycerides	SPME	GC–MS	Guo et al. (2021)
Liquid/ aerosol	Caffeine, tributylphosphine oxide catechol, isophorone, vanillin, 2,4-diaminotoluene harmaline, skatole, triethyl citrate difenzoquat and methyl sulfate, tributyl O-acetylcitrate	Dilution	LC–HRMS	Tehrani et al. (2021)
Spectroscopy techniques				
Liquid	Nicotine/PG:VG ratio	Direct injection	Raman	Richardson et al. (2023)
Liquid	Nicotine, Vitamin E, Cannabinoids	Direct injection	MIR/NIR	Akhtar et al. (2024)
Liquid	Heavy metal; aluminum, chromium, copper, iron, lead, nickel, selenium, silicon, tin, zinc	Acid digestion	ICP–OES	Williams et al. (2022)
Liquid	SC; JWH-210, 5 F-AKB48, and 5 F-MDMB-PICA	Dilution	NMR	Wu et al. (2021)
Aerosol	VOC	Direct injection	NMR	Kerber et al. (2022)
Aerosol	VOC	Direct injection	rtCRDS	Rajapaksha et al. (2021)
Aerosol	Heavy metal; aluminum, chromium, lead, tin, etc.	Cold trap/Impinger collection	ICP-OES	Williams et al. (2019)
Aerosol	Heavy metal; chromium, copper, manganese, nickel, and zinc	Acid digestion	ICP–MS	Kapiamba et al. (2022)
Aerosol	Nicotine fraction	Direct injection	X-ray	Berkeley and Ahmed (2023)
Liquid/ aerosol	Nicotine fraction	Direct injection	NMR	Duell et al. (2018)
Liquid/ aerosol	Heavy metal; chromium, nickel, copper, zinc, cadmium, tin, and lead	Dilution	ICP-QQQ-MS	Gray et al. (2020)

(continued)

Table 13.1 (continued)

Sample	Analyte	Sample preparation	Analytical method	References
Other techniques				
Liquid	Nicotine/cannabinoids	Dilution	DART-MS	Falconer and Morales-Garcia (2023)
Aerosol	Methamphetamine	SPME-LLE	DART-MS	Krakowiak et al. (2019)
Aerosol	Nicotine, Δ9-THC, methamphetamine	Direct injection	DART-MS	Forbes and Krauss (2021)

labeled products was determined to be lower than the specified amount, suggesting possible degradation to nicotine N-oxide.

13.3.2 Flavors

Complex interactions between nicotine and flavors can influence the formation of HPHCs (Kerber et al. 2022). More than 200 flavors have been reported, present at variable concentrations. Hence, continuous research is necessary to determine the implications of using emerging flavors (Havermans et al. 2021). Chromatographic methods continue to be the preferred method for the analysis of flavors in e-liquids. Flavor chemical concentrations in 277 e-liquids were determined utilizing a GC–MS methodology (Omaiye et al. 2019). The study determined that the majority of the samples contained flavor levels exceeding 1 mg/mL, with some exceeding 10 mg/mL. Cinnamaldehyde, a chemical classified as harmful, was determined at a notable elevated concentration of 343 mg/mL in a sample, surpassing its cytotoxic level by over 100,000 times. In another study, a simple dilution method utilizing GC–MS was developed for the determination of 79 flavorings in 320 e-liquids; most prevalent is the sweet or fruity flavors of vanillin, ethyl butyrate, and cis-3-hexenol (Krüsemann et al. 2020). The concentration of flavors was found to vary and may undergo changes during storage. High concentrations (32.49 mg/mL ethyl maltol, 23.15 mg/mL triacetin, 19.07 mg/mL ethyl vanillin, and 19.01 mg/mL l-menthol) of flavors were detected in another study utilizing GC–TOFMS, fruity flavors remained the most predominant flavors (Page and Goniewicz 2021). Pulegone, a suspected carcinogen at high concentrations, was found in menthol- or mint-flavored e-liquids in the 0.02–0.32 mg/mL range. The use of TOFMS allowed for the determination of flavors of concern even at low concentrations with lower limit of quantitation (LLOQ) values ranging from 0.02 to 0.63 mg/mL. In a screening of 129 e-liquids, two different GC–MS methodologies were employed to address the diverse volatility levels of flavor compounds (Barhdadi et al. 2021). Headspace analysis of samples diluted in water was performed for highly volatile compounds, while semi-volatile compounds were extracted using liquid-liquid extraction with hexane and a "freeze-and-pour" technique, differing in temperature gradient for the GC conditions.

13.3.3 Volatile Organic Compounds (VOC)

Thermal degradation of PG, VG, and flavors is known to produce harmful VOC, such as formaldehyde, acetaldehyde, and acrolein, among others, from the carbonyl class compounds. In particular, α-dicarbonyl compounds are potentially produced during the vaping process. Glyoxal, methylglyoxal, diacetyl, and acetyl propionyl in e-liquid and aerosol samples were examined using an UPLC–MS approach, providing LOQ in the range of 12–40 ng/g (Melvin et al. 2020). Recoveries ranged from 76% to 116%, and inter-day precision with RSD < 16% for both sample types was attained. The formation pathway via aldol condensation reaction to yield diacetyl, an HPHC, was explored using a microwave reaction system, suggesting PG or VG as precursors. The diacetyl concentration in tested aerosols was found to be below the established exposure limits, markedly lower than that found in tobacco smoke. This suggests reduced exposure to this HPHC from vaping compared to smoking cigarettes. Uchiyama et al. (2020) utilized GC–MS, GC–TCD, and HPLC–PDA approaches to identify dicarbonyls and oxides generated from third-generation devices, with power levels being adjusted accordingly. Minimal chemical compounds were produced at low power levels (10 W), but there was an exponential increase with higher power (> 40 W). Thermal decomposition products, notably formaldehyde, were exceedingly high, with levels reaching 4400 μg/15 puffs at 50 W and 6200 μg/15 puffs at 60 W, despite the supplier's recommended power range of 15–60 W. In another study, a cold trap technique was applied to collect vaping emissions. PFBHA derivatization and SPME extraction were implemented prior to analysis with GC–MS for the identification of carbonyls (Chen et al. 2021). VOC in commercially available strawberry-flavored e-liquids marketed in the UK was investigated using a headspace GC–MS approach (Haworth-duff et al. 2023). Among the VOC determined, 5 were classified as health hazards, and 17 were categorized as irritants.

13.3.4 Tobacco-Specific Nitrosamines (TSNA)

TSNA, which occurs naturally in tobacco leaves, is a recognized carcinogen and has been linked to lung, esophageal, and pancreatic cancers (Edwards et al. 2017). Given that the nicotine in e-liquids originates from tobacco, the analysis of TSNA in both e-liquids and aerosols remains a relevant area of interest. Lee et al. (2018) developed a rapid methodology utilizing LC–MS/MS to quantitate N′-nitrosonornicotine (NNN), 4-(Methylnitrosamino) – 1-(3-pyridyl) – 1-butanone (NNK), N′-nitrosoanabasine (NAB) and N′-nitrosoanatabine (NAT) in e-liquid and aerosol. The estimated detection limits for NNN, NNK, NAT, and NAB were 4.40, 4.47, 3.71, and 3.28 ng/mL. Two aerosol capture filters, a Cambridge filter pad (CFP) and quartz wool (QW), were utilized in the study, with QW exhibiting greater efficiency. Fix et al. (2021) developed a methodology using LC–MS/MS with improved detection capabilities, yielding LOQ values of 1.0 ng/mL for NNK and 0.5 ng/mL for NNN, NAT, and NAB. The study reported higher levels of nitrosamines observed in tobacco-flavored liquids, and notably low levels of nitrosamines were unexpectedly found in nicotine-free liquids. A novel approach utilizing ultrasound-assisted solvent extraction with a polypropylene (PP) porous membrane was developed for the determination of NNN, NNK, NAT, and NAB in e-liquids (Paweł Kubica 2019). The proposed method achieved good recoveries ranging from 87% to 105%, with coefficients of variation <4.9%. In all nine samples examined, at least one TSNA was detected, with NNK and NNN being the most prevalent. NAB was not detected in any of the samples.

13.3.5 Polycyclic Aromatic Hydrocarbons (PAH)

PAH, formed during the combustion of tobacco, are recognized as potential carcinogens to humans. These pollutants have been detected in aerosols generated by e-cigs (Beauval et al. 2017). Benz[*a*]anthracene, chrysene, benzo [*b*]fluoranthene, benzo [*k*]fluoranthene, benzo [*a*]pyrene, dibenz [*a,h*]anthracene, and benzo [*g,h,i*]perylene (B*ghi*P) in smoking cessation aids have been quantitatively determined using solid phase extraction (SPE) and HPLC with fluorimetric detection (Kim et al. 2021). The method LODs ranged from 0.04 to 0.18 ng/mL in e-liquid and from 0.06 to 0.17 ng/mL in aerosol. Low levels of PAH were detected in the aerosol samples, confirming the formation of PAH through the vaporization process. Doosti et al. (2023) synthesized two coordination polymers featuring nanorod and microsphere structures, proposing their utilization as extractive phases in a needle trap device for PAH analysis. When paired with GC–MS, the nanorod-prepared device demonstrated good linearity (0.2–750 ng/L), low limits of detection (0.05–5 ng/L), precision (RSD < 12%), and prolonged durability (> 100 usages). Acenaphthylene, fluorene, and pyrene were detected in trace levels, while naphthalene and anthracene were below the limit of detection.

13.3.6 Heavy Metals

The metal content in e-liquids and aerosols, which may originate from the corrosion of heating elements and atomizing units, or transfer from heated metal filaments, is of concern to consumers because of reported levels of lead, chromium, nickel, copper, zinc, aluminum, tin, and other metals (Omaiye et al. 2021; Williams et al. 2019; Kapiamba et al. 2022). Gray et al. (2020) investigated the metal contents in e-liquids and aerosols emitted from pod-type e-cigs utilizing inductively coupled plasma triple quadrupole mass spectrometry (ICP–MS/MS). They observed significant variability in metal content between devices, which was attributed to variations in design, components, and materials used. Additionally, the pH of the e-liquid may affect the corrosion of the metal components. Williams et al. (2019) examined two aerosol collection methods for metal analysis, the cold trap and impinger methods, using inductively coupled plasma optical emission spectroscopy (ICP–OES). They found that elemental concentrations varied depending on the collection method employed. The cold trap technique was found to yield a total concentration of metals 3.5 times higher than that obtained using the impinger method. The authors concluded that the cold trap method was more efficient in recovering alkali (sodium and potassium) and alkaline earth metals (magnesium and calcium) as well as metalloids (silicon, boron). However, this method was less effective than the impinger approach for collecting transition metals (chromium, copper, iron, nickel, and zinc).

A study utilizing an inductively coupled plasma quadrupole mass spectrometer (ICP–MS) was conducted by Kapiamba et al. (2022) to examine metal contents in primary and second-hand aerosols released by e-cigs, reporting levels of several metals (Cr, Ni, and Mn) above the regulated limits. The researchers deduced that metals were primarily generated from the aerosolization of e-liquids, as supported by the positive correlation between metal content, nicotine content, and puff duration. This concept was supported by a study conducted by Williams et al. (2022), which utilized an ICP–OES approach to investigate the levels of metals, nicotine, and flavors in unused and used first-generation e-cigs that have been stored for a specific period. The study noted higher concentrations of metals in e-liquids from devices after repeated use compared to unused liquids. Additionally, e-liquids with lower pH values were found to contain multiple acids and exhibited the highest total metal concentrations.

13.3.7 Illicit Drugs

Recently, there have been concerns regarding the use of e-cigs as a tool for illicit drug delivery. Notably, drugs derived from the cannabis and synthetic cannabinoids class are of particular interest (Breitbarth et al. 2018). The legalization of marijuana in some countries has marked a shift in the cannabis market, potentially changing e-cig consumer preferences with the incorporation of cannabinoids in e-liquids. Grafinger et al. (2020) examined 20 e-liquid samples to detect CBD, THC, and various other phytocannabinoids (compounds from the cannabis plant), GC–MS and HPLC coupled to a diode array detector (HPLC–DAD) approaches were developed for qualitative and quantitative purposes, respectively. With LLOQ values of 0.001 mg/mL for total THC and CBD, the THC amount for all samples was determined to be below 0.11%, which is lower than the EU legal threshold of 0.2%. Half of the analyzed samples were found to exhibit a discrepancy for the labeled CBD content, which is more than the acceptable 10% deviation (45.9–117.9%) in comparison to the actual total CBD concentration.

The emergence of SC as new psychoactive substances (NPS) presents complexities in their detection, primarily due to challenges in their characterization. Apirakkan et al. (2020) reported a study that utilized HPLC–DAD, HRMS, NMR, and GC–MS techniques to identify and characterize SCs, successfully identifying MDMB-CHMICA, 5F-CUMYL-PINACA, and 5F-APINACA in the e-liquids. Prior to analysis, the samples were subjected to a semipreparative LC–DAD-MS technique for the purification of SCs. The structural identification of ADB-BRINACA, a novel bromoindazole SC analog, was accomplished through the combined use of GC–MS, LC–QTOFMS, and NMR techniques (Choi et al. 2022). A quantitative thermal-assisted carbon fiber ionization mass spectrometry technique was proposed for the detection of three SCs (MDMB-4en-PINACA, ADB-BUTINACA, and ADB-4en-PINACA). This approach mitigated the matrix effect in the presence of heat (Dou et al. 2023). The desorption of SCs directly added to the carbon fiber was facilitated by applying a metal-ceramic heater. During this process, impurities were retained on the carbon fiber, enabling ionization of the volatilized compounds in a corona discharge process and subsequent detection by the mass spectrometer. According to the study findings, a decrease in matrix effect was observed compared to the LC–MS/MS approach. Due to the sophisticated design of e-cig devices, consumers are able to modify the devices for efficient delivery of illicit drugs. Krakowiak et al. (2019) reported methodological approaches for analyzing aerosolized methamphetamine using SPME-DART–MS and SPME-GC–MS. Characterizing the vapors produced by e-liquids with methamphetamine provides insights into the dose-delivery mechanisms of e-cigs, thereby assessing the efficacy of these devices as illicit drug delivery systems.

13.4 Conclusion

The use of e-cigs to quit smoking remains controversial, with different countries imposing different regulatory standards for vaping. Despite this, global regulatory guidelines on e-cigs are notably limited. We are of the opinion that additional research on e-cigs is warranted to reduce their potential health risks for users. For instance, the determination of the chemical constituents of e-cigs and their resulting by-products during vaping necessitates further studies. This information may shed light on the toxicity or carcinogenicity of the generated aerosols and other chemical entities based on their corresponding transfer efficiency and reactivity during vaping (Huimin et al. 2023). The availability of this information can be useful in setting guidelines or regulations to control additives and potential toxicants in e-cigs. Following the analytical techniques for the targeted and/or untargeted analysis of chemical constituents in e-cigs described in this chapter, chromatography approaches (typically HPLC and GC) are still by far the most widely used techniques compared to spectroscopy methods. This might be attributed to the higher resolving power of chromatography techniques to untangle many chemical constituents within a single analysis, and with hyphenation to a MS, identities of the resolved peaks can be tentatively determined. It can be envisioned that the progression toward the use of chromatographic techniques with HRMS, especially the newer Orbitrap MS technology that offers high mass resolving power and mass accuracy, will be appreciated. These powerful hyphenations will greatly facilitate the trace analysis and identification of "known unknowns" in e-liquids and the generated aerosols. Thus far, the applicability of HPLC or GC with Orbitrap MS for high-resolution analysis of chemical entities in e-cig aerosols has not been reported. The continuous advances in separation science will also witness the emergence of multidimensional chromatography techniques (e.g., comprehensive two dimensional LC and GC × GC) to chromatographically deconvolute e-cig constituents. Recent application of GC × GC−TOFMS for the profiling of VOCs produced from heated e-cig liquids, as highlighted in Sect. 13.3.1, affords the capability of this high-resolution chromatographic approach to enhance the separation and detection of >1000 VOCs, which is difficult to achieve via one-dimensional GC analysis. These high-resolution multidimensional chromatography approaches should be further explored due to their unprecedented peak capacity and higher resolving power compared to one-dimensional GC and LC methods, offering promises to separate and identify many more compounds in e-cig aerosols.

References

Akhtar Z, Barhdadi S, De Braekeleer K, Delporte C, Adams E, Deconinck E (2024) Spectroscopy and chemometrics for conformity analysis of E-liquids: illegal additive detection and nicotine characterization. Chemosensors 12(1):1–15. https://doi.org/10.3390/chemosensors12010009

Almeida-da-silva C, Matshik H (2021) Effects of electronic cigarette aerosol exposure on oral and systemic health. Biom J 44:252–259. https://doi.org/10.1016/j.bj.2020.07.003

Apirakkan O, Frinculescu A, Denton H, Shine T, Cowan D, Abbate V, Frascione N (2020) Isolation, detection and identification of synthetic cannabinoids in alternative formulations or dosage forms. Forensic Chem 18:100227. https://doi.org/10.1016/j.forc.2020.100227

Augustini AL, Sielemann S, Telgheder U (2021) Strategy for the identification of flavor compounds in E-liquids by correlating the analysis of GCxIMS and GC-MS. Talanta 230:122318. https://doi.org/10.1016/j.talanta.2021.122318

Augustini AL, Sielemann S, Telgheder U (2022) Quantitation of flavor compounds in refill solutions for electronic cigarettes using HS-GCxIMS and internal standards. Molecules 27. https://doi.org/10.3390/molecules27228067

Augustini AL, Borg C, Sielemann S, Telgheder U (2023) Making every single puff count – simple and sensitive E-cigarette aerosol sampling for GCxIMS and GC-MS analysis. Molecules 28(18):1–22. https://doi.org/10.3390/molecules28186574

Barhdadi S, Mertens B, Van Bossuyt M, Van De Maele J, Anthonissen R, Canfyn M, Courselle P, Rogiers V, Deconinck E, Vanhaecke T (2021) Identification of flavouring substances of genotoxic concern present in E-cigarette refills. Food Chem Toxicol 147:111864. https://doi.org/10.1016/j.fct.2020.111864

Barhdadi S, Courselle P, Deconinck E, Vanhee CCC, Deconinck É, Vanhee CCC (2023) The analysis of cannabinoids in E-cigarette liquids using LC-HRAM-MS and LC-UV. J Pharm Biomed Anal 230(April):115394. https://doi.org/10.1016/j.jpba.2023.115394

Beauval N, Antherieu S, Soyez M, Gengler N, Grova N, Howsam M, Hardy EM et al (2017) Chemical evaluation of electronic cigarettes: multicomponent analysis of liquid refills and their corresponding aerosols. J Anal Toxicol 41(8):670–678. https://doi.org/10.1093/jat/bkx054

Berkeley L, Ahmed M (2023) Fraction of free-base nicotine in simulated vaping aerosol particles determined by X-ray spectroscopies. J Phys Chem Lett 14:1279–1287. https://doi.org/10.1021/acs.jpclett.2c03748

Breitbarth AK, Morgan J, Jones AL (2018) E-cigarettes – an unintended illicit drug delivery system. Drug Alcohol Depend 192:98–111. https://doi.org/10.1016/j.drugalcdep.2018.07.031

Breland A, Soule E, Lopez A, Ramôa C, El-Hellani A, Eissenberg T (2017) Electronic cigarettes: what are they and what do they do? Ann N Y Acad Sci 1394(1):5–30. https://doi.org/10.1111/nyas.12977

Budzyńska E, Sielemann S, Puton J, Surminski ALRM (2020) Analysis of E-liquids for electronic cigarettes using GC-IMS/MS with headspace sampling. Talanta 209(November 2019):120594. https://doi.org/10.1016/j.talanta.2019.120594

Cao YY, Dezhen Daming W, Cao YY, Ma Y, Dezhen Daming W, Ma Y, Ma X et al (2021) Toxicity of electronic cigarettes: a general review of the origins, health hazards, and toxicity mechanisms. Sci Total Environ 772:145475. https://doi.org/10.1016/j.scitotenv.2021.145475

Chen JY, Canchola A, Lin Y-H (2021) Carbonyl composition and Electrophilicity in vaping emissions of flavored and unflavored E-liquids. Toxics 9. https://doi.org/10.3390/toxics9120345

Cheng T (2014) Chemical evaluation of electronic cigarettes. Tob Control 23(SUPPL. 2):ii11. https://doi.org/10.1136/tobaccocontrol-2013-051482

Chien JY, Gu YC, Liu CH, Tsai HM, Lee CN, Yang AC, Huang J, Wang YL, Wang JK, Lin CH (2023) Rapid detection of nicotine and benzoic acid in E-liquids with surface-enhanced Raman scattering and artificial intelligence-assisted Spectrum interpretation. J Pharm Biomed Anal 233(155):115456. https://doi.org/10.1016/j.jpba.2023.115456

Choi H, Kim S, Jang H, Kim HJ, Choe S (2022) Identification of a novel Bromoindazole synthetic cannabinoid analogue in seized E-cigarette liquid: ADB-BRINACA. Forensic Sci Int 338:111385. https://doi.org/10.1016/j.forsciint.2022.111385

Clapp PW, Jaspers I (2017) Electronic cigarettes: their constituents and potential links to asthma. Curr Allergy Asthma Rep 17(11):79. https://doi.org/10.1007/s11882-017-0747-5

Crenshaw MD, Tefft ME, Buehler SS, Brinkman MC, Clark PI, Gordon SM (2017) Determination of nicotine, glycerol, propylene glycol and water in electronic cigarette fluids using quantitative 1 H NMR. Magn Reson Chem 54(11):901–904. https://doi.org/10.1002/mrc.4498. Determination

Dagla I, Gikas E, Tsarbopoulos A (2023) Two fast GC-MS methods for the measurement of nicotine, propylene glycol, vegetable glycol, Ethylmaltol, Diacetyl, and Acetylpropionyl in refill liquids for E-cigarettes. Molecules 28(4):1902. https://doi.org/10.3390/molecules28041902

Doosti M, Soufi G, Bagheri H (2023) Nanorod coordination polymers for needle trap extraction of E-cigarettes and cigarettes smoke. Microchem J 187:108439. https://doi.org/10.1016/j.microc.2023.108439

Dou Q, Liu W, Xiang P, Zhao J (2023) Quantitative analysis of three synthetic cannabinoids MDMB-4en-PINACA, ADB-BUTINACA, and ADB-4en-PINACA by thermal-assisted carbon fiber ionization mass spectrometry. J Am Soc Mass Spectrom 34(10):2316–2322. https://doi.org/10.1021/jasms.3c00229

Du Y, Liu B, Guifeng X, Rong S, Sun Y, Yuxiao W, Snetselaar LG, Wallace RB, Bao W (2020) Association of Electronic Cigarette Regulations with electronic cigarette use among adults in the United States. JAMA Netw Open 3(1):1–12. https://doi.org/10.1001/jamanetworkopen.2019.20255

Duell AK, Pankow JF, Peyton DH (2018) "Free-Base nicotine determination in electronic cigarette liquids by 1H NMR spectroscopy." rapid-communication. Chem Res Toxicol 31(6):431–434. https://doi.org/10.1021/acs.chemrestox.8b00097

Duffy B, Li L, Shijun L, Durocher L, Dittmar M, Delaney-Baldwin E, Panawennage D, LeMaster D, Navarette K, Spink D (2020) Analysis of cannabinoid-containing fluids in illicit vaping cartridges recovered from pulmonary injury patients: identification of vitamin E acetate as a major diluent. Toxics 8(1):1–19. https://doi.org/10.3390/toxics8010008

Edwards SH, Rossiter LM, Taylor KM, Holman MR, Zhang L, Ding YS, Watson CH (2017) Tobacco-specific nitrosamines in the tobacco and mainstream smoke. Chem Res Toxicol 30(2):540–551. https://doi.org/10.1021/acs.chemrestox.6b00268

Eshraghian EA, Al-Delaimy WK (2021) A review of constituents identified in E-cigarette liquids and aerosols. Tob Prev Cessat 7:1–15. https://doi.org/10.18332/tpc/131111

Falconer TM, Morales-Garcia F (2023) Rapid screening of vaping liquids by DART-MS. J AOAC Int 106(2):436–444. https://doi.org/10.1093/jaoacint/qsac103

Fan Y, Zong X, Liu J, Ke X, Huang Z, Yu X (2021) Development of a fragmentation pattern of synthetic cannabinoids based on electrospray ionization mass spectrometry in positive ion mode to screen synthetic cannabinoids in illicit products. J Pharm Biomed Anal 193:113723. https://doi.org/10.1016/j.jpba.2020.113723

Fix BV, Oconnor RJ, Goniewicz ML, Leigh NL, Cummings M, Hitchman SC, Fong GT et al (2021) Characterisation of vaping liquids used in vaping devices across four countries :

results from an analysis of selected vaping liquids reported by users in the 2016 ITC four country smoking and vaping survey. Tob Control 32(1):51–59. https://doi.org/10.1136/tobaccocontrol-2020-056338

Forbes TP, Krauss ST (2021) Confined DART-MS for rapid chemical analysis of electronic cigarette aerosols and spiked drugs. J Am Soc Mass Spectrom 32(8):2274–2280. https://doi.org/10.1021/jasms.1c00103

Gilbert HA (1965) Smokeless non-tobacco cigarette. US Patent No. 3,200,819, issued 1965

Grafinger KE, Krönert S, Broillet A, Weinmann W (2020) Cannabidiol and tetrahydrocannabinol concentrations in commercially available CBD E-liquids in Switzerland. Forensic Sci Int 310:110261. https://doi.org/10.1016/j.forsciint.2020.110261

Gray N, Halstead M, Valentin-Blasini L, Watson C, Steven Pappas R (2020) Toxic metals in liquid and aerosol from pod-type electronic cigarettes. J Anal Toxicol 46(1):69–75. https://doi.org/10.1093/jat/bkaa185

Guo W, Vrdoljak G, Liao VC, Moezzi B (2021) Major constituents of cannabis vape oil liquid, vapor and aerosol in California vape oil cartridge samples. Front Chem 9(June):1–14. https://doi.org/10.3389/fchem.2021.694905

Halstead M, Gray N, Gonzalez-Jimenez N, Fresquez M, Valentin-Blasini L, Watson C, Steven Pappas R (2019) Analysis of toxic metals in electronic cigarette aerosols using a novel trap design. J Anal Toxicol 44(2):149–155. https://doi.org/10.1093/jat/bkz078

Harris CC (2018) Tobacco smoking, E-cigarettes, and nicotine harm. Proc Natl Acad Sci USA 115(7):1406–1407. https://doi.org/10.1073/pnas.1722636115

Harvanko AM, Havel CM, Jacob P, Benowitz NL (2020) Characterization of nicotine salts in 23 electronic cigarette refill liquids. Nicotine Tob Res 22(7):1239–1243. https://doi.org/10.1093/ntr/ntz232

Havermans A, Krüsemann EJZ, Pennings J, De Graaf K, Boesveldt S, Talhout R (2021) Nearly 20 000 E-liquids and 250 unique flavour descriptions: an overview of the Dutch market based on. Tob Control 30:57–62. https://doi.org/10.1136/tobaccocontrol-2019-055303

Haworth-duff A, Parkes GMB, Reed NJ (2023) Profiling flavourings in strawberry-flavoured e-liquid using headspace-gas chromatography – mass spectrometry. Drug Test Anal 15:1077–1083. https://doi.org/10.1002/dta.3451

Helen GS, Eaton DL (2018) Public health consequences of E-cigarette use. JAMA Intern Med 178:984. https://doi.org/10.1001/jamainternmed.2018.1600

Holt AK, Poklis JL, Peace MR (2021) A retrospective analysis of chemical constituents in regulated and unregulated E-cigarette liquids. Front Chem 9. https://doi.org/10.3389/fchem.2021.752342

Hon L (2003).Flameless electronic atomizing cigarette. US Patent No. US 2006/0196518A1, issued 2003

Huerta TR, Walker DM, Mullen D, Johnson TJ, Ford EW (2017) Trends in E-cigarette awareness and perceived harmfulness in the U.S. Am J Prev Med 52(3):339–346. https://doi.org/10.1016/j.amepre.2016.10.017

Huimin D, Tang S, Yang F, Chen D, Bian Z, Wang Y, Tang G, Lee HK (2023) 'Recent advances in the analysis of electronic cigarette liquids and aerosols: sample preparation and chromatographic characterization'. J Chromatogr A 1712(October):464495. https://doi.org/10.1016/j.chroma.2023.464495

Jabba SV, Erythropel HC, Torres DG, Delgado LA, Woodrow JG, Anastas PT, Zimmerman JB, Jordt S-E (2022) Synthetic cooling agents in US-marketed E-cigarette refill liquids and popular disposable Ecigarettes: chemical analysis and risk assessment. Nicotine Tob Res 24(7):1037–1046. https://doi.org/10.1093/ntr/ntac046

Jo SH, Kim KH (2016) Development of a sampling method for carbonyl compounds released due to the use of electronic cigarettes and quantitation of their conversion from liquid to aerosol. J Chromatogr A 1429:369–373. https://doi.org/10.1016/j.chroma.2015.12.061

Johne S, van der Toorn M, Iskandar AR, Majeed S, Torres LO, Hoeng J, Peitsch MC (2023) An in vitro evaluation of E-vapor products: the contributions of chemical adulteration, concentration, and device power. Food Chem Toxicol 175(January):113708. https://doi.org/10.1016/j.fct.2023.113708

Kapiamba KF, Hao W, Adom S, Liu W, Huang YW, Wang Y (2022) Examining metal contents in primary and secondhand aerosols released by electronic cigarettes. Chem Res Toxicol 35(6):954–962. https://doi.org/10.1021/acs.chemrestox.1c00411

Kerber PJ, Duell AK, Powers M, Peyton DH (2022) Effects of common E-liquid Flavorants and added nicotine on toxicant formation during vaping analyzed by 1H NMR spectroscopy. Chem Res Toxicol 35(7):1267–1276. https://doi.org/10.1021/acs.chemrestox.2c00110.Effects

Khlystov A, Samburova V (2016) Flavoring compounds dominate toxic aldehyde production during E-cigarette vaping. Environ Sci Technol 50(23):13080–13085. https://doi.org/10.1021/acs.est.6b05145

Kim D-Y, Lee B-E, Shin H-S (2021) Determination of polycyclic aromatic hydrocarbons (PAHs) in smoking cessation aids by using high-performance liquid chromatography. Anal Biochem 617:114119. https://doi.org/10.1016/j.ab.2021.114119

Kong G, Chaffee BW, Wu R, Krishnan-Sarin S, Liu F, Leventhal AM, McConnell R, Barrington-Trimis J (2022) E-cigarette device type and combustible tobacco use: results from a pooled analysis of 10,482 youth. Drug Alcohol Depend 232(November 2021):109279. https://doi.org/10.1016/j.drugalcdep.2022.109279

Kovach AL, Carter RR, Thornburg JW, Wiethe R, Fennell TR, Wiley JL (2022) Thermal degradants identified from the vaping of vitamin E acetate. J Anal Toxicol 46(7):750–756. https://doi.org/10.1093/jat/bkab109

Krakowiak RI, Poklis JL, Peace MR (2019) The analysis of aerosolized methamphetamine from E-cigarettes using high resolution mass spectrometry and gas chromatography mass spectrometry. J Anal Toxicol 43(8):592–599. https://doi.org/10.1093/jat/bkz067

Krüsemann EJZ, Pennings JLA, Cremers JWJM, Bakker F, Boesveldt S, Talhout R (2020) GC – MS analysis of e-cigarette refill solutions: a comparison of flavoring composition between flavor categories. J Pharm Biomed Anal 188:113364. https://doi.org/10.1016/j.jpba.2020.113364

Kubica P (2019) Ultrasound-assisted solvent extraction of a porous membrane packed sample for the determination of tobacco-specific nitrosamines in the replacement. Molecules 24. https://doi.org/10.3390/molecules24244618

Kubica P (2023) Determination of glycerol, propylene glycol, and nicotine as the main components in refill liquids for electronic cigarettes. Molecules 28(11). https://doi.org/10.3390/molecules28114425

Lechasseur A, Jubinville É, Routhier J, Bérubé JC, Hamel-Auger M, Talbot M, Lamothe J et al (2017) Exposure to electronic cigarette vapors affects pulmonary and systemic expression of circadian molecular clock genes. Physiol Rep 5(19):1–13. https://doi.org/10.14814/phy2.13440

Lee YS, Kim KH, Lee SS, Brown RJ, Jo SH (2018) Analytical method for measurement of tobacco-specific nitrosamines in E-cigarette liquid and aerosol. Appl Sci 8:2699. https://doi.org/10.3390/app8122699

Lee YJ, Na CJ, Botao L, Kim KH, Son YS (2020) Quantitative insights into major constituents contained in or released by electronic cigarettes: propylene glycol, vegetable glycerin, and nicotine. Sci Total Environ 703:134567. https://doi.org/10.1016/j.scitotenv.2019.134567

Melvin MS, Avery KC, Ballentine RM, Flora JW, Gardner W, Karles GD, Pithawalla YB, Smith DC, Ehman KD, Wagner KA (2020) Formation of Diacetyl and other α – Dicarbonyl compounds during the generation of E – vapor product aerosols. ACS Omega 5:17565–17575. https://doi.org/10.1021/acsomega.0c02018

Omaiye EE, Mcwhirter KJ, Wentai L, Tierney PA, Pankow JF, Talbot P (2019) High concentrations of flavor chemicals are present in electronic cigarette refill fluids. Sci Rep 9:2468. https://doi.org/10.1038/s41598-019-39550-2

Omaiye EE, Williams M, Bozhilov KN, Talbot P (2021) Design features and elemental/metal analysis of the atomizers in pod-style electronic cigarettes. PLoS One 16(3):1–19. https://doi.org/10.1371/journal.pone.0248127

Omaiye EE, Luo W, McWhirter KJ, Pankow JF, Talbot P (2022) Ethyl maltol, vanillin, corylone and other conventional confectionery-related flavour chemicals dominate in some e-cigarette liquids labelled 'tobacco' Flavoured. Tob Control 31:S238–S244. https://doi.org/10.1136/tc-2022-057484

Orellana-Barrios MA, Payne D, Mulkey Z, Nugent K (2015) Electronic cigarettes – a narrative review for clinicians. Am J Med 128(7):674–681. https://doi.org/10.1016/j.amjmed.2015.01.033

Page MK, Goniewicz ML (2021) New analytical method for quantifying flavoring chemicals of potential respiratory health risk concerns in E-cigarette liquids. Front Chem 9(October):1–12. https://doi.org/10.3389/fchem.2021.763940

Palazzolo D, Nelson JM, Hudson Z (2019) The use of HPLC-PDA in determining nicotine and nicotine-related alkaloids from E-liquids: a comparison of five E-liquid brands purchased locally. Int J Environ Res Public Health 16:8–15. https://doi.org/10.3390/ijerph16173015

Palmisani J, Abenavoli C, Famele M, Di Gilio A, Palmieri L, de Gennaro G, Draisci R (2020) Chemical characterization of electronic cigarette (e-cigs) refill liquids prior to EU tobacco product directive adoption: evaluation of BTEX contamination by HS-SPME-GC-MS and identification of flavoring additives by GC-MS-O. Atmos 11(374). https://doi.org/10.3390/ATMOS11040374

Patel D, Verena Taudte R, Nizio K, Herok G, Cranfield C, Shimmon R (2021) Headspace analysis of E-cigarette fluids using comprehensive two dimensional GC×GC-TOF-MS reveals the presence of volatile and toxic compounds. J Pharm Biomed Anal 196:113930. https://doi.org/10.1016/j.jpba.2021.113930

Rajapaksha RD, Tehrani MW, Rule AM, Harb CC (2021) A rapid and sensitive chemical screening method for E – cigarette aerosols based on runtime cavity Ringdown spectroscopy. Environ Sci Technol 55:8090–8096. https://doi.org/10.1021/acs.est.0c07325

Richardson PIC, Burke A, Gotts N, Goodacre R (2023) Quantifying PG: VG ratio and nicotine content in commercially available e-liquids using handheld Raman spectroscopy. Analyst 148(17):4002–4011. https://doi.org/10.1039/d3an00888f

Sleiman M, Logue JM, Nahuel Montesinos V, Russell ML, Litter MI, Gundel LA, Destaillats H (2016) Emissions from electronic cigarettes: key parameters affecting the release of harmful chemicals. Environ Sci Technol 50(17):9644–9651. https://doi.org/10.1021/acs.est.6b01741

Son Y, Weisel C, Wackowski O, Schwander S, Delnevo C, Meng Q (2020) The impact of device settings, use patterns, and flavorings on carbonyl emissions from electronic cigarettes. Int J Environ Res Public Health 17(16):1–14. https://doi.org/10.3390/ijerph17165650

Stephens WE (2018) Comparing the cancer potencies of emissions from vapourised nicotine products including E-cigarettes with those of tobacco smoke. Tob Control 27(1):10–17. https://doi.org/10.1136/tobaccocontrol-2017-053808

Strongin RM (2019) E-cigarette chemistry and analytical detection. Annu Rev Anal Chem 12:23–39. https://doi.org/10.1146/annurev-anchem-061318-115329

Tehrani MW, Newmeyer MN, Rule AM, Prasse C (2021) Characterizing the chemical landscape in commercial E-cigarette liquids and aerosols by liquid chromatography-high-resolution mass spectrometry. Chem Res Toxicol 34(10):2216–2226. https://doi.org/10.1021/acs.chemrestox.1c00253

Tierney PA, Karpinski CD, Brown JE, Luo W, Pankow JF (2016) Flavour chemicals in electronic cigarette fluids. Tob Control 25(E1):e10–e15. https://doi.org/10.1136/tobaccocontrol-2014-052175

Uchiyama S, Noguchi M, Sato A, Ishitsuka M, Inaba Y, Kunugita N (2020) Determination of thermal decomposition products generated from E-cigarettes. Chem Res Toxicol 33(2):576–583. https://doi.org/10.1021/acs.chemrestox.9b00410

Valentine G, Jatlow P, Coffman M, Nadim H, Gueorguieva R, Sofuoglu M (2016) The effects of alcohol-containing E-cigarettes on young adult smokers. Drug Alcohol Dependence 159:272–276. https://doi.org/10.1016/j.drugalcdep.2015.12.011

Wagner KA, Flora JW, Melvin MS, Avery KC, Ballentine RM, Brown AP, McKinney WJ (2018) An evaluation of electronic cigarette formulations and aerosols for harmful and potentially harmful constituents (HPHCs) typically derived from combustion. Regul Toxicol Pharmacol 95:153–160. https://doi.org/10.1016/j.yrtph.2018.03.012

Walley SC, Wilson K (2021) Electronic cigarettes and vape devices: a comprehensive guide for clinicians and health professionals. In: Electronic cigarettes and vape devices: a comprehensive guide for clinicians and health professionals. https://doi.org/10.1007/978-3-030-78672-4

Williams M, Li J, Talbot P (2019) Effects of model, method of collection, and topography on chemical elements and metals in the aerosol of tank-style electronic cigarettes. Sci Rep 9:13969. https://doi.org/10.1038/s41598-019-50441-4

Williams M, Luo W, McWhirter K, Ikegbu O, Talbot P (2022) Chemical elements, flavor chemicals, and nicotine in unused and used electronic cigarettes aged 5–10 years and effects of PH. Int J Environ Res Public Health 19(24). https://doi.org/10.3390/ijerph192416931

Wu N, Danoun S, Balayssac S, Malet-Martino M, Lamoureux C, Gilard V (2021) Synthetic cannabinoids in E-liquids: a proton and fluorine NMR analysis from a conventional spectrometer to a compact one. Forensic Sci Int 324. https://doi.org/10.1016/j.forsciint.2021.110813

Yassine A, Antossian C, El-Hage R, Saliba NA (2023) A quick method for the determination of the fraction of freebase nicotine in electronic cigarettes. Chem Res Toxicol 36(7):1021–1027

Zhao D, Navas-Acien A, Ilievski V, Slavkovich V, Olmedo P, Adria-Mora B, Domingo-Relloso A et al (2019) Metal concentrations in electronic cigarette aerosol: effect of open-system and closed-system devices and power settings. Environ Res 174(February):125–134. https://doi.org/10.1016/j.envres.2019.04.003

Bibliometric Analysis of E-cigarette's Emissions

14

Maria Makrygianni, Zoe Gareiou, Niki Matsouki, Chara Tsipa, Evi Bekou, Sofia Gortsilas, and Efthimios Zervas

Abstract

Electronic cigarettes (e-cigs) have been commercially introduced in recent years as an alternative to conventional cigarette smoking. Due to the high concern for human health, scientific production dealing with e-cigs is constantly increasing. Bibliometric analysis can be used to reveal the trends of scientific research. The aim of this chapter is to investigate and analyze the trends in the production of scientific documents on e-cig's emissions by conducting a bibliometric analysis. Scopus is used for this work and the time span of the research starts from 2002 and goes up to 2023. The trends concerning the Year, Source type, Country, Document type, Subject Area, Institutions, and Number of citations are analyzed. An analysis of keywords of the published documents was also performed. The number of publications constantly increases, with a probable saturation during the last few years. The majority of the published documents are articles in journals. The most productive journal is Nicotine and Tobacco Research, while the most productive institution is the University of California, San Francisco. Also, the United States and the UK are at the top in this research, considering the total number of publications and the predominant subject area is Medicine, followed by Pharmacology, Toxicology and Pharmaceutics, and Environmental Science. The keywords with the highest frequency are associated with the categories of humans, studies and methods, devices and chemical compounds, and smoking habits.

M. Makrygianni · Z. Gareiou · N. Matsouki · C. Tsipa · E. Bekou · S. Gortsilas
E. Zervas (✉)
Laboratory of Technology and Policy of Energy and the Environment, School of Applied Arts and Sustainable Design, Hellenic Open University, Patra, Greece
e-mail: zervas@eap.gr

Y.-H. Leong, A. S. Mohd Samin (eds.), *E-Cigarettes*,
https://doi.org/10.1007/978-981-96-5087-3_14

Keywords

Electronic cigarettes · Bibliometrics · Research trends · Emissions · Particulate matter · Vaping · Smoking

14.1 Introduction

Electronic cigarettes (e-cigs) are a relatively new tobacco product, alternative to conventional cigarettes (CCs), aiming to reduce health risks attributed to the use of CCs (Lu et al. 2022), even if this statement is not fully proven today (Pearce et al. 2020). E-cigs were initially designed by Herbert A. Gilbert in the early 1960s, but did not manage to enter the market, until 2000s. In 2004, they were launched in China (Qu et al. 2018) and, during the same period, in the United States and Europe as well (Giovacchini et al. 2022). They were promoted as a tool to quit conventional smoking; and their popularity has been constantly growing, especially among adolescents (Qu et al. 2018; Giovacchini et al. 2022). A large variety of products is now available in the market, under different brand names and extend to products that use a disposable cartridge (cigalike), others where the tank can be filled in with e-liquid by the user (vape pens) or most advanced types (pod-mods), where the user can control several parameters, such as heating temperature (Yingst et al. 2021). Moreover, thousands of e-liquids can be purchased by the user (Behar et al. 2018). The health impacts of e-cigs still attract the interest of all involved parties (Ting et al. 2020). In terms of scientific research, in order to estimate e-cigs safety or, even their possible reduced health risks, the scientific community is focused on the analysis of their emissions, which are found to be affected by the diversity of e-cigs and the difference in use conditions (Matsouki et al. 2021; Zervas et al. 2018, 2020; Zhao et al. 2019).

This work deals with a cartography of the publications related to e-cig emissions by using a bibliometric analysis. Through bibliometric analysis, it is possible to investigate and analyze a scientific field (Donthu et al. 2021). Bibliometric analysis is a popular method in recent years, as the availability and accessibility of scholarly bases, such as Scopus and Web of Sciences, has helped to analyze large volumes of scholarly data (Donthu et al. 2021).

More specifically, researchers use bibliometrics for various reasons, either to study general topics, such as how to make a bibliometric analysis (Donthu et al. 2021), or to explore a specific field that concerns the global scientific community (Drimili et al. 2019; Iliopoulou et al. 2022; Kalyvas and Zervas 2021; Mandalas et al. 2023; Rejeb et al. 2022).

This chapter explores and analyzes the production of scientific documents and trends regarding e-cigarette emissions using bibliometric analysis.

14.2 Methods

Elsevier's Scopus is the search engine for this work, as is one of the main scientific databases used today. The search was conducted using two combinations of keywords. The first one concerns the "electronic cigarette": ("ecig" OR e-cig OR "electronic cigarette*" OR e-cigarette* OR "Vape pen*" OR "Vaping device" OR "electronic nicotine delivery system*") and the second one denotes "emission": (emission* OR smoke OR vapor OR aerosol OR voc OR pollutant* OR particle* OR "particulate matter" OR "pm"). The search option of having these keywords in the title, abstract, and keywords was chosen.

The search was limited to documents published from 2002 until 2023, as 2024 is still ongoing. The selected language of the documents was English (which corresponds to more than 94% of the total documents), while the other languages were excluded. In addition, "trade journal" and "undefined" of the "source type" were excluded (three trade journals and two undefined) because they are in a minority compared to the other source types. The research was performed from the 19th to the 25th of April 2024.

The documents were extracted and analyzed according to Year of publication, Document type, Source title, Country, Subject area, Institution (worldwide), and Number of Citations in Scopus. Furthermore, the keywords used were also analyzed. The impact factors (IFs) which are an important tool for scientific evaluation and it is used as a measure for judging academic success, come from the Journal Citation Reports (JCR) Science Edition 2022.

14.3 Results and Discussion

14.3.1 Annual Number of Documents and Citations

The final search resulted in 3927 documents. In the first step, the publication and citation number of each year is analyzed, where the blue line corresponds to the number of documents published per year and the red line refers to the number of citations that these documents received each year from 2002 to 2023 (Fig. 14.1). The first document indexed in Scopus dealing with e-cig emissions was published in 2002 and only 6 documents were published until 2009. Figure 14.1 shows a large increase of published documents from 2010 to 2020, while the number of documents remained relatively constant from 2020 to 2023. As can be observed from Fig. 14.1 the period with the highest growth rate is from 2014 until 2020.

The number of citations that the articles published each specific year receive shows a quite different pattern. The total number of citations of those 3927 documents is 102,248. Three periods can be distinguished. From the beginning until 2014, where a high increase in the number of citations is observed starting from 0 to reach 13,104 citations in 2014. This trend is expected, as the number of articles increases every year, and the general scientific interest also increases. For the documents published from 2014 until 2019, the number of citations has slight variations

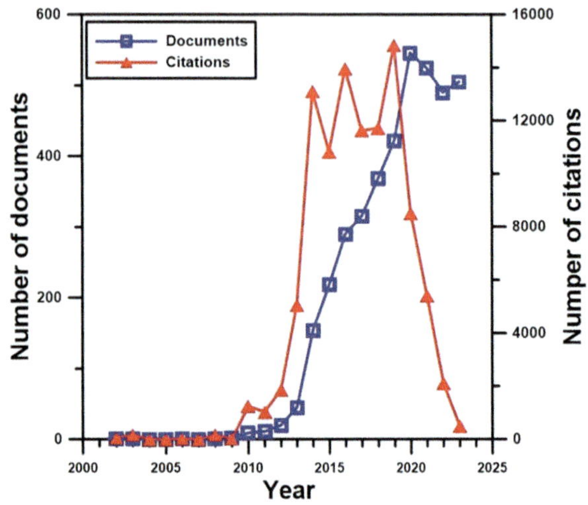

Fig. 14.1 Annual number of documents; number of citations received until the end of 2023 for the articles published each year

Table 14.1 Total number of documents and citations/document type

Document type	Total number of document type (2002–2023)	Total citations of document type (2002–2023)	Total citations/total number of document type
Article	2911	76,677	26.34
Review	513	21,942	42.77
Letter	113	1568	13.87
Editorial	110	604	5.49
Note	104	904	8.69
Conference paper	65	232	3.56
Book chapter	62	85	1.37
Short survey	23	178	7.73
Erratum	19	15	0.78
Book	4	24	6.00
Conference review	3	0	-

with an average number of about 12,500 citations/year. This shows a relative saturation of global interest because the number of citations remains constant despite the increasing number of documents published. As expected, the number of citations decreases for the documents published after 2020 due to the limited time between the publication of the initial articles and those cited them.

14.3.2 Document and Source Type

Eleven different document types are identified. Table 14.1 shows the number of documents and also of citations/document type and the percentage of the ratio

"Total citations/Total number of document type" for each document type. Table 14.1 shows that the most common document type is "Article", followed by "Review", "Letter", "Editorial", and "Note". Other document types include "Conference paper", "Book chapter", "Short survey", "Erratum", "Book", and "Conference review" with the corresponding results as shown in Table 14.1. As expected, "Review" is the type which is the most cited, with approximately 43 citations/document, followed by articles with 26 and letters with 14 citations/document.

14.3.3 Main Sources of Publication

There is a high number of journals that have published the documents corresponding to our research. However, there is an export limit of 160 terms to Scopus. Thus, the analysis is focused on the first 157 journals that are enlisted in Scopus which cover the most published documents when compared with the total publication document number (2511 documents of the 3927 in total).

Figure 14.2 shows the (cumulative) percentage of the 2511 documents as a function of the number of publication sources (journals) that have published them, arranged in decreasing order of the number of articles published per publication source. It can be observed that the first 100 journals have published only 57% of the total documents. This observation is confirmed in Fig. 14.3 which shows the percentage of total documents for the top 10 most productive Journals from 2002 to 2023. Those statements show that there is a concentration of relevant publications only in a few journals.

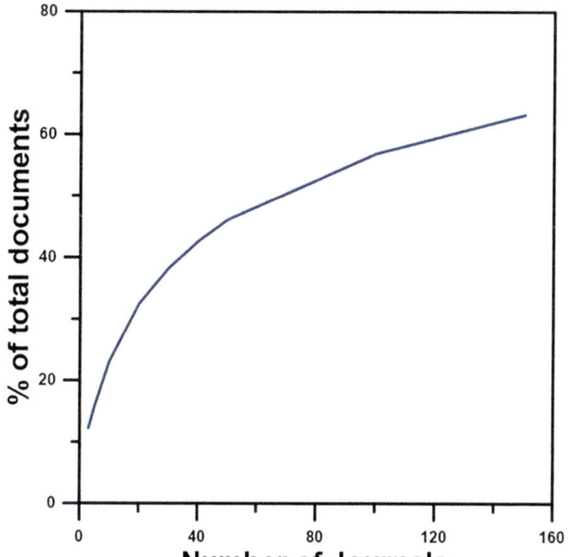

Fig. 14.2 Cumulative percentage of documents published as a function of the number of publication sources (journals) that have published them

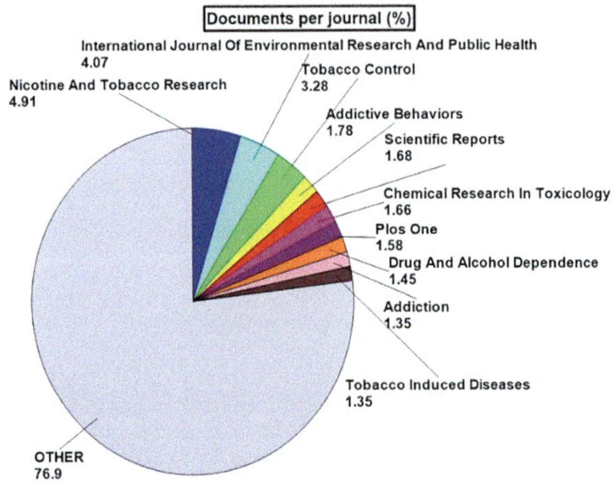

Fig. 14.3 Pie chart of the percentage of total documents for the top 10 most productive journals (2002–2023)

Nicotine and Tobacco Research, which ranks first in the number with 193 published documents, owns only 4.91% of the total documents. The second journal is the International Journal of Environmental Research and Public Health with 160 publications (4.07%), followed by Tobacco Control (129 documents, 3.28%). The journals after the tenth position, which are presented in the pie chart as "others" cover 76.9% of the total articles (Fig. 14.3).

The first 20 sources of publication (journals) by number of documents published and the corresponding citations are shown in Table 14.2. These sources correspond to 32.5% of the total publications. The columns "Rank Citations" and "Rank Publications" correspond to the classification of the publication sources based on the number of publications and citations received.

Taking those 20 journals into consideration, we searched for a possible correlation between the number of citations and the number of publications. The r^2 of this correlation is 0.73, indicating a positive correlation between the number of citations and the number of publications. Average citation/document is a parameter that can express the influence/impact of a publication source (Jin et al. 2018). Figure 14.4 shows that the medium impact factor (IF) journals have published the highest number of publications. Also, this figure shows that high IF journals have a high number of citations/document, but it is not true that the low IF journals have a low number of citations/document, as, in our search, there is not a clear trend between the citations/documents and the journal's IF.

14.4 Countries Classification Comparison

Figure 14.5 shows a pie chart of the first ten countries in terms of the number of documents published. The calculations are based on the total number of documents that resulted from the research in Scopus for the countries analysis, which is 5027 articles, as several articles are counted in more than one country. The first country is

Table 14.2 Ranking 20 journals in terms of publications and citations

Journal name	Documents	Document ranking	Citations	Citation ranking	Citations/ documents	Impact factor 2022
Nicotine and Tobacco Research	193	1	7525	1	39.0	4.7
International Journal of Environmental Research and Public Health	160	2	3560	3	22.3	4.6
Tobacco Control	129	3	7180	2	55.7	5.2
Addictive Behaviors	70	4	1359	14	19.4	4.4
Scientific Reports	66	5	1445	11	21.9	4.6
Chemical Research in Toxicology	65	6	1648	8	25.4	4.1
Plos One	62	7	3331	4	53.7	3.7
Drug and Alcohol Dependence	57	8	1467	10	25.7	4.2
Addiction	53	9	2778	5	52.4	6
Tobacco Induced Diseases	53	10	543	18	10.2	3.7
Regulatory Toxicology and Pharmacology	52	11	2521	6	48.5	3.4
American Journal of Physiology Lung Cellular and Molecular Physiology	49	12	1778	7	36.3	4.9
BMC Public Health	43	13	1159	15	27.0	4.5
Preventive Medicine	43	14	1438	12	33.4	5.1
Toxics	34	15	321	20	9.4	4.6
Inhalation toxicology	33	16	1415	13	42.9	2.1
BMJ Open	30	17	338	19	11.3	2.9
Food and Chemical Toxicology	29	18	837	17	28.9	4.3
Environmental Research	28	19	845	16	30.2	8.3
Pediatrics	27	20	1631	9	60.4	8

the United States with 44.5% of the total publications, either from single-country articles or international collaborations (2249 documents), followed by the UK with 9.14% (462 documents), Australia 3.6% (182 documents), Italy 3.3% (170 documents), Canada 3.3% (167 documents), Germany 2.75% (139 documents), China

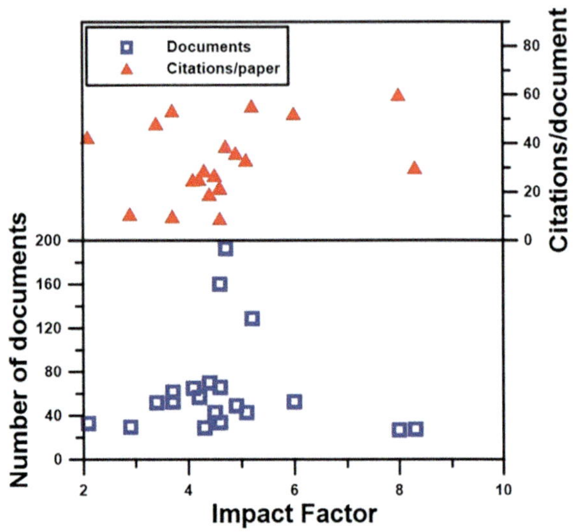

Fig. 14.4 Number of documents and citations/document as a function of the impact factor of each journal (for the first 20 journals)

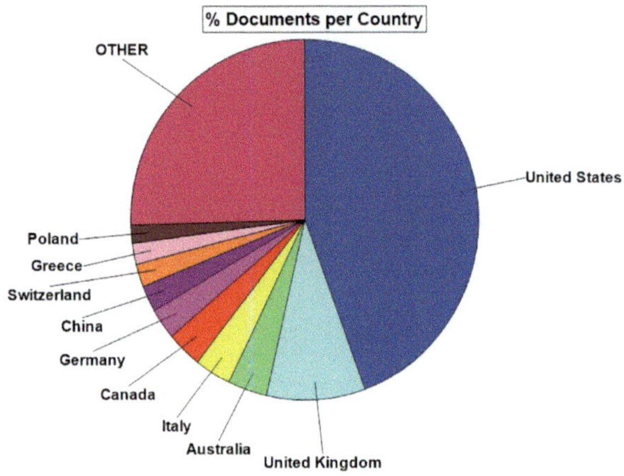

Fig. 14.5 Pie chart of the % of total documents for the top 10 most productive countries (2002–2023)

2.35% (119 documents), Switzerland 2.06% (104 documents), Greece 1.88% (95 documents), and Poland 1.66% (84 documents).

The comparison of the evolution of the relative contribution of these countries for the period 2010–2023 is presented in Fig. 14.6, where most of the countries have started publishing their documents. Some fluctuations are observed during the first years due to the low number of publications. The United States definitely occupies the first place in publications during this period; however, with a sharp decrease during last year. It should be noted that China started to publish in this field only in 2014; however, China's relative contribution has been in constant increase since that date.

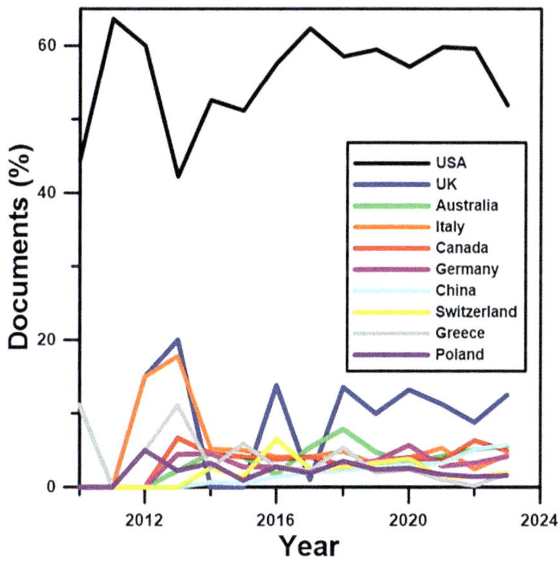

Fig. 14.6 Evolution of the relative contribution of the first ten countries (2010–2023)

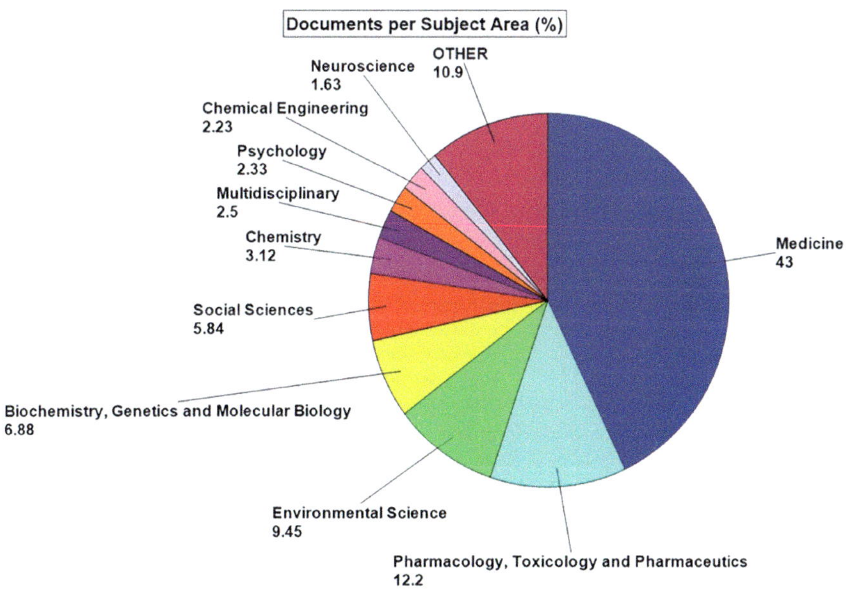

Fig. 14.7 Pie chart of the % of total documents for the top 10 productive subject areas (2002–2023)

14.5 Subject Domain Comparison

The ten most frequent subject areas based on the percentage of total publications are represented in Fig. 14.7 for each subject area. The calculations are based on the total

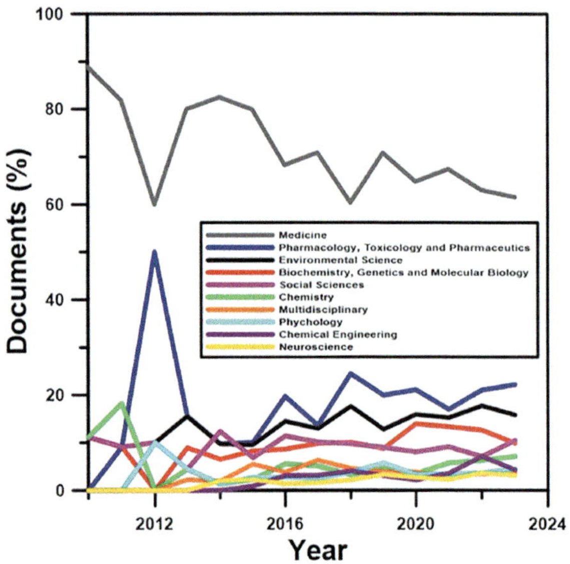

Fig. 14.8 Evolution of the relative contribution for the top 10 productive subject areas (2010–2023)

number of documents that resulted from the research in Scopus for the subject areas analysis, which is 6150 documents, as several documents are counted in more than one subject area. Medicine is the area with the highest percentage of total documents (43%) across the years 2002–2023. The next two subject areas show a percentage of 12.3% for Pharmacology, Toxicology and Pharmaceutics and 9.45% for Environmental Science.

Furthermore, Fig. 14.8 shows the comparison of the evolution of the relative contribution of the first 10 subject areas. The results before 2010 are not presented because of the small number of documents. This is also the reason for the fluctuations observed until 2013. After that year, Medicine had a constant decrease during those years, from 89% of published documents in 2010 to 62% in 2022. Pharmacology, Toxicology and Pharmaceutics, and Environmental Science, which are in the second and third position respectively, show a constant increase; the same increase is observed for Social Sciences and Chemistry.

14.6 Institution Comparison

Among the first 160 institutions in the number of publications extracted by Scopus, the first 10 most productive are shown in Table 14.3. University of California occupies the first place with 127 documents, followed by the Virginia Commonwealth University (113 documents), the UCSF School of Medicine (90 documents), the University of North Carolina at Chapel Hill (84 documents), the Roswell Park Cancer Institute (83 documents), the University of California, San Diego (74 documents), the British American Tobacco (71), the Centers for Disease Control and

Table 14.3 Contribution of the top 10 productive institutions

Institute	Documents	% of the total documents	Document ranking	Citations	Citation ranking	Citations/ documents
University of California, San Francisco	127	2.69	1	8318	1	65.5
Virginia Commonwealth University	113	2.40	2	3839	4	34.0
UCSF School of Medicine	90	1.91	3	5686	3	63.2
The University of North Carolina at Chapel Hill	84	1.78	4	2948	5	35.1
Roswell Park Cancer Institute	83	1.76	5	6132	2	73.9
University of California, San Diego	74	1.57	6	1988	9	26.9
British American Tobacco	71	1.51	7	2016	7	28.4
Centers for Disease Control and Prevention	66	1.40	8	2002	8	30.3
Johns Hopkins Bloomberg School of Public Health	60	1.27	9	2491	6	41.5
UNC School of Medicine	57	1.20	10	1946	10	34.1

Prevention (66 documents), the Johns Hopkins Bloomberg School of Public Health (60 documents), and the UNC School of Medicine (57 documents).

The columns "Document ranking" and "Citation ranking" (Table 14.3) show the ranking of those institutions according to the number of publications and the citations received. Taking into consideration those 10 institutions we searched for a possible correlation between the number of citations and the number of published documents. The r^2 of this correlation is 0.62, indicating that there is a moderate correlation between the two variables.

These top 10 affiliations cover just 18.5% of the total articles from 2002 to 2023, while 82.5% belong to the rest of the institutions.

14.7 Keywords

The keywords provided by Scopus are derived from the keywords defined by the authors and those assigned to the documents by Scopus. All the keywords mentioned in the articles are analyzed using VOSviewer, which creates a map based on the *.ris data files format of the articles exported from Scopus, presenting the keywords' co-occurrence and links. This process penetrates the interactions between the different aspects of the examined topic. For the construction of the map, the minimum number of occurrences of a keyword was set to 20, and out of the 16,813 total keywords, 777 were found to satisfy this threshold. The 100 keywords with the greatest link strength are selected and the analysis is performed following the association strength method. Figure 14.9 shows this visualization map of the obtained results.

Also, the first cluster consists of 160 keywords (this is the limit of Scopus), with the highest occurrence being extracted and clustered. According to the most common keywords, seven main clusters can be created as second cluster (Tables 14.4 and 14.5). Table 14.4 consists of the following clusters: Humans, Studies and methods, Devices and compounds, and Smoking habits, while Table 14.5 includes the clusters of Emissions, Health and Plants. The third cluster includes words related to studies and medicine, including words such as controlled studies, procedure, animal, human cell, toxicity, etc. (57 keywords appeared 14,758 times). The second cluster includes words, related to humans, such as human(s), male, female, child, adolescent, young adult, adult, middle aged, aged, etc. (19 keywords appeared 12,913 times). The third cluster includes words related to the devices and the chemical compounds, such as electronic cigarettes, electronic equipment, liquid, glycerol,

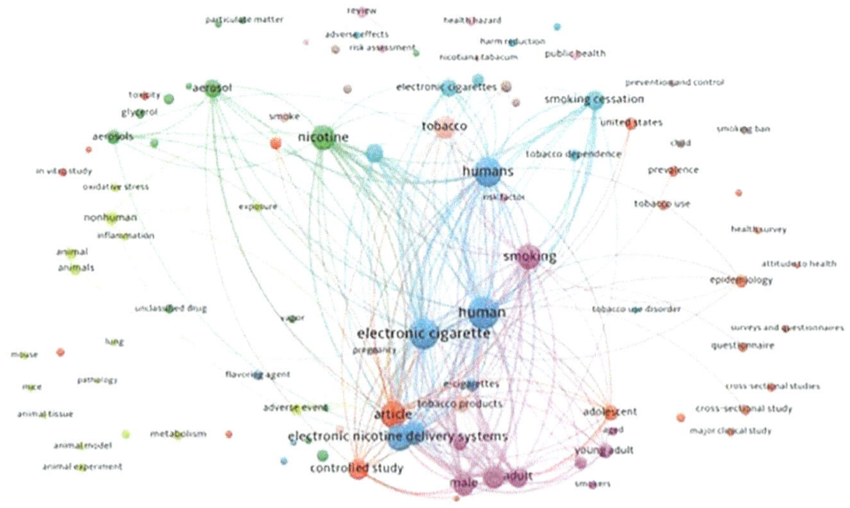

Fig. 14.9 The VOSviewer visualization map. (C1: purple, C2: yellow, C3: blue, C4: light blue, C5: green, C6: orange, C7: light brown)

Table 14.4 Keywords clusters (1–4)

Humans (C1)	No.	Studies and methods (C2)	No.	Devices and compounds (C3)	No.	Smoking habits (C4)	No.
Human	2868	Article	1987	Electronic cigarette	2633	Smoking	1717
Humans	2396	Controlled study	1185	Electronic nicotine delivery systems	2047	Vaping	1460
Male	1404	Priority journal	909	Nicotine	1744	Smoking cessation	1006
Female	1361	United States	496	Tobacco	1556	Cigarette smoking	644
Adult	1209	Procedures	454	Electronic cigarettes	809	Passive smoking	448
Adolescent	712	Animals	419	Tobacco products	796	Tobacco dependence	374
Young adult	590	Animal	413	E-cigarettes	509	Tobacco use	337
Nonhuman	573	Review	411	E-cigarette	451	Psychology	297
Middle aged	422	Cross-sectional study	343	Propylene glycol	368	Nicotine replacement therapy	192
Smokers	227	Questionnaire	339	Glycerol	312	Tobacco use disorder	146
Aged	212	Risk assessment	296	Flavoring agent	253	Perception	123
Child	208	Comparative study	292	Flavoring agents	190	Tobacco smoking	123
Pregnancy	154	Oxidative stress	283	Flavor	152	Smoking habit	113
Sex difference	128	Risk factor	279	Tobacco industry	137	Marketing	111
Student	116	Animal experiment	270	Menthol	122	Nicotine gum	105
Age	114	Unclassified drug	270	Smokeless tobacco	120	Inhalation	104
Students	113	Drug effect	267	Electronic equipment	118	TOTAL	7300
Normal human	106	Cross-sectional studies	253	Liquid	108		

(continued)

Table 14.4 (continued)

Humans (C1)	No.	Studies and methods (C2)	No.	Devices and compounds (C3)	No.	Smoking habits (C4)	No.
TOTAL	12,913	Major clinical study	242	Heating	101		
		Mouse	235	TOTAL	12,526		
		Surveys and questionnaires	230				
		Human cell	228				
		Inflammation	215				
		In vitro study	212				
		Toxicity	212				
		Statistics and numerical data	194				
		Animal model	178				
		Cotinine	173				
		Mice	173				
		Cytotoxicity	168				
		Animal tissue	166				
		Biological marker	158				
		Randomized controlled trial	136				
		Analysis	135				
		Pathology	134				
		Pathophysiology	125				
		Rat	124				
		Follow up	119				
		Gene expression	119				
		Cohort analysis	118				
		Protein expression	118				
		Biomarkers	117				

	Reactive oxygen metabolite	116				
	Human experiment	113				
	Cell viability	112				
	Epithelium cell	108				
	Self report	108				
	Clinical article	107				
	Animal cell	106				
	Editorial	102				
	Food and drug administration	102				
	Interleukin 6	101				
	Mass fragmentography	101				
	Drug effects	97				
	Letter	97				
	Systematic review	97				
	Mass spectrometry	96				
	TOTAL	14,758				

Table 14.5 Keywords clusters (5–7)

Emissions (C5)	No.	Health (C6)	No.	Plants (C7)	No.
Aerosol	936	Epidemiology	470	Nicotiana tabacum	148
Aerosols	628	Adverse event	434	Cannabis	124
Cigarette smoke	438	Prevalence	362	TOTAL	272
Smoke	388	Metabolism	335		
Tobacco smoke pollution	291	Public health	329		
Exposure	271	Health hazard	288		
Tobacco smoke	258	Smoking ban	248		
Environmental exposure	254	Harm reduction	241		
Vapor	246	Lung	200		
Chemistry	237	Adverse effects	187		
Particulate matter	207	Health survey	187		
Electronic cigarette vapor	197	Prevention and control	181		
Formaldehyde	178	Legislation and jurisprudence	151		
Acrolein	153	Attitude to health	142		
Particle size	143	Smoke-free policy	142		
Chemical composition	140	Cardiovascular disease	136		
Carbon monoxide	139	Health care policy	131		
E-cigarette vapor	126	Asthma	127		
Chemical analysis	123	Lung injury	127		
Acetaldehyde	122	Chronic obstructive lung disease	123		
Concentration (parameter)	114	Risk factors	122		
Volatile organic compound	108	Risk reduction	97		
Indoor air pollution	106	Safety	96		
Carcinogen	103	TOTAL	4856		
Nitrosamine	97				
TOTAL	6003				

etc. (19 keywords appeared 12,529 times). The fourth cluster includes words related to smoking habits, such as smoking, vaping, etc. (16 keywords appeared 7300 times). The fifth cluster is related to the emissions, including words such as aerosol, cigarette smoke, particular matter, etc. (25 keywords appeared 6003 times). The sixth cluster includes the keywords related to health and the corresponding keywords are epidemiology, public health, etc. (23 keywords appeared 4856 times). The seventh category is related to the plants and includes the words nicotiana tabacum and cannabis (2 keywords appeared 272 times).

14.8 Conclusion

This chapter shows a bibliometric analysis of the emissions of e-cigs worldwide. A total of 3927 publications were obtained from Scopus for the period from 2002 to 2023. The annual number of publications presents a high increase from 2010 until 2020, while it remains quite constant from 2021 until 2023, indicating a probable saturation of the scientific production in this field. Also, probably this stagnation is a result of the COVID-19 pandemic. 64% of the total documents were published in the first 160 different journals. "Article" is the most commonly used document type, while reviews have the highest ratio of citations/document. Nicotine And Tobacco Research, International Journal of Environmental Research And Public Health, and Tobacco Control are the top three journals based on the document number, while Nicotine And Tobacco Research also had the highest number of citations. The most productive country is the United States (44.5% of the total publications), followed by the UK and Australia. The three predominant subject areas are Medicine, Pharmacology, Toxicology and Pharmaceutics, and Environmental Science. The most productive institutions were the University of California, San Francisco followed by the Virginia Commonwealth University, and the UCSF School of Medicine. According to the classification of the institutions based on the number of citations and publications, the University of California, San Francisco ranked first.

In addition, the keywords are clustered according to their content in seven main clusters, related to humans, studies and methodology, devices and chemical compounds, smoking conditions, emissions, health effects, and the plants (tobacco and cannabis).

References

Behar RZ, Luo W, McWhirter KJ, Pankow JF, Talbot P (2018) Analytical and toxicological evaluation of flavor chemicals in electronic cigarette refill fluids. Sci Rep 8(1):8288. https://doi.org/10.1038/s41598-018-25575-6

Donthu N, Kumar S, Mukherjee D, Pandey N, Lim WM (2021) How to conduct a bibliometric analysis: an overview and guidelines. J Bus Res 133:285–296. https://doi.org/10.1016/j.jbusres.2021.04.070

Drimili E, Gareiou Z, Zervas E (2019) Analysis of the world scientific production on public's opinion on environmental issues. Environ Res Technol 2(4):171–182. https://doi.org/10.35208/ERT.582337

Giovacchini CX, Crotty Alexander LE, Que LG (2022) Electronic cigarettes: a pro–con review of the current literature. J Allergy Clin Immunol Pract 10(11):2843–2851. https://doi.org/10.1016/j.jaip.2022.07.009

Iliopoulou Z, Rapsomanikis N, Zervas E (2022) Analysis of the world scientific production on energy poverty, fuel poverty and energy vulnerability. IOP Conf Ser Earth Environ Sci 1123(1):012059. https://doi.org/10.1088/1755-1315/1123/1/012059

Jin R, Gao S, Cheshmehzangi A, Aboagye-Nimo E (2018) A holistic review of off-site construction literature published between 2008 and 2018. J Clean Prod 202:1202–1219. https://doi.org/10.1016/j.jclepro.2018.08.195

Kalyvas T, Zervas E (2021) Bibliometric analysis of scientific production on energy, sustainability, and climate change. In: Low carbon energy technologies in sustainable energy systems. Elsevier, pp 287–307. https://doi.org/10.1016/B978-0-12-822897-5.00011-0

Lu F, Yu M, Chen C, Liu L, Zhao P, Shen B, Sun R (2022) The emission of VOCs and CO from heated tobacco products, electronic cigarettes, and conventional cigarettes, and their health risk. Toxics 10(1). https://doi.org/10.3390/toxics10010008

Mandalas I, Gizani N, Zervas E (2023) New trends on environmental Kuznets curve research—a bibliometric analysis. E3S Web Conf 436. https://doi.org/10.1051/e3sconf/202343605005

Matsouki N, Konstantinidis E, Tsipa C, Zervas E (2021) Comparison of carbonyls emitted from conventional cigarettes, electronic cigarettes and heated tobacco products. IOP Conf Ser Earth Environ Sci 899(1):012007. https://doi.org/10.1088/1755-1315/899/1/012007

Pearce K, Gray N, Gaur P, Jeon J, Suarez A, Shannahan J, Pappas RS, Watson-Wright C (2020) Toxicological analysis of aerosols derived from three electronic nicotine delivery systems using normal human bronchial epithelial cells. Toxicol In Vitro 69:104997. https://doi.org/10.1016/j.tiv.2020.104997

Qu Y, Kim KH, Szulejko JE (2018) The effect of flavor content in e-liquids on e-cigarette emissions of carbonyl compounds. Environ Res 166:324–333. https://doi.org/10.1016/j.envres.2018.06.013

Rejeb A, Rejeb K, Abdollahi A, Al-Turjman F, Treiblmaier H (2022) The interplay between the Internet of Things and agriculture: a bibliometric analysis and research agenda. Internet Things (Netherlands) 19:100580. https://doi.org/10.1016/j.iot.2022.100580

Ting CY, Ahmad Sabri NA, Tiong LL, Zailani H, Wong LP, Agha Mohammadi N, Anchah L (2020) Heavy metals (Cr, Pb, Cd, Ni) in aerosols emitted from electronic cigarettes sold in Malaysia. J Environ Sci Health A Tox Hazard Subst Environ Eng 55(1):55–62. https://doi.org/10.1080/10934529.2019.1665950

Yingst JM, Bordner CR, Hobkirk AL, Hoglen B, Allen SI, Krebs NM, Houser KR, Livelsberger C, Foulds J (2021) Response to flavored cartridge/pod-based product ban among adult juul users: "you get nicotine however you can get it". Int J Environ Res Public Health 18(1):1–13. https://doi.org/10.3390/ijerph18010207

Zervas E, Litsiou E, Konstantopoulos K, Poulopoulos S, Katsaounou P (2018) Physical characterization of the aerosol of an electronic cigarette: impact of refill liquids. Inhal Toxicol 30(6):218–223. https://doi.org/10.1080/08958378.2018.1500662

Zervas E, Matsouki N, Kyriakopoulos G, Poulopoulos S, Ioannides T, Katsaounou P (2020) Transfer of metals in the liquids of electronic cigarettes. Inhal Toxicol 32(6):240–248. https://doi.org/10.1080/08958378.2020.1776801

Zhao D, Navas-Acien A, Ilievski V, Slavkovich V, Olmedo P, Adria-Mora B, Domingo-Relloso A, Aherrera A, Kleiman NJ, Rule AM, Hilpert M (2019) Metal concentrations in electronic cigarette aerosol: effect of open-system and closed-system devices and power settings. Environ Res 174:125–134. https://doi.org/10.1016/j.envres.2019.04.003

Emission of Metals from E-cigarettes 15

Niki Matsouki, Chara Tsipa, Evangelia Bekou, Zoe Gareiou, Maria Makrygianni, Sofia Gortsilas, and Efthimios Zervas

Abstract

The presence of metals in the emissions of electronic cigarettes (e-cigs) has been reported in several studies. Their presence may be attributed to metal filaments used for evaporating e-liquids, the composition of these liquids, or their interaction with metal parts. Many of the detected metals are known to be highly toxic. There is a high scarcity of their concentrations in the emissions due to the uncontrolled parameters during e-cigs use, such as smoking regime, battery voltage, e-liquids aging, etc., but also due to the different consumables used. This chapter presents the results of extensive literature research concerning the presence of metals in e-cig emissions. This search was performed on the Scopus database using specific keywords and comprises the papers published from 2010 until 2023. The initial query resulted in 3927 articles; after a detailed analysis, 41 articles report the detection of metals in e-cig emissions. Fifty-eight (58) different metals were reported, sixteen (16) of them below the detection limit. The concentrations of the metals, the generation of vapors, their collection, the analytical method used, and any other information available in the published studies, such as the type of device used, are reported in this chapter.

Keywords

Electronic cigarettes · E-cigs · Metals · Emissions · Vapors · Health

N. Matsouki · C. Tsipa · E. Bekou · Z. Gareiou · M. Makrygianni · S. Gortsilas
E. Zervas (✉)
Laboratory of Technology and Policy of Energy and the Environment, School of Applied Arts and Sustainable Design, Hellenic Open University, Patra, Greece
e-mail: zervas@eap.gr

Y.-H. Leong, A. S. Mohd Samin (eds.), *E-Cigarettes*,
https://doi.org/10.1007/978-981-96-5087-3_15

15.1 Introduction

Smoking conventional cigarettes (CCs) is associated with many respiratory and cardiovascular diseases and a lot of premature deaths globally would have been avoided by quitting smoking (WHO 2021). For several years, there has been a continuous effort for the development of novel tobacco products, which resulted in the introduction of electronic cigarettes (e-cigs) into the market. E-cigs, which entered the market dynamically during the last years, have been promoted as healthier alternatives compared to CCs (Stratton et al. 2018; Yayan et al. 2024), and even as a way to quit smoking (Etter and Bullen 2011; Grana and Ling 2014). In order to estimate their safety, several researchers performed a chemical analysis of e-cig emissions, since health impacts are related to the composition of the inhaled vapor. A significant number of chemical compounds, such as nitrosamines, carbonyls, metals, etc., many of them having adverse health effects, are detected in the emissions of e-cigs (Li et al. 2020; Pinto et al. 2022; Talih et al. 2023; Tang et al. 2022).

Many metals are toxic to humans and can bioaccumulate causing severe health problems (Mitra et al. 2022). As it is already known, the general operation process of an e-cig device involves the heating of a liquid (usually a mixture of glycerol, propylene glycol, and nicotine), called e-liquid, through a heating element/atomizer. The heating of this e-liquid produces vapor, which is inhaled by the user. The transfer of metals from the heating element in the e-liquid is considered one of the main sources of metals in the inhaled vapors (Prokopowicz et al. 2019; Zervas et al. 2020). Other sources can be the e-liquids themselves since they may contain small quantities of metals (Kapiamba et al. 2022; Zervas et al. 2017), or other metallic components of the device that come in contact with the e-liquid (Gray et al. 2022).

The scope of this chapter is to show in a comprehensive way the presence of metals in e-cigs' emissions, as published in the scientific literature. The ultimate scope is to contribute to the determination of the health effects of e-cigs through an appropriate toxicological analysis.

15.2 Methodology

This work is part of a larger work concerning the chemical compounds detected in the emissions of e-cigs. An extended bibliographic search was conducted in order to collect the available scientific data concerning the presence of metals. Scopus database is used and the papers published from 2010 until 2023 are taken into consideration. Only the papers published in English were included. The keywords used for this search were: ((ecig OR e-cig OR "electronic cigarette*" OR e-cigarette* OR "Vape pen*" OR "Vaping device" OR "electronic nicotine delivery system*") AND (emission* OR smoke OR vapor OR aerosol OR voc OR pollutant* OR particle* OR "particulate matter" OR pm")). The selected option was the presence of those keywords in the title or in the abstract or in the keywords section of the papers.

From the publications found using the above criteria, only those dealing with the presence of metals in the emissions of e-cigs were further processed. The information included in the publications was subsequently recorded in a table. This information comprises the name of the metal reported and its CAS number, three pairs of columns with the average, minimum, and maximum concentration and their standard deviation (where applicable), the units of the measured concentration, the puff volume, the limit of detection (LOD), and the limit of quantification (LOQ) of the method used (if reported), as well as the collection method and the analytical method used. Other relevant information, that can facilitate the comparison of data, such as the puffing regime or the device model, is recorded in a separate column.

15.3 Results

The results of the aforementioned analysis will focus on the following parameters: e-liquids, e-cig devices, puff generation, puff regime, puff number per sample, sample collection method, and sample analytical method.

15.3.1 E-liquids

A wide variety of e-liquids, both market available and laboratory made, were used. They consist of a mixture of propylene glycol and glycerine in percentages that can possibly vary from 0:100 to 100:0; water may also be added in several cases. E-liquids also contain nicotine in concentrations varying from 0 to 24 mg/mL and they may be flavored or not (Beauval et al. 2017; Belushkin et al. 2020; Kim et al. 2017; Schober et al. 2014). The exact synthesis of the e-liquids used (commercially available or laboratory made) was not described in most of the articles (Kim et al. 2017; Beauval et al. 2017; Belushkin et al. 2020; Schober et al. 2014). Two more parameters are the aging of e-liquids and the collection method (Jameson et al. 2023). Each device contains a specific amount of e-liquid, delivering a specific number of puffs. As a consequence, the analysis of the vapor can be performed either to the whole number of puffs or a part of them (first, middle, or last puffs). The collection and analysis of the first puff refers to a fresh e-liquid, while the collection and analysis of the last puffs to an aged one. An example is the work of Kim et al. (2017) who collected vapors from a fresh e-liquid and from an e-liquid after 20 hrs of usage (Kim et al. 2017).

15.3.2 E-cig Device

A large variety of different e-cig brands, each being available on the market (Belushkin et al. 2020; Jameson et al. 2023; Saffari et al. 2014), several e-cigs testing devices (Kim et al. 2018) were used for the production of vapor. Some researchers give detailed information about the e-cig device type and brand name (Lerner et al. 2015), while others do not report this information (Olmedo et al. 2018). Due to the high number of e-cig devices available and the high rate of new entries in the market, it is not possible to test all of them. The e-cig devices tested include both disposable and rechargeable devices, pods, mods, and disposable pods (Aherrera et al. 2023; Talih et al. 2022). A large variety is also reported in the heating coils used, which are complex metal alloys (Olmedo et al. 2018). The most commonly used coils are Kanthal consisting of 70–73% Fe, 20–24% Cr, 5.8% Al, and < 1.2% other metals, Nichrome a mixture of 80% Ni and 20% Cr and Stainless steel coils consisting of 65–70% Fe, 16–18% Cr, 10–12% Ni, and 4–5% other metals (Haworth-Duff et al. 2023; Olmedo et al. 2018).

15.3.3 Puff Generation

Some of the publications refer to the metals determined in e-cig vapors (mainstream emissions), while others refer to the metals in the indoor air environment (room or chamber) during or after the use of e-cigs (sidestream emissions).

Mainstream Emissions

Some of the publications refer to the metals determined in the mainstream emissions of e-cigs (Kim et al. 2018). Mainstream vapors are produced with the use of a pump (Haworth-Duff et al. 2023), or, more specifically, a peristaltic pump (Tehrani et al. 2023). Alternatively, several types of smoking machines are used (Jameson et al. 2023; Margham et al. 2021), such as a Palaczbot, which is a one-port linear piston-like smoking machine with adjustable puffing regimes (Goniewicz et al. 2014), a CETI 8 MK3 smoking machine (Xu et al. 2023), etc. In some cases, the Coresta CRM81 method was reported as the e-cig vapor generation method (Belushkin et al. 2020).

Sidestream Emissions

Some other publications refer to the metals in the indoor air environment (room or chamber) during or after the use of e-cigs (sidestream emissions) (Oldham et al. 2017). In the case of indoor environment measurements, vapors are also produced by volunteers using their e-cigs in a specific room and under specific conditions (frequency and number of puffs, etc.) (Saffari et al. 2014; Schober et al. 2014).

15.3.4 Puffing Regime

Two often used puffing profiles are: 3 s puff, 30 s puff interval (Beauval et al. 2017; Gonzalez-Jimenez et al. 2021) and rectangular puff profile (Gray et al. 2022; Halstead et al. 2020), or 4 s puff, 30 s puff interval and square shaped puff profile (Oldham et al. 2017). Four seconds (4 s) puff was also applied, with a puff interval of 18 s (Kim et al. 2018). In several publications a 4.3 s puff was also applied (Williams et al. 2013). In general, puff duration varies from 1.8 to 7 s and puff interval from 17 to 60 s (Pearce et al. 2020; Prokopowicz et al. 2019; Williams et al. 2015).

Puff Number per Sample

The number of puffs collected per sample differs among the authors. However, it should be noticed that all authors collected more than one puff per sample. Fifty (50) puffs are often collected (Belushkin et al. 2020; Gray et al. 2022; Halstead et al. 2020; Jeon et al. 2023; Palazzolo et al. 2017); however, 60 puffs (Prokopowicz et al. 2019; Williams et al. 2017) or 75 puffs (Mikheev et al. 2016) or 99 puffs (Tayyarah and Long 2014) per sample are also reported. The maximum number of puffs collected for one sample was 150 puffs (Kim et al. 2017, 2018) and the minimum was 4 puffs (Lerner et al. 2015).

Sample Collection Method

Various collection methods were reported in the publications. They were based either on the use of glass impingers (one or more often two) (Ji et al. 2016; Williams et al. 2019), round bottom flasks (Williams et al. 2019), filters (quartz fiber, glass fiber, teflon, methyl-cellulose, or cotton filters) (Mikheev et al. 2016; Palazzolo et al. 2017), condensation tubes (fluorinated ethylene propylene, FEP) (Gonzalez-Jimenez et al. 2021) or tubes with nitric acid as a trapping solution (Ting et al. 2020). In the case of impingers and round bottom flasks as a trapping solution, all researchers used nitric acid (Haworth-Duff et al. 2023; Prokopowicz et al. 2019; Williams et al. 2019). For example, Beauval et al. used two midget impingers connected in series, containing a 20 mL aqueous nitric acid (5%v/v) solution each (Beauval et al. 2017). Nitric acid solution was also used for the collection of the sample when the vapor was driven through a condenser (Kim et al. 2018). In the case of filters, metals are analyzed after the dissolution of the filters in nitric acid (Kapiamba et al. 2022; Oldham et al. 2017) and extraction through vortexing (Miyashita et al. 2018).

Analytical Methods

The dominant analytical method used for the determination of the metals is Inductively Coupled Plasma Mass Spectroscopy (ICP-MS) (Aherrera et al. 2023; Kapiamba et al. 2022; Margham et al. 2021; Mikheev et al. 2016; Miyashita et al. 2018; Olmedo et al. 2018; Pearce et al. 2020; Saffari et al. 2014; Talih et al. 2022; Tayyarah and Long 2014; Tehrani et al. 2023; Ting et al. 2020; Xu et al. 2023; Zhao et al. 2019). In one article, this method was connected to a High-Performance Liquid Chromatography (HPLC-ICP-MS) system (Liu et al. 2020). High-Resolution Magnetic-Sector (SF-ICP-MS) (Zhao et al. 2018), Triple Quadrupole ICP (ICP-QQQ-MS) (Gonzalez-Jimenez et al. 2021; Gray et al. 2022; Halstead et al. 2020) and Inductively Coupled Plasma Optical Emission Spectroscopy (ICP-OES) (Ji et al. 2016; Kim et al. 2018; Williams et al. 2019, 2020) are also used. Atomic Absorption (Lerner et al. 2015), is also used, especially when a very small volume of sample is available. Atomic Absorption can be used either as Graphite Furnace Atomic Absorption (GFAA) (Rastian et al. 2022) or Electrothermal Atomic Absorption Spectrometry (Prokopowicz et al. 2019). Energy Dispersive X-ray spectroscopy (EDX) is also used for the determination of the elemental composition of the sample collected onto the grid of a Transmission Electron Microscope (TEM)

(Wilson et al. 2019). Alpha Spectrometry was used only in the case of Polonium 210, while a Delayed Neutron Counting Analysis was used in the case of Uranium 235 and Uranium 238 (Margham et al. 2016).

Metals Detected

The bibliographic search, under the criteria described in the methodology section, resulted in 3927 documents. After a detailed analysis, 41 papers report the presence of metals in the emissions of e-cigs. Those 41 publications report the presence of 58 different metals. Uranium is counted as one metal, though reported as two different isotopes (Uranium-238 and Uranium-235). Haworth-Duff et al. also determined different isotopes of the same metal (as ^{50}Cr, ^{52}Cr, and ^{53}Cr) (Haworth-Duff et al. 2023); however, the data of the most abundant isotope is reported. Liu et al. found six arsenic species in the vapor, three of them not fully characterized (Liu et al. 2020); those authors give the median concentration of their results. The list of metals detected and their corresponding concentrations are shown in Table 15.1. Sixteen of the examined metals were found below the analytical method's limit of detection (<LOD). More specifically, Saffari et al. examined the presence of metals in indoor air after the use of e-cigs by three volunteers, in conditions simulating real-life conditions, and they found that the Cesium, Dysprosium, Europium, Holmium, Lutetium, Neodymium, Niobium, Phosphorus, Platinum, Praseodymium, Rhodium, Samarium, Scandium, Thorium, Ytterbium, and Yttrium were detected, however, in concentrations below LOD (Saffari et al. 2014). Haworth-Duff et al., in their work, also examined the presence of Rhodium, and their results are in agreement with Saffari et al. (concentration < LOD).

Some of the publications refer to the metals determined in e-cig vapors (mainstream emissions), while others refer to the metals in the indoor air environment (room or chamber) during or after the use of e-cigs (sidestream emissions).

During the construction of Table 15.1, and in order to facilitate comparison between the concentrations, the reported units, when feasible, were converted: ng/m^3 is converted to μg/L, ppb to ng/kg, etc. The lowest and highest concentration values reported in the examined papers are included in Table 15.1.

15.4 Discussion

The review of the published works dealing with the emissions of e-cigs revealed 58 different metals. Three of them, Arsenic, Beryllium, and Cadmium, are classified in the Group 1 of IARC (agents carcinogenic to humans) (IARC 2024). In addition, though metallic Nickel and Chromium do not impose this kind of risk, Nickel compounds and Chromium (VI) compounds are also classified in Group 1. Several other metals, such as Lead, are classified in Groups 2A and 2B, as possible or probable carcinogenic. Metals are generally bioaccumulating (Mitra et al. 2022); thus, their health effects depend on the dose received (depending on the emissions and also from the inhalation of metals from other sources) and the exposure period (depending on the years of vaping and the daily consumption). As the use of e-cigs may be

Table 15.1 Concentration of the metals detected in the e-cigs mainstream and side-stream emissions

Metal	Cas number	Concentration	References
Aluminum	7429-90-5	ND—394 ng/10 puffs	Gonzalez-Jimenez et al. (2021), Palazzolo et al. (2017), and Williams et al. (2013)
		ND—134 µg/L	Jeon et al. (2023), Saffari et al. (2014), Schober et al. (2014), and Williams et al. (2019, 2020)
		0.90–29.6 µg/kg	Haworth-Duff et al. (2023) and Zhao et al. (2019)
Antimony	7440-36-0	ND—27.3 ng/puff	Talih et al. (2022), Williams et al. (2017), and Xu et al. (2023)
		<LOQ—0.74 µg/L	Beauval et al. (2017), Miyashita et al. (2018), Saffari et al. (2014), and Schober et al. (2014)
		0.08–6.23 µg/kg	Mikheev et al. (2016) and Zhao et al. (2019)
Arsenic	7440-38-2	ND—1.33 ng/puff	Belushkin et al. (2020), Goniewicz et al. (2014), Jameson et al. (2023), Margham et al. (2016, 2021), Palazzolo et al. (2017), Talih et al. (2022), Tayyarah and Long (2014), Williams et al. (2017), and Xu et al. (2023)
		ND—0.052 µg/L	Beauval et al. (2017), Jeon et al. (2023), Kim et al. (2018), Liu et al. (2017), Liu et al. (2020), Miyashita et al. (2018), Oldham et al. (2017), Olmedo et al. (2018), Schober et al. (2014), Tehrani et al. (2023), and Williams et al. (2020)
		0.05–2.39 µg/kg	Mikheev et al. (2016) and Zhao et al. (2019)
Barium	7440-39-3	ND—0.012 µg/10 puffs	Goniewicz et al. (2014), Gonzalez-Jimenez et al. (2021), Williams et al. (2013, 2017)
		ND—22.06 µg/L	Jeon et al. (2023), Miyashita et al. (2018), and Saffari et al. (2014)
Beryllium	7440-41-7	ND—0.047 ng/puff	Jameson et al. (2023), Margham et al. (2016, 2021), and Tayyarah and Long (2014)
		ND (ng/mg of TPM)	Mikheev et al. (2016)

(continued)

Table 15.1 (continued)

Metal	Cas number	Concentration	References
Bismuth	7440-69-9	ND—0.003 μg/10 puffs	Williams et al. (2017)
		0.1–1.2 ng/m^3	Schober et al. (2014)
Boron	7440-42-8	ND—3.83 μg/10 puffs	Williams et al. (2013, 2017)
		ND—300 μg/L	Miyashita et al. (2018), Saffari et al. (2014), Williams et al. (2019), and Zhao et al. (2018)—qualitative determination
Cadmium	7440-43-9	ND—1.47 ng/puff	Belushkin et al. (2020), Goniewicz et al. (2014), Gonzalez-Jimenez et al. (2021), Gray et al. (2022), Halstead et al. (2020), Jameson et al. (2023), Margham et al. (2016, 2021), Palazzolo et al. (2017), Prokopowicz et al. (2019), Talih et al. (2022), Tayyarah and Long (2014), Williams et al. (2017), and Xu et al. (2023)
		ND—0.3 μg/L	Beauval et al. (2017), Jeon et al. (2023), Kim et al. (2017, 2018), Liu et al. (2017), Miyashita et al. (2018), Oldham et al. (2017), Saffari et al. (2014), Schober et al. (2014), Ting et al. (2020), and Williams et al. (2019)
		ND—0.26 μg/kg	Mikheev et al. (2016) and Zhao et al. (2019)
Calcium	7440-70-2	0.073–1.03 μg/10 puffs	Williams et al. (2013, 2019)
		10.81 × 10^{-6}—3630 μg/L	Ji et al. (2016), Kim et al. (2018), Saffari et al. (2014), Schober et al. (2014), and Williams et al. (2019, 2020)
Cerium	7440-45-1	ND—2.4 ng/m^3	Saffari et al. (2014) and Schober et al. (2014)

(continued)

Table 15.1 (continued)

Metal	Cas number	Concentration	References
Chromium	7440-47-3	ND—80.3 ng/puff	Aherrera et al. (2023), Belushkin et al. (2020), Goniewicz et al. (2014), Gonzalez-Jimenez et al. (2021), Gray et al. (2022), Halstead et al. (2020), Jameson et al. (2023), Kapiamba et al. (2022), Margham et al. (2016, 2021), Pearce et al. (2020), Tayyarah and Long (2014), Williams et al. (2013, 2015, 2017), and Xu et al. (2023)
		ND—120 μg/L (the value of 120 mg/L is retrieved from a figure)	Beauval et al. (2017), Jeon et al. (2023), Kim et al. (2017, 2018), Liu et al. (2017), Miyashita et al. (2018), Oldham et al. (2017), Olmedo et al. (2018), Rastian et al. (2022), Saffari et al. (2014), Schober et al. (2014), Ting et al. (2020), and Williams et al. (2019, 2020)
		0.05–54.5 μg/kg	Mikheev et al. (2016) and Zhao et al. (2019)
Cobalt	7440-48-4	ND—0.3 ng/puff	Aherrera et al. (2023), Goniewicz et al. (2014), Gonzalez-Jimenez et al. (2021), Jameson et al. (2023), Margham et al. (2016, 2021), Mikheev et al. (2016), Tayyarah and Long (2014), and Williams et al. (2017)
		ND—1.62 μg/L	Jeon et al. (2023), Kim et al. (2017, 2018), Miyashita et al. (2018), Saffari et al. (2014), and Schober et al. (2014)
Copper	7440-50-8	ND—224.7 ng/puff	Goniewicz et al. (2014), Gonzalez-Jimenez et al. (2021), Gray et al. (2022), Halstead et al. (2020), Jameson et al. (2023), Kapiamba et al. (2022), Lerner et al. (2015), Margham et al. (2016, 2021), Palazzolo et al. (2017), Pearce et al. (2020), Talih et al. (2022), Williams et al. (2013, 2015, 2017), and Zhao et al. (2018)
		ND—201 μg/L	Jeon et al. (2023), Kim et al. (2017, 2018), Miyashita et al. (2018), Rastian et al. (2022), Saffari et al. (2014), Schober et al. (2014), and Williams et al. (2019, 2020)
		0.12–913 μg/kg	Mikheev et al. (2016) and Zhao et al. (2019)
Germanium	7440-56-4	ND—0.014 μg/10 puffs	Williams et al. (2017)

(continued)

Table 15.1 (continued)

Metal	Cas number	Concentration	References
Indium	7440-74-6	ND—0.005 µg/10 puffs	Williams et al. (2017)
Iridium	7439-88-5	ND—0.008 µg/10 puffs	Williams et al. (2017)
Iron	7439-89-6	ND—246.9 ng/puff	Gonzalez-Jimenez et al. (2021), Jameson et al. (2023), Margham et al. (2016, 2021), Palazzolo et al. (2017), Talih et al. (2022), and Williams et al. (2013, 2017)
		ND—828 µg/L	Jeon et al. (2023), Ji et al. (2016), Kim et al. (2018), Miyashita et al. (2018), Saffari et al. (2014), Schober et al. (2014), and Williams et al. (2019, 2020)
		0.30–229 µg/kg	Haworth-Duff et al. (2023) and Zhao et al. (2019)
Lanthanum	7439-91-0	ND—0.005 µg/10 puffs	Williams et al. (2017)
		0.2–1.9 ng/m^3	Saffari et al. (2014) and Schober et al. (2014)
Lead	7439-92-1	ND—46.3 ng/puff	Aherrera et al. (2023), Belushkin et al. (2020), Goniewicz et al. (2014), Gonzalez-Jimenez et al. (2021), Gray et al. (2022), Halstead et al. (2020), Jameson et al. (2023), Margham et al. (2016, 2021), Palazzolo et al. (2017), Pearce et al. (2020), Prokopowicz et al. (2019), Tayyarah and Long (2014), Williams et al. (2013, 2015, 2017), and Xu et al. (2023)
		ND—1133 µg/L	Beauval et al. (2017), Jeon et al. (2023), Kim et al. (2017, 2018), Miyashita et al. (2018), Olmedo et al. (2018), Rastian et al. (2022), Saffari et al. (2014), Schober et al. (2014), Ting et al. (2020), and Williams et al. (2019)
		ND—749 µg/kg	Mikheev et al. (2016) and Zhao et al. (2019)
Lithium	7439-93-2	ND (ng/m^3)	Saffari et al. (2014)
		0.008 µg/10 puffs	Williams et al. (2017)
Magnesium	7439-95-4	ND—0.066 µg/10 puffs	Williams et al. (2013, 2020)
		$13.3 \times 10^{-6} - 115$ µg/L	Ji et al. (2016), Kim et al. (2018), Saffari et al. (2014), Schober et al. (2014), and Williams et al. (2019, 2020)

(continued)

Table 15.1 (continued)

Metal	Cas number	Concentration	References
Manganese	7439-96-5	ND—30.3 ng/puff	Aherrera et al. (2023), Goniewicz et al. (2014), Kapiamba et al. (2022), Palazzolo et al. (2017), Talih et al. (2022), Tayyarah and Long (2014), and Williams et al. (2013, 2017)
		ND—19.4 μg/L	Jeon et al. (2023), Kim et al. (2017), Miyashita et al. (2018), Olmedo et al. (2018), Saffari et al. (2014), and Schober et al. (2014)
		0.2–90.6 μg/kg	Zhao et al. (2019)
Mercury	7439-97-6	ND—0.029 ng/puff	Margham et al. (2016, 2021), Tayyarah and Long (2014), Williams et al. (2017), and Xu et al. (2023)
		60 μg/L	Williams et al. (2020)
Molybdenum	7439-98-7	0.4 ng/puff	Williams et al. (2017)
		1.36×10^{-6} – 1.67 μg/L	Miyashita et al. (2018), Saffari et al. (2014), Schober et al. (2014), and Williams et al. (2017)
Nickel	7440-02-0	ND—37.3 ng/puff	Aherrera et al. (2023), Belushkin et al. (2020), Goniewicz et al. (2014), Gonzalez-Jimenez et al. (2021), Gray et al. (2022), Halstead et al. (2020), Jameson et al. (2023), Kapiamba et al. (2022), Margham et al. (2016, 2021), Palazzolo et al. (2017), Pearce et al. (2020), Talih et al. (2022), Tayyarah and Long (2014), Williams et al. (2013, 2015, 2017), and Xu et al. (2023)
		ND—2600 μg/L (the value of 2600 μg/L is retrieved from a figure)	Jeon et al. (2023), Kim et al. (2017, 2018), Liu et al. (2017), Miyashita et al. (2018), Oldham et al. (2017), Olmedo et al. (2018), Rastian et al. (2022), Saffari et al. (2014), Schober et al. (2014), Ting et al. (2020), and Williams et al. (2019, 2020)
		0.09–2148 μg/kg	Haworth-Duff et al. (2023), Mikheev et al. (2016), and Zhao et al. (2019)
Palladium	7440-05-3	ND—0.003 μg/10 puffs	Kim et al. (2017, 2018), Saffari et al. (2014), and Williams et al. (2017)
Polonium-210	13981-52-7	2.5×10^{-5} Bq/puff	Margham et al. (2016)
Potassium	7440-09-7	ND—40.4 ng/puff	Williams et al. (2013, 2017) and Zhao et al. (2018)
		ND—213 μg/L	Saffari et al. (2014), Schober et al. (2014), and Williams et al. (2019, 2020)

(continued)

Table 15.1 (continued)

Metal	Cas number	Concentration	References
Rubidium	7440-17-7	ND—8.8 ng/puff	Goniewicz et al. (2014) and Williams et al. (2017)
		0.33×10^{-6} – 1.21 µg/L	Jeon et al. (2023), Miyashita et al. (2018), and Saffari et al. (2014)
Selenium	7782-49-2	ND—7.1 ng/puff	Goniewicz et al. (2014), Jameson et al. (2023), Margham et al. (2016, 2021), Tayyarah and Long (2014), and Williams et al. (2017)
		0.000–347 µg/L	Williams et al. (2020)
		ND (ng/mg of TPM)	Mikheev et al. (2016)
Silicon	7440-21-3	224–683 ng/puff	Williams et al. (2013, 2017)
		13–3826 µg/L	Ji et al. (2016), Kim et al. (2018), and Williams et al. (2019, 2020)
Silver	7440-22-4	ND—1.8 ng/puff	Goniewicz et al. (2014), Jameson et al. (2023), and Williams et al. (2015, 2017)
		ND—0.46×10^{-6} µg/L	Jeon et al. (2023) and Saffari et al. (2014)
Sodium	7440-23-5	ND—418 ng/puff	Williams et al. (2015, 2017) and Zhao et al. (2018)
		ND—2289 µg/L	Ji et al. (2016), Schober et al. (2014), and Williams et al. (2017)
Strontium	7440-24-6	ND—1.9 ng/puff	Goniewicz et al. (2014), Talih et al. (2022), and Williams et al. (2013, 2017)
		ND—40.47 µg/L	Jeon et al. (2023), Miyashita et al. (2018), and Saffari et al. (2014)
Thallium	7440-28-0	ND—< LOQ	Beauval et al. (2017), Goniewicz et al. (2014), Jeon et al. (2023), Saffari et al. (2014), and Schober et al. (2014)
Tin	7440-31-5	ND—1130 ng/puff	Gonzalez-Jimenez et al. (2021), Gray et al. (2022), Halstead et al. (2020), Jameson et al. (2023), Margham et al. (2016, 2021), Pearce et al. (2020), Talih et al. (2022), Tayyarah and Long (2014), and Williams et al. (2013, 2015, 2017)
		0.95×10^{-6} – 345 µg/L	Saffari et al. (2014), Schober et al. (2014), and Williams et al. (2019, 2020)
		0.05–391 µg/kg	Mikheev et al. (2016) and Zhao et al. (2019)
Titanium	7440-32-6	0.02 ng/puff	Williams et al. (2013)
		ND—52.2×10^{-6} µg/L	Saffari et al. (2014), Schober et al. (2014), and Williams et al. (2019, 2020)

(continued)

Table 15.1 (continued)

Metal	Cas number	Concentration	References
Tungsten	7440-33-7	ND—1.8 ng/puff	Williams et al. (2017)
		0.67×10^{-6} µg/L	Saffari et al. (2014)
		0.04–0.33 µg/kg	Zhao et al. (2019)
Uranium	7440-61-1	ND—0.06 µg/kg	Margham et al. (2016), Saffari et al. (2014), and Zhao et al. (2019)
Vanadium	7440-62-2	ND—0.3 ng/puff	Goniewicz et al. (2014) and Williams et al. (2017)
		ND—0.4 µg/L	Jeon et al. (2023), Miyashita et al. (2018), Saffari et al. (2014), Schober et al. (2014), and Williams et al. (2019)
Zinc	7440-66-6	ND—110 ng/puff	Goniewicz et al. (2014), Gray et al. (2022), Halstead et al. (2020), Jameson et al. (2023), Kapiamba et al. (2022), Margham et al. (2016, 2021), Palazzolo et al. (2017), Pearce et al. (2020), Williams et al. (2013, 2015, 2017), and Zhao et al. (2018)
		56.01×10^{-6} – 789 µg/L	Jeon et al. (2023), Miyashita et al. (2018), Saffari et al. (2014), Schober et al. (2014), and Williams et al. (2019, 2020)
		3–4915 µg/kg	Mikheev et al. (2016) and Zhao et al. (2019)
Zirconium	7440-67-7	ND—0.246 µg/10 puffs	Williams et al. (2013, 2017)

ND not detected or below limits of detection, *LOQ* limits of quantification, *TPM* total particulate matter

a long-term habit, more detailed information about the intake of metals from inhalation is necessary.

Metals presence in the e-vapors is attributed to the metallic filament that is used to heat the e-liquid (Prokopowicz et al. 2019; Zervas et al. 2020) and to the composition of the e-liquids (Kapiamba et al. 2022), that may contain some metals, or due to the contact of the e-liquid with other metallic parts of the device (Gray et al. 2022). The metallic filaments are mainly made of Nickel, Iron, and Chromium in different concentrations (Fowles et al. 2020), the origin of e-liquid determines its metal content (Olmedo et al. 2018) and the design of the device and the materials used determine the metals coming from the contact of the e-liquid with the metallic parts of the device (Omaiye et al. 2021).

It should be noticed that the set of parameters used by each research team varies significantly, affecting the concentration of the emitted metals. The concentration of the most detected metals varies from below the detection limit (mainly when the determination concerns sidestream air) (Saffari et al. 2014) up to a maximum level, and is correlated with the selected parameters of the experimental method used. It is evident that the accurate exposure of users to the metals emitted from e-cigs cannot be determined.

The mechanism behind the generation of the metals in the emissions and the impact of the studied parameters is sometimes analyzed in the relevant publications and a number of correlations are suggested. It is proposed that the used or ceramic-coated metal coils generate fewer particles compared to the new coils (Wilson et al. 2019). The concentration of metals in the vapor increases with the nicotine content of the e-liquid, while the impact of flavors in the e-liquid is not clear. Moreover, in general, the concentration of the metals emitted is positively correlated with device power and puff duration; however, the opposite trends are also reported (Kapiamba et al. 2022; Zhao et al. 2019). Device power affects heating temperature and the concentration of the metal emitted increases with heating temperature (Williams et al. 2019). This statement is in agreement with the finding that mod systems, having higher heating temperatures of metal filaments than pod systems, emit higher concentration of metals than the latter (Kapiamba et al. 2022). Aging of the e-liquids also affects metals emission, which seems to increase with time (Rastian et al. 2022).

15.5 Conclusion

The presence of metals in the emissions of e-cigs, as reported in the already published works, is studied here. For this, a bibliographic search was performed using the Scopus database and specific keywords, under specified criteria. The number of published articles from 2010 to 2023, related to e-cigs emissions is 3927; 41 of them study the presence of metals in the emissions. Fifty-eight (58) different metals were detected; however, 16 of them were found below the detection limit (LOD). A high scattering in the concentrations of metals is reported, which can be attributed to the high diversity of e-cig devices used, and to the heterogeneity of the used parameters, such as the composition and aging of the e-liquids, the puff generation method, the puffing regime, the puff number per sample, the vapor collection and the sample analysis method. Several of the detected metals, such as Arsenic, Beryllium, and Cadmium, are carcinogenic, and others, such as Chromium, Nickel, and Lead, have probable or possible carcinogenic effects. Since real-use conditions cannot be controlled, it is not possible to accurately determine the intake of metals by the users and their health impact. Further research concerning the emissions of metals from e-cigs is more than necessary in order to establish standard methods for metals determination in e-cig emissions and to monitor short and long-term health impacts on e-cig users attributed to metals' presence in the inhaled e-cig vapors.

References

Aherrera A, Lin JJ, Chen R, Tehrani M, Schultze A, Borole A, Tanda S, Goessler W, Rule AM (2023) Metal concentrations in E-cigarette aerosol samples: a comparison by device type and flavor. Environ Health Perspect 131(12):027010. https://doi.org/10.1289/EHP2175

Beauval N, Antherieu S, Soyez M, Gengler N, Grova N, Howsam M, Hardy EM, Fischer M, Appenzeller BMR, Goossens J-F, Allorge D, Garçon G, Lo-Guidice J-M, Garat A (2017) Chemical evaluation of electronic cigarettes: multicomponent analysis of liquid refills and their corresponding aerosols. J Anal Toxicol 41(8):670–678. https://doi.org/10.1093/jat/bkx054

Belushkin M, Tafin Djoko D, Esposito M, Korneliou A, Jeannet C, Lazzerini M, Jaccard G (2020) Selected harmful and potentially harmful constituents levels in commercial e-cigarettes. Chem Res Toxicol 33(2):657–668. https://doi.org/10.1021/acs.chemrestox.9b00470

Etter J-F, Bullen C (2011) Electronic cigarette: users profile, utilization, satisfaction and perceived efficacy: E-cigarette survey. Addiction 106(11):2017–2028. https://doi.org/10.1111/j.1360-0443.2011.03505.x

Fowles J, Barreau T, Wu N (2020) Cancer and non-cancer risk concerns from metals in electronic cigarette liquids and aerosols. Int J Environ Res Public Health 17(6):2146. https://doi.org/10.3390/ijerph17062146

Goniewicz ML, Knysak J, Gawron M, Kosmider L, Sobczak A, Kurek J, Prokopowicz A, Jablonska-Czapla M, Rosik-Dulewska C, Havel C, Jacob P, Benowitz N (2014) Levels of selected carcinogens and toxicants in vapour from electronic cigarettes. Tob Control 23(2):133–139. https://doi.org/10.1136/tobaccocontrol-2012-050859

Gonzalez-Jimenez N, Gray N, Pappas R, Halstead M, Lewis E, Valentin-Blasini L, Watson C, Blount B (2021) Analysis of toxic metals in aerosols from devices associated with electronic cigarette, or vaping, product use associated lung injury. Toxics 9(10):240. https://doi.org/10.3390/toxics9100240

Grana RA, Ling PM (2014) Smoking revolution. Am J Prev Med 46(4):395–403. https://doi.org/10.1016/j.amepre.2013.12.010

Gray N, Halstead M, Valentin-Blasini L, Watson C, Pappas RS (2022) Toxic metals in liquid and aerosol from pod-type electronic cigarettes. J Anal Toxicol 46(1):69–75. https://doi.org/10.1093/jat/bkaa185

Halstead M, Gray N, Gonzalez-Jimenez N, Fresquez M, Valentin-Blasini L, Watson C, Pappas RS (2020) Analysis of toxic metals in electronic cigarette aerosols using a novel trap design. J Anal Toxicol 44(2):149–155. https://doi.org/10.1093/jat/bkz078

Haworth-Duff A, Parkes GMB, Reed NJ (2023) A simple approach to analysing trace metals in aerosols produced by e-cigarettes. Drug Test Anal 15(10):1084–1090. https://doi.org/10.1002/dta.3416

IARC (2024) IARC monographs on the identification of carcinogenic hazards to humans. Retrieved July 24, 2024, from https://monographs.iarc.who.int/agents-classified-by-the-iarc/

Jameson JB, Wang J, Bailey PC, Oldham MJ, Smith CR, Jeong LN, Cook DK, Bates AL, Ullah S, Pennington ASC, Gillman IG (2023) Determination of chemical constituent yields in e-cigarette aerosol using partial and whole pod collections, a comparative analysis. Front Chem 11:1223967. https://doi.org/10.3389/fchem.2023.1223967

Jeon J, Zhang Q, Chepaitis PS, Greenwald R, Black M, Wright C (2023) Toxicological assessment of particulate and metal hazards associated with vaping frequency and device age. Toxics 11(2):155. https://doi.org/10.3390/toxics11020155

Ji EH, Sun B, Zhao T, Shu S, Chang CH, Messadi D, Xia T, Zhu Y, Hu S (2016) Characterization of electronic cigarette aerosol and its induction of oxidative stress response in oral keratinocytes. PLoS One 11(5):e0154447. https://doi.org/10.1371/journal.pone.0154447

Kapiamba KF, Hao W, Owusu SY, Liu W, Huang Y-W, Wang Y (2022) Examining metal contents in primary and secondhand aerosols released by electronic cigarettes. Chem Res Toxicol 35(6):954–962. https://doi.org/10.1021/acs.chemrestox.1c00411

Kim JJ, Sabatelli N, Tutak W, Giuseppetti A, Frukhtbeyn S, Shaffer I, Wilhide J, Routkevitch D, Ondov JM (2017) Universal electronic-cigarette test: physiochemical characterization of reference e-liquid. Tob Induc Dis 15(1):14. https://doi.org/10.1186/s12971-017-0119-x

Kim SA, Smith S, Beauchamp C, Song Y, Chiang M, Giuseppetti A, Frukhtbeyn S, Shaffer I, Wilhide J, Routkevitch D, Ondov JM, Kim JJ (2018) Cariogenic potential of sweet flavors in electronic-cigarette liquids. PLoS One 13(9):e0203717. https://doi.org/10.1371/journal.pone.0203717

Lerner CA, Sundar IK, Watson RM, Elder A, Jones R, Done D, Kurtzman R, Ossip DJ, Robinson R, McIntosh S, Rahman I (2015) Environmental health hazards of e-cigarettes and their components: oxidants and copper in e-cigarette aerosols. Environ Pollut 198:100–107. https://doi.org/10.1016/j.envpol.2014.12.033

Li L, Lin Y, Xia T, Zhu Y (2020) Effects of electronic cigarettes on indoor air quality and health. Annu Rev Public Health 41(1):363–380. https://doi.org/10.1146/annurev-publhealth-040119-094043

Liu J, Liang Q, Oldham M, Rostami A, Wagner K, Gillman I, Patel P, Savioz R, Sarkar M (2017) Determination of selected chemical levels in room air and on surfaces after the use of cartridge- and tank-based E-vapor products or conventional cigarettes. Int J Environ Res Public Health 14(9):969. https://doi.org/10.3390/ijerph14090969

Liu Q, Huang C, Chris Le X (2020) Arsenic species in electronic cigarettes: determination and potential health risk. J Environ Sci 91:168–176. https://doi.org/10.1016/j.jes.2020.01.023

Margham J, McAdam K, Forster M, Liu C, Wright C, Mariner D, Proctor C (2016) Chemical composition of aerosol from an E-cigarette: a quantitative comparison with cigarette smoke. Chem Res Toxicol 29(10):1662–1678. https://doi.org/10.1021/acs.chemrestox.6b00188

Margham J, McAdam K, Cunningham A, Porter A, Fiebelkorn S, Mariner D, Digard H, Proctor C (2021) The chemical complexity of e-cigarette aerosols compared with the smoke from a tobacco burning cigarette. Front Chem 9:743060. https://doi.org/10.3389/fchem.2021.743060

Mikheev VB, Brinkman MC, Granville CA, Gordon SM, Clark PI (2016) Real-time measurement of electronic cigarette aerosol size distribution and metals content analysis. Nicotine Tob Res 18(9):1895–1902. https://doi.org/10.1093/ntr/ntw128

Mitra S, Chakraborty AJ, Tareq AM, Emran TB, Nainu F, Khusro A, Idris AM, Khandaker MU, Osman H, Alhumaydhi FA, Simal-Gandara J (2022) Impact of heavy metals on the environment and human health: novel therapeutic insights to counter the toxicity. J King Saud Univ Sci 34(3):101865. https://doi.org/10.1016/j.jksus.2022.101865

Miyashita L, Suri R, Dearing E, Mudway I, Dove RE, Neill DR, Van Zyl-Smit R, Kadioglu A, Grigg J (2018) E-cigarette vapour enhances pneumococcal adherence to airway epithelial cells. Eur Respir J 51(2):1701592. https://doi.org/10.1183/13993003.01592-2017

Oldham MJ, Wagner KA, Gene Gilman I, Beach JB, Liu J, Rostami AA, Sarkar MA (2017) Development/verification of methods for measurement of exhaled breath and environmental e-vapor product aerosol. Regul Toxicol Pharmacol 85:55–63. https://doi.org/10.1016/j.yrtph.2017.01.006

Olmedo P, Goessler W, Tanda S, Grau-Perez M, Jarmul S, Aherrera A, Chen R, Hilpert M, Cohen JE, Navas-Acien A, Rule AM (2018) Metal concentrations in e-cigarette liquid and aerosol samples: the contribution of metallic coils. Environ Health Perspect 126(2):027010

Omaiye EE, Williams M, Bozhilov KN, Talbot P (2021) Design features and elemental/metal analysis of the atomizers in pod-style electronic cigarettes. PLoS One 16(3):e0248127. https://doi.org/10.1371/journal.pone.0248127

Palazzolo DL, Crow AP, Nelson JM, Johnson RA (2017) Trace metals derived from electronic cigarette (ECIG) generated aerosol: potential problem of ECIG devices that contain nickel. Front Physiol 7. https://doi.org/10.3389/fphys.2016.00663

Pearce K, Gray N, Gaur P, Jeon J, Suarez A, Shannahan J, Pappas RS, Watson-Wright C (2020) Toxicological analysis of aerosols derived from three electronic nicotine delivery systems using normal human bronchial epithelial cells. Toxicol In Vitro 69:104997. https://doi.org/10.1016/j.tiv.2020.104997

Pinto MI, Thissen J, Hermes N, Cunningham A, Digard H, Murphy J (2022) Chemical characterisation of the vapour emitted by an e-cigarette using a ceramic wick-based technology. Sci Rep 12(1):16497. https://doi.org/10.1038/s41598-022-19761-w

Prokopowicz A, Sobczak A, Szuła-Chraplewska M, Ochota P, Kośmider L (2019) Exposure to cadmium and lead in cigarette smokers who switched to electronic cigarettes. Nicotine Tob Res 21(9):1198–1205. https://doi.org/10.1093/ntr/nty161

Rastian B, Wilbur C, Curtis DB (2022) Transfer of metals to the aerosol generated by an electronic cigarette: influence of number of puffs and power. Int J Environ Res Public Health 19(15):9334. https://doi.org/10.3390/ijerph19159334

Saffari A, Daher N, Ruprecht A, De Marco C, Pozzi P, Boffi R, Hamad SH, Shafer MM, Schauer JJ, Westerdahl D, Sioutas C (2014) Particulate metals and organic compounds from electronic and tobacco-containing cigarettes: comparison of emission rates and secondhand exposure. Environ Sci Process Impacts 16(10):2259–2267. https://doi.org/10.1039/C4EM00415A

Schober W, Szendrei K, Matzen W, Osiander-Fuchs H, Heitmann D, Schettgen T, Jörres RA, Fromme H (2014) Use of electronic cigarettes (e-cigarettes) impairs indoor air quality and increases FeNO levels of e-cigarette consumers. Int J Hyg Environ Health 217(6):628–637. https://doi.org/10.1016/j.ijheh.2013.11.003

Stratton KR, Kwan LY, Eaton DL, National Academies of Sciences and Medicine (U.S.), E (2018) Public health consequences of e-cigarettes, Consensus study report. The National Academies Press, Washington, DC

Talih S, Salman R, Soule E, El-Hage R, Karam E, Karaoghlanian N, El-Hellani A, Saliba N, Shihadeh A (2022) Electrical features, liquid composition and toxicant emissions from 'pod-mod'-like disposable electronic cigarettes. Tob Control 31(5):667–670. https://doi.org/10.1136/tobaccocontrol-2020-056362

Talih S, Karaoghlanian N, Salman R, Hilal E, Patev A, Bell A, Fallah S, El-Hage R, Saliba NA, Cobb C, Barnes A, Shihadeh A (2023) Effects of aftermarket electronic cigarette pods on device power output and nicotine, carbonyl, and ROS emissions. Chem Res Toxicol 36(12):1930–1937. https://doi.org/10.1021/acs.chemrestox.3c00213

Tang M, Lee H-W, Weng M, Wang H-T, Hu Y, Chen L-C, Park S-H, Chan H, Xu J, Wu X-R, Wang H, Yang R, Galdane K, Jackson K, Chu A, Halzack E (2022) DNA damage, DNA repair and carcinogenicity: tobacco smoke versus electronic cigarette aerosol. Mutat Res Rev Mutat Res 789:108409. https://doi.org/10.1016/j.mrrev.2021.108409

Tayyarah R, Long GA (2014) Comparison of select analytes in aerosol from e-cigarettes with smoke from conventional cigarettes and with ambient air. Regul Toxicol Pharmacol 70(3):704–710. https://doi.org/10.1016/j.yrtph.2014.10.010

Tehrani MW, Aherera AD, Tanda S, Chen R, Borole A, Goessler W, Rule AM (2023) Arsenic and arsenic species in MOD, POD, and disposable POD electronic cigarette aerosols: a pilot study. J Environ Expo Assess 2(2):9. https://doi.org/10.20517/jeea.2023.03

Ting CY, Ahmad Sabri NA, Tiong LL, Zailani H, Wong LP, Agha Mohammadi N, Anchah L (2020) Heavy metals (Cr, Pb, Cd, Ni) in aerosols emitted from electronic cigarettes sold in Malaysia. J Environ Sci Health A 55(1):55–62. https://doi.org/10.1080/10934529.2019.1665950

WHO (2021) WHO report on the global tobacco epidemic 2021: addressing new and emerging products. World Health Organization, Geneva

Williams M, Villarreal A, Bozhilov K, Lin S, Talbot P (2013) Metal and silicate particles including nanoparticles are present in electronic cigarette cartomizer fluid and aerosol. PLoS One 8(3):e57987. https://doi.org/10.1371/journal.pone.0057987

Williams M, To, A, Bozhilov K, Talbot P (2015) Strategies to reduce tin and other metals in electronic cigarette aerosol. PLoS One 10(9):e0138933. https://doi.org/10.1371/journal.pone.0138933

Williams M, Bozhilov K, Ghai S, Talbot P (2017) Elements including metals in the atomizer and aerosol of disposable electronic cigarettes and electronic hookahs. PLoS One 12(4):e0175430. https://doi.org/10.1371/journal.pone.0175430

Williams M, Li J, Talbot P (2019) Effects of model, method of collection, and topography on chemical elements and metals in the aerosol of tank-style electronic cigarettes. Sci Rep 9(1):13969. https://doi.org/10.1038/s41598-019-50441-4

Williams M, Ventura J, Loza A, Wang Y, Talbot P (2020) Chemical elements in electronic cigarette solvents and aerosols inhibit mitochondrial reductases and induce oxidative stress. Nicotine Tob Res 22(Supplement_1):S14–S24. https://doi.org/10.1093/ntr/ntaa193

Wilson MD, Prasad KA, Kim JS, Park JH (2019) Characteristics of metallic nanoparticles emitted from heated Kanthal e-cigarette coils. J Nanopart Res 21(7):156. https://doi.org/10.1007/s11051-019-4598-y

Xu T, Niu Z, Xu J, Li X, Luo Q, Luo A, Huang Y, Jiang X, Wu Z (2023) Chemical analysis of selected harmful and potentially harmful constituents and in vitro toxicological evaluation of leading flavoured e-cigarette aerosols in the Chinese market. Drug Test Anal 15(10):1156–1163. https://doi.org/10.1002/dta.3337

Yayan J, Franke K-J, Biancosino C, Rasche K (2024) Comparative systematic review on the safety of e-cigarettes and conventional cigarettes. Food Chem Toxicol 185:114507. https://doi.org/10.1016/j.fct.2024.114507

Zervas E, Ioannides T, Athanasopoulou S, Litsiou E, Gratziou C, Zakynthinos S, Katsaounou P (2017) Metals found in the e-cigarette liquids. 7th ECTOH conference on tobacco or health, 22 March

Zervas E, Matsouki N, Kyriakopoulos G, Poulopoulos S, Ioannides T, Katsaounou P (2020) Transfer of metals in the liquids of electronic cigarettes. Inhal Toxicol 32(6):240–248. https://doi.org/10.1080/08958378.2020.1776801

Zhao J, Nelson J, Dada O, Pyrgiotakis G, Kavouras IG, Demokritou P (2018) Assessing electronic cigarette emissions: linking physico-chemical properties to product brand, e-liquid flavoring additives, operational voltage and user puffing patterns. Inhal Toxicol 30(2):78–88. https://doi.org/10.1080/08958378.2018.1450462

Zhao D, Navas-Acien A, Ilievski V, Slavkovich V, Olmedo P, Adria-Mora B, Domingo-Relloso A, Aherrera A, Kleiman NJ, Rule AM, Hilpert M (2019) Metal concentrations in electronic cigarette aerosol: effect of open-system and closed-system devices and power settings. Environ Res 174:125–134. https://doi.org/10.1016/j.envres.2019.04.003

Assessment of Prevalence, Patterns, and Characteristics of E-cigarette Use

16

Norny Syafinaz Ab Rahman, Syahrir Zaini, and Mohamad Haniki Nik Mohamed

Abstract

This chapter describes the prevalence of e-cigarette (e-cig) usage globally and in various nations, age groups, sociodemographic traits linked to e-cig use, and variables that encourage e-cig use. The use of e-cigs is on the rise everywhere, particularly among young individuals. What is most concerning is that e-cigs are starting to have a "gateway effect" on those who are cigarette-naïve and eventually switch to conventional cigarettes after using them. It has been determined that a variety of variables, such as tastes, price, accessibility, laws, and others, encourage the use of e-cigs. To develop methods to reduce this serious public health concern, it is essential to comprehend the present prevalence, features, and factors that affect e-cig usage.

Keywords

Electronic cigarette · Prevalence · Patterns · Characteristics

16.1 Introduction

Electronic cigarettes (e-cigs), also known as e-cigarettes or vapes, have emerged as a popular alternative to conventional cigarette smoking. These battery-powered devices heat a liquid solution, typically containing nicotine, flavorings, and other chemicals to produce an aerosol, which is then inhaled by the user. While initially marketed as a potentially safer alternative to combustible tobacco, the prevalence of

N. S. A. Rahman (✉) · S. Zaini · M. H. N. Mohamed
Department of Pharmacy Practice, Kulliyyah of Pharmacy, International Islamic University Malaysia (IIUM), Kuantan, Pahang, Malaysia
e-mail: norny@iium.edu.my

Y.-H. Leong, A. S. Mohd Samin (eds.), *E-Cigarettes*,
https://doi.org/10.1007/978-981-96-5087-3_16

e-cig use has surged in recent years. The trend of e-cig use showed a rapid rise while there was a slight decline in conventional cigarette consumption (Fadus et al. 2019), prompting concerns about its potential health impacts and regulatory challenges.

Globally, e-cigs use has now become a major public health highlight due to its harmful effects on social and health, especially among young adults. Although e-cigs was first introduced as a method to quit smoking among conventional cigarette users, it is actually an impediment to reducing smoking. Increasing numbers of evidence have shown that e-cigs are less likely to help smokers to reduce or stop smoking, but in fact, the former smokers' relapse rate was higher if they used e-cigs (Gomajee et al. 2019).

Despite the beneficial effect of e-cigs as an option for conventional cigarette cessation, they also can become a great threat, which can lead to addiction and eventually increase the prevalence of e-cigs users. This is more worrisome, especially among conventional cigarette-naïve adolescents, because initial use of e-cigs led them to subsequent use of conventional cigarettes, suggesting a "gateway effect", even when controlling for demographic, behavioral, and smoking risk factors for conventional cigarette use (Soneji et al. 2017; Epstein et al. 2020).

The presence of nicotine in e-cigs liquid was seen as another threat whereby it has been demonstrated that the frequency and severity of subsequent use of e-cigs and conventional cigarettes were correlated with nicotine dosages in e-cigs (Goldenson et al. 2017). Moreover, adolescents who tried conventional cigarettes were at high risk of established smoking when taking e-cigs (Chaffee et al. 2018). Research demonstrates that teenagers who have friends or parents who use e-cigs are more likely to use e-cigs themselves in the future (Wang et al. 2018).

E-cigs users can have adverse effects from e-cigs used due to their compounds and materials such as volatile organic compounds, carcinogenic carbonyls, carcinogenic tobacco-specific nitrosamines, cytotoxic components, fine/ultrafine particles, and propylene glycol (Pisinger and Dossing 2014). Therefore, the risk of respiratory and cardiovascular illnesses is increased with e-cigs use (Wagener et al. 2017, Bhatta and Glantz 2020).

This chapter details the prevalence of e-cigs use worldwide and across different countries, age groups, the sociodemographic characteristics associated with the use of e-cigs, and factors promoting the use of e-cigs.

16.2 Global Prevalence and Trends

Understanding the global prevalence of e-cigs use is essential for policymakers, healthcare professionals, and researchers to assess the scope of this phenomenon and its potential consequences. Various factors influence prevalence rates, including cultural norms, regulatory frameworks, marketing strategies, and socioeconomic factors.

Over the years, e-cigs have become popular worldwide, and their use gradually increased when it was first introduced in 2003 as a smoking cessation aid in adults but the usage has drastically increased in recent years. Globally, the overall trend of current e-cig use has shown an upward trend from 2011 to 2019, whereby the trend of e-cig use increased from 2011 to 2014, but then it showed a constant trend from 2014 to 2019 (Tehrani et al. 2022).

Tehrani et al. conducted a systematic review and meta-analysis study investigating the global prevalence and trends of e-cigs use for lifetime or ever e-cig users and current e-cigs users. Lifetime e-cig users refer to e-cigs used by an individual during his/her life, while current e-cig users refer to e-cigs used during the last 12 months (Tehrani et al. 2022). Meanwhile, in many other studies, different definitions were used to represent current e-cig users. Most studies involving adolescents and adults refer to current e-cig user as e-cigarettes user in the past 30 days (Jane Ling et al. 2023; Driezen et al. 2022).

Overall, according to Tehrani et al. (2022), the prevalence of e-cigs use globally from 2011 to 2019 showed an upward trend. Similar trends were also observed for the prevalence of lifetime e-cig users, but not for current e-cig users. The prevalence of e-cigs use also varies across genders, ages, and continents. The prevalence of current e-cig users among women from 2014 to 2019 showed a declining tendency that followed the increasing trend from 2011 to 2014, in contrast to men, whose prevalence first showed an upward trend before becoming nearly constant. Adolescents have seen an increase in the prevalence of e-cigs users throughout, whereas adults are seeing a decline in this demographic (Tehrani et al. 2022). Across the four continents, the America and Asia continents showed an increasing trend in the beginning but then almost a constant trend. On the contrary, the Europe continent's current e-cigs use rates are consistently increasing (Tehrani et al. 2022).

Table 16.1 illustrates the global prevalence of lifetime and current e-cig users with prevalence according to gender, age, and continent, as reported by Tehrani et al. (2022). The global prevalence of lifetime e-cigs was 23%, with 22% and 16% of men and women using cigarettes, respectively. Meanwhile, the global prevalence of current e-cigs was 11%, with men and women being 12% and 8%, respectively.

Table 16.1 Prevalence of lifetime and current e-cig users (based on data reported by Tehrani et al. 2022)

		Lifetime users, %	Current users, %
Global		23	11
	Male	22	12
	Female	16	8
	Lifetime cigarette smoker	–	39
	Current cigarette smoker	–	43
	Adults	19	11
	Adolescents & school students	25	11
	College students	26	14
	Patients	29	10
Continents			
	Europe	26	14
	America	24	10
	Asia	16	11
	Oceania	25	6

By age group, the prevalence of lifetime and current e-cigs among adolescents was 25% and 11%, respectively. Meanwhile, the prevalence of lifetime and current e-cigs among adults was 19% and 11%, respectively.

The prevalence of lifetime e-cigs use varies by region; Europe has the highest rate (26%), followed by Oceania (25%), America (24%), and Asia (16%). As for the prevalence of current e-cigs use, it spans from 6% to 14%, with Europe having the highest rate at 14%, followed by Asia at 11%, America at 10%, and Oceania at 6%. There was an increasing trend from 2011 to 2019 in the lifetime prevalence of e-cigs smoking across the Americas, Asia, Europe, and Oceania. The prevalence of lifetime e-cigs smoking in the region of America, Asia, Europe, and Oceania showed an upward trend. Different trends were seen for the current e-cigs users in the continents of America and Asia, which first showed an upward trend and then presented a steady trend. However, in the European continent, it showed an upward trend between 2011 and 2019.

16.2.1 North America

In North America, particularly in the United States (US) and Canada, e-cigs use has seen significant growth, especially among adolescents and young adults. E-cigs use became particularly popular among North American adolescents, with rates often exceeding rates of conventional cigarette use (Hammond et al. 2019). In the US, the current prevalence has increased rapidly from 1.5% to 20.8% in the years 2011 and 2018, respectively (Cullen et al. 2018). It was also reported by Chapman and Wu (2014) that in the US, young adults aged 20 to 28 were the highest ever e-cig users, followed by adults above 18 and adolescents aged between 11 and 19 years old. In America, the prevalence of lifetime e-cigs among males and females was 9.6% and 7.4%, respectively, and the prevalence of current e-cigs among males and females was reported to be 2.6% and 2.1%, respectively (Zhu et al. 2017). According to research done in the US, 9% and 30% of college students were lifetime and current e-cig users, respectively (Sutfin et al. 2013). In Canada, the percentage of adolescents who have ever used an e-cigs is 37%, and those who are current users were 14.6% in 2018 (Hammond et al. 2019).

16.2.2 Europe

Across Europe, the prevalence of e-cigs use varies considerably among countries due to differences in tobacco control policies and cultural attitudes toward smoking. While some countries have embraced e-cigs as a harm reduction tool and have relatively high prevalence rates, others have adopted more cautious approaches, citing concerns about the potential normalization of smoking behaviors and the lack of long-term safety data. The prevalence of lifetime e-cig use among college students was found to be 34.4%, 42.7%, 45%, 33.4%, and 42.7% in a study conducted in five European countries: Slovakia, Belarus, Poland, Russia, and Lithuania. The prevalence of current e-cig use was found to be 2.3%, 2.7%, 2.8%, 4%, and 3.5% in these same countries (Brozek et al. 2019).

Among adolescents in seven Western European cities, a study conducted in 2016–2017 found that 35.5% were lifetime e-cig users (from 23% in Germany to nearly 50% in Italy and Belgium). Of those, 6.6% reported using e-cigs currently (i.e., monthly or more frequently), compared to 17.2% who currently use conventional cigarettes (Kinnunen et al. 2021). In 2017–2018, 43.7% of university students in multiple countries in Central and Eastern Europe reported lifetime use of e-cigs; however, only 2.9% were current e-cig users, which is much less than the 12.3% of current conventional cigarette users (Brozek et al. 2019). According to Fiippidis et al. (2017), during the course of 4 years (2012–2016), the prevalence of lifetime e-cig users in 27 European nations increased from 7.2% to 11.6%. The percentage of people who have ever used e-cigs ranged from 17% to 62% across European nations (Kwon et al. 2018; Geidne et al. 2017; Goniewicz and Zielinska-Danch 2012). Many European countries (such as Iceland, Monaco, Lithuania, Poland, Ireland, Germany, Czechia, Hungary, France, and Norway) have reported a higher prevalence of e-cigs use than conventional cigarettes use in which the e-cig users were smokers previously or began smoking at an early age (Cerrai et al. 2022). Young people in Europe, particularly the UK, are using e-cigs less often now that the EU Tobacco Products Directive restrictions have gone into effect (Moore et al. 2020; Vardavas 2022). Nonetheless, in contrast to other nations, England continues to see a rise in both lifetime and current e-cig usage between 2014 and 2016.

16.2.3 Asia

In Asia, where tobacco smoking rates have traditionally been high, the emergence of e-cigs presents both opportunities and challenges. Countries like Japan and South Korea have seen a rapid uptake of e-cigs, particularly among smokers looking to quit or reduce their tobacco consumption. However, regulatory responses in Asia have been mixed, with some countries imposing strict bans on e-cigs sales and advertising, while others have adopted more lenient regulations. Concerns over the rising use of e-cigs, as in other nations, are due to the absence of regulations governing the sale of e-cigs and the easy availability of tobacco in Chinese retail establishments (Xiao et al. 2019).

In Korea, the prevalence of lifetime e-cig use among young adults showed an increase of 40% over 9 years, that is from 0.5% in 2008 to 42% in 2017 (Park et al. 2017; Cho et al. 2011). Almost similar rates of the increment (about 4% per year) in the prevalence of current e-cig use were seen in young adults in Taiwan (2.5% in 2014 to 6.4% in 2017) (Gao et al. 2021).

Various studies involving college students found varying prevalences of e-cigs usage. According to reports, the percentage of Saudi Arabian health sciences students who have used e-cigs during their lifetime is 27.7% (Qanash et al. 2019), whereas the percentage among Malaysian college students is 20.4% (Puteh et al. 2018). According to Iqbal et al. (2018), 13.9% of Pakistani medical students reported using e-cigs, while Brozek et al. (2019) found that the prevalence of current e-cig usage among medical students and non-medical students was 4.4% and 12.4%, respectively.

According to Malaysian research, the proportion of lifetime conventional cigarette users and current conventional cigarette smokers who also use e-cigs was 4.3% and 8%, respectively (Ho et al. 2019). The current rates of usage among South Korean adolescents, which are considerably lower at 3% and 1% among young men and women, respectively, have remained consistent between 2016 and 2018 (Lee and Paek 2020). It has been determined that one element influencing regional variations in the adoption of e-cigs is the stringency of laws pertaining to mass marketing and sales to adolescents (Cho et al. 2018; Hammond et al. 2020).

16.2.4 Oceania (Australia)

A comprehensive report on the prevalence of e-cigs use in Australia showed that the prevalence of lifetime e-cig users was the highest among the young adults aged between 18 and 24 years (48%) and the older age group with a lesser percentage (Greenhalgh et al. 2024). According to Greenhalgh et al. (2024), between 2019 and 2022–2023, the percentage of adults who had ever used e-cigs rose dramatically across all age categories with the exception of the oldest smokers and non-smokers. The percentage of people who had never smoked before using an e-cig increased significantly between 2019 and 2022–2023, especially among young adults. Most Australians (57.6%) who used e-cigs for the first time between the ages of 18 and 24 in 2022–2023 had never smoked previously.

The percentage of current e-cig users rose dramatically from 5.3% to 20.6% among users aged 18 to 24 between 2019 and 2022–2023. Among those below 30, the percentage of current e-cig users was nearly 60% in 2022–2023. Age-wise comparison of the daily e-cig users for 2019 and 2022–2023 revealed that the largest group of users in both years was 18–24 years old, with a significant rise from 9% in 2019 to 10% in 2022–2023. Other age groups, however, displayed declining trends during the same time span (Greenhalgh et al. 2024).

16.3 Age Groups and Gender

The prevalence of e-cigs use among adolescents varies widely between countries. As reported in the Global Youth Tobacco Survey (GYTS) 2012–2019, across 68 countries, about 9.2% of current e-cig users are male, higher than female (11.7% and 6.6%, respectively) and aged 15 to 16 years than aged 12 to 14 years (11.2% and 8%, respectively). Across WHO regions, the current e-cig use among 12–16 years ranges from 3.3% in Southeast Asia to 10.8% in the Western Pacific (7.8% in the Americas, 9.3% in Europe, 9.9% in Africa, and 10.6% in Eastern Mediterranean) (Sun et al. 2022). The prevalence was higher in youths who smoked cigarettes, in those who smoked other tobacco products, and in low-income and high-income countries (Sun et al. 2022). The use of e-cigs was highly associated with parental smoking and second-hand smoking exposure (Yoong et al. 2021).

16.4 Factors of Changes in Prevalence and Trends of E-cigs Use

Men's positive expectations of e-cigs, such as their good taste, improved social performance, and increased energy, are contributing to the current prevalence trend in comparison to women. In contrast, women only have one positive expectation of using e-cigs: weight loss from smoking them (Pineiro et al. 2016). According to a study, pregnant women are less likely to use e-cigs than young women because of the harmful impacts of these products (Ashford et al. 2016). The inability of e-cigs to aid in fitness and weight reduction may be the cause of the decline in e-cigs smoking among women. Furthermore, women tend to be more health-conscious than men, therefore the decrease in use might be attributable to a higher knowledge of the risks associated with smoking e-cigs (Tehrani et al. 2022). It has also been reported that medical students may have a low prevalence since they were well aware of the risks associated with smoking e-cigs during their education (Brozek et al. 2019).

Generally speaking, e-cig usage is rising in all nations and areas, maybe as a result of insufficient e-cig taxes. Furthermore, the regulation might not have been passed yet considering the rise in e-cigs in recent years. The passage of regulations to decrease advertising, prohibit sales, raise taxes, and launch awareness efforts in the industry may be the cause of the variations in the prevalence of e-cigs among nations (Tehrani et al. 2022).

16.5 Factors of E-cigs Use

There were many reasons associated with the increasing prevalence of e-cigs use as reported in various studies. Reasons for e-cigs uptake include taste and entertainment, experimentation, to replace conventional cigarettes, curiosity, to quit smoking, and as a complement to tobacco use, and peer influence (Evans-Polce et al. 2018; Eichler et al. 2016; Kinouani et al. 2020). The availability of a wide range of flavors and sleek designs, coupled with aggressive marketing tactics, has contributed to the popularity of e-cigs.

Youth-targeted marketing on e-cigs or vaping, social media e-cigs promotion, and peer pressure are all potential contributing factors to the rising use of e-cigs among adolescents (Becker and Rice 2022). Adolescents' intention to use e-cigs has always been linked to advertising (Amin et al. 2020). Adolescents are left open to advertising since certain nations have been sluggish to restrict the marketing of e-cigs (Chen et al. 2019). Additionally, it has been discovered that youngsters are more likely to use e-cigs if they receive larger monthly allowances from their parents (Zhu ct al. 2020; Park et al. 2017; Czoli et al. 2015; Cui et al. 2019).

A systematic review discovered that social media platforms had been extensively used for public discussion of e-cigs in a way that yields generally favorable or neutral e-cig depictions, in addition to traditional marketing, which frequently uses youth-directed strategies (Collins et al. 2019). Adolescent e-cigs use is also impacted by the attitudes and behaviors of others in their immediate vicinity regarding conventional cigarette smoking. Adolescent e-cigs usage is influenced by parents' views toward e-cigs use, and youth e-cigs use is at risk due to parents' conventional cigarette consumption (Trucco et al. 2021). It might be difficult for certain family systems to put reasonable restrictions on adolescents' access to and usage of e-cigs. According to Watkins and Ohannessian (2020), teenagers with divorced parents had a 51% higher likelihood of using e-cigs than adolescents with married parents.

E-cigs were used by both current and former users of conventional cigarettes because they were cheaper than conventional cigarettes, tasted better due to flavor, helped people quit smoking, could be used in places where smoking is prohibited or were used due to their belief in e-cigs for harm reduction (Kinouani et al. 2020). The fact that smokers often use e-cigs to cut back on or quit conventional cigarettes but eventually develop a continuous e-cigs use habit might be one of the causes of the rise in e-cigs consumption among smokers of conventional cigarettes in the European continent (Tehrani et al. 2022). According to Ho et al. (2019), current conventional cigarette smokers may use e-cigs out of curiosity to try them, help them quit, or cut back on their conventional cigarette usage.

Legal limitations by themselves do not completely explain cross-national diversity in e-cigs usage, even while policy contexts appear to explain part of it (Cho et al. 2018; Hammond et al. 2020). For example, the Tobacco Control Scale indicates that Germany has one of the least restrictive settings for nicotine products (Joossens et al. 2020). Nonetheless, Germany has lower rates of youth e-cigs usage than two of the most restricted European countries, Finland and Ireland (Joossens et al. 2020). These results highlight the need for more research on other variables causing cross-country variations (Kinnunen et al. 2021).

The US high school student population has been using e-cigs at worrisome rates in recent years, according to statistics from the Centers for Disease Control and Prevention (CDC), which has prompted public health initiatives and legislative steps. This suggests that families ought to talk to their children and adolescents more about smoking e-cigs. Health experts might also be a key source of information on e-cig usage, particularly for teenagers and schoolchildren. Furthermore, e-cig usage among teenagers and schoolchildren may be decreased by outlawing the sale of e-cigs to anybody less than 18 years old.

16.6 Regulatory Environment

Regulatory policies and frameworks governing e-cigs vary widely across countries and regions, impacting accessibility, marketing practices, and public perceptions. Understanding the regulatory landscape is essential for contextualizing patterns of e-cigs use. Evidence has shown that e-cigs regulatory policy influenced e-cigs use, where e-cigs use increased significantly in countries with fewer restrictions on e-cigs sales and marketing, for instance, the US and England, as compared to countries that have more restrictive e-cigs regulations, such as Australia, Uruguay, and Brazil (Gravely et al. 2019).

Even though many places forbid teenagers from purchasing e-cigs (Kennedy et al. 2017), minors nonetheless manage to get their hands on the devices. According to Pepper et al. (2019), the majority of American teenagers who use e-cigs possess their own device, although they also commonly share it with others. According to one survey, many young people in the US bought e-cigs via physical stores or online; informal transactions, purchases made on behalf of others, and receipts as presents were less prevalent (Pepper et al. 2019). According to different research, American teenagers most often acquired e-liquids through social media (McKeganey et al. 2019). Most American teenagers reported reducing their e-cigs usage because they were having trouble finding e-cig goods, according to a recent survey conducted during the COVID-19 epidemic after the US tightened restrictions around e-cigs marketing. However, young people who kept smoking e-cigs started purchasing goods online, including accepting delivery without having their age verified (Gaiha et al. 2020).

16.7 Characteristics of E-cigs User

16.7.1 Smoking Status

E-cigs use patterns often differ based on smoking status. Current conventional smokers may use e-cigs as smoking cessation aids or harm reduction tools, while never-smokers may experiment with e-cigs out of curiosity or social influences. Between current and former smokers, current smokers among adults have mainly used e-cigs compared to former smokers. According to Ho et al. (2019), the primary motivations for beginning to use e-cigs among current smokers were the desire to give them a try, followed by the desire to reduce their smoking and stop using conventional cigarettes. Among current smokers, it was found that e-cigs use was significantly higher among younger smokers (18–44 years old), urban residents, those with secondary education or above, students, those who were single, divorced, or widowed, and those who began smoking tobacco at an age of less than 20 (Ho et al. 2019).

16.7.2 Age

E-cigs use spans across age groups, although prevalence rates often vary. While young adults and adolescents are commonly associated with e-cigs use, older adults may also use these devices, particularly as smoking cessation aids. According to Wang et al. (2020), of the teenagers who now use e-cigs in the US, 38.9% reported using them at least 20 days a month, and 83% reported using flavored e-cigs. E-cigs are a vehicle for drug use, which may or may not contain nicotine, despite the fact that nicotine is most commonly associated with them. In a nationally representative 2017 poll of US teenagers, 25% of current e-cig users reported that they only vaped flavoring; the remaining respondents indicated that they vaped nicotine (45%), marijuana (13%), or both substances (17%) (Dai and Siahpush 2020). Adolescents in Europe who had ever used e-cigs reported smoking non-nicotine liquids at a rate of 43%, nicotine at 37%, and not knowing what they were vaping at 19.6% (Kinnunen et al. 2021). A lot of young people can misjudge or be unaware of what they are vaping. Teenagers in the US frequently believe that nicotine is a tobacco derivative, and those who claim to merely be flavoring vapers have been shown to know very little about using nicotine (Pepper et al. 2018). According to Boykan et al. (2019), 40% of young people who reported using nicotine-free products exhibited nicotine traces in their urine.

16.7.3 Flavor Preferences

For those who use e-cigs, having a large variety of flavors is a big lure, especially for younger audiences. Variety in flavor is essential for influencing consuming habits and preferences. Flavors encourage young people to start using e-cigs and keep them using them. When provided by a friend, American teenagers expressed a greater desire to try menthol, fruit, or candy-flavored e-cigs than tobacco (Pepper et al. 2016). According to studies conducted by Pepper et al. (2016) and Ford et al. (2016), teenagers in the US and the UK have said that they think fruit flavors in e-cigs are less dangerous than tobacco flavors.

In comparison to peers whose initial exposure to e-cigs was unflavored, adolescents who experience flavored e-cigs progress to current and more frequent use more quickly (Audrain-McGovern et al. 2019). Additionally, e-cigs use by adolescents with nonconventional flavors (such as candy and fruit) was linked to the vaping continuation and puffs per use 6 months later compared to conventional flavors (such as menthol, tobacco, and flavorless) (Leventhal et al. 2019).

16.8 Conclusion

The characteristics of e-cig users globally are multifaceted, reflecting a complex interplay of demographics, behaviors, motivations, and environmental factors. As e-cigs use continues to evolve, ongoing research is needed to track trends, identify emerging patterns, and inform targeted interventions aimed at promoting public health and reducing potential harms associated with these devices.

References

Amin S, Dunn AG, Laranjo L (2020) Social influence in the uptake and use of electronic Cigarettes: a systematic review. Am J Prev Med 58:129–141. https://doi.org/10.1016/j.amepre.2019.08.023

Ashford K, Wiggins A, Butler K, Ickes M, Rayens MK, Hahn E (2016) E-cigarette use and perceived harm among women of childbearing age who reported tobacco use during the past year. Nurs Res 65(5):408–414

Audrain-McGovern J, Rodriguez D, Pianin S, Alexander E (2019) Initial e-cigarette flavoring and nicotine exposure and e-cigarette uptake among adolescents. Drug Alcohol Depend 202:149–155. https://doi.org/10.1016/j.drugalcdep.2019.04.037

Becker TD, Rice TR (2022) Youth vaping: a review and update on global epidemiology, physical and behavioral health risks, and clinical considerations. Eur J Pediatr 181:453–462

Bhatta DN, Glantz SA (2020) Association of e-cigarette use with respiratory disease among adults: a longitudinal analysis. Am J Prev Med 58(2):182–190

Boykan R, Messina CR, Chateau G et al (2019) Self-reported use of tobacco, E-cigarettes, and marijuana versus urinary biomarkers. Pediatrics 143:e20183531. https://doi.org/10.1542/peds.2018-3531

Brożek GM, Jankowski M, Lawson JA, Shpakou A, Poznański M, Zielonka TM et al (2019) The prevalence of cigarette and E-cigarette smoking among students in Central and Eastern Europe—results of the YUPESS Study. Int J Environ Res Public Health 16(13):2297

Cerrai S, Benedetti E, Colasante E, Scalese M, Gorini G, Gallus S, Molinaro S (2022) E-cigarette use and conventional cigarette smoking among European students: findings from the 2019 ESPAD survey. Addiction 117:2918

Chaffee BW, Watkins SL, Glantz SA (2018) Electronic cigarette use and progression from experimentation to established smoking. Pediatrics 141:e20173594. https://doi.org/10.1542/peds.2017-3594

Chapman SLC, Wu L (2014) E-cigarette prevalence and correlates of use among adolescents versus adults: a review and comparison. J Psychiatr Res 54:43–54

Chen PC, Chang LC, Hsu C, Lee YC (2019) Dual use of e-cigarettes and traditional cigarettes among adolescents in Taiwan, 2014–2016. Nicotine Tob Res 21:48–54. https://doi.org/10.1093/ntr/nty003

Cho JH, Shin E, Moon S (2011) Electronic-cigarette smoking experience among adolescents. J Adolesc Health 49:542–546

Cho H-J, Dutra LM, Glantz SA (2018) Differences in adolescent e-cigarette and cigarette prevalence in two policy environments: South Korea and the United States. Nicotine Tob Res 20:949–953. https://doi.org/10.1093/ntr/ntx198

Collins L, Glasser AM, Abudayyeh H et al (2019) E-cigarette marketing and communication: how e-cigarette companies market e-cigarettes and the public engages with e-cigarette information. Nicotine Tob Res 21:14–24

Cui Y, Forget EL, Zhu Y, Torabi M, Oguzoglu U (2019) The effects of cigarette price and the amount of pocket money on youth smoking initiation and intensity in Canada. Can J Public Health 110(1):93–102

Cullen KA, Ambrose BK, Gentzke AS, Apelberg BJ, Jamal A, King BA (2018) Notes from the field: use of electronic cigarettes and any tobacco product among middle and high school students—United States, 2011–2018. Morb Mortal Wkly Rep 67(45):1276

Czoli CD, Hammond D, Reid JL, Cole AG, Leatherdale ST (2015) Use of conventional and alternative tobacco and nicotine products among a sample of Canadian youth. J Adolesc Health 57(1):123–125

Dai H, Siahpush M (2020) Use of e-cigarettes for nicotine, marijuana, and just flavoring among U.S. youth. Am J Prev Med 58:244–249. https://doi.org/10.1016/j.amepre.2019.09.006

Driezen T, Amer Nordin AS, Mohd Hairi F, Yee N, Ahmad Tajuddin NA, Hasan SI, Danaee M, Kamaludin IS, Kaai SC, Yan M, Grey M, Quah ACK, Thompson ME, Fong GT (2022) E-Cigarette prevalence among Malaysian adults and types and flavours of e-cigarette products used by cigarette smokers who vape: findings from the 2020 ITC Malaysia survey. Tob Induc Dis 20(March):32

Eichler M, Blettner M, Singer S (2016) The use of e-cigarettes. Dtsch Aerzteblatt Online 113:847–854. https://doi.org/10.3238/arztebl.2016.0847

Epstein M, Bailey JA, Kosterman R et al (2020) E-cigarette use is associated with subsequent cigarette use among young adult non-smokers, over and above a range of antecedent risk factors: a propensity score analysis. Addiction add.15317. 10.1111/add.15317

Evans-Polce RJ, Patrick ME, Lanza ST et al (2018) Reasons for vaping among U.S. 12th graders. J Adolesc Health 62:457–462. https://doi.org/10.1016/j.jadohealth.2017.10.009

Fadus MC, Smith TT, Squeglia LM (2019) The rise of e-cigarettes, pod mod devices, and JUUL among youth: factors influencing use, health implications, and downstream effects. Drug Alcohol Depend 201:85–93

Filippidis FT, Laverty AA, Gerovasili V, Vardavas CI (2017) Two-year trends and predictors of e-cigarette use in 27 European Union member states. Tob Control 26(1):98–104. https://doi.org/10.1136/tobaccocontrol-2015-052771. Epub 2016 May 24. PMID: 27220621; PMCID: PMC5256312

Ford A, MacKintosh AM, Bauld L et al (2016) Adolescents' responses to the promotion and flavouring of e-cigarettes. Int J Public Health 61:215–224. https://doi.org/10.1007/s00038-015-0769-5

Gaiha SM, Lempert LK, Halpern-Felsher B (2020) Underage youth and young adult e-cigarette use and access before and during the coronavirus disease 2019 pandemic. JAMA Netw Open 3:e2027572. https://doi.org/10.1001/jamanetworkopen.2020.27572

Gao W, Sanna M, Chuluunbaatar E, Tsai M, Levy DT, Wen CP (2021) Are e-cigarettes reviving the popularity of conventional smoking among Taiwanese male adolescents? A time-trend population-based analysis for 2004–2017. Tob. Control 30,132–136

Geidne S, Beckman L, Edvardsson I, Hulldin J (2017) Prevalence and risk factors of electronic cigarette use among adolescents: data from four Swedish municipalities. Nordic Stud Alcohol Drugs 33:225–240

Goldenson NI, Leventhal AM, Stone MD et al (2017) Associations of electronic cigarette nicotine concentration with subsequent cigarette smoking and vaping levels in adolescents. JAMA Pediatr 171:1192. https://doi.org/10.1001/jamapediatrics.2017.3209

Gomajee R, El-Khoury F, Goldberg M, Zins M, Lemogne C, Wiernik E, Lequy-Flahault E, Romanello L, Kousignian I, Melchior M (2019) Association between electronic cigarette use and smoking reduction in France. JAMA Intern Med 179(9):1193–1200. https://doi.org/10.1001/jamainternmed.2019.1483

Goniewicz ML, Zielinska-Danch W (2012) Electronic cigarette use among teenagers and young adults in Poland. Pediatrics 130:e879–e885

Gravely S, Driezen P, Ouimet J et al (2019) Prevalence of awareness, ever-use and current use of nicotine vaping products (NVPs) among adult current smokers and exsmokers in 14 countries with differing regulations on sales and marketing on NVPs: cross-sectional findings from the ITC Project. Addiction 114:1060–1073. https://doi.org/10.1111/add.14558

Greenhalgh EM, Jenkins S, Bain E, Scollo MM (2024) 18.3 Prevalence of e-cigarette use. In Greenhalgh, EM, Scollo, MM andWinstanley, MH [editors]. Tobacco in Australia: Facts and issues. Melbourne: Cancer Council Victoria; Available from: https://www.tobaccoinaustralia.org.au/chapter-18-e-cigarettes/18-3-extent

Hammond D, Reid JL, Rynard VL et al (2019) Prevalence of vaping and smoking among adolescents in Canada, England, and the United States: repeat national cross sectional surveys. BMJ 365:l2219. 10.1136/BMJ.l2219

Hammond D, Reid JL, Burkhalter R, Rynard VL (2020) E-cigarette marketing regulations and youth vaping: crosssectional surveys, 2017–2019. Pediatrics 146:e20194020. https://doi.org/10.1542/peds.2019-4020

Ho B, Haniki NM, Jamalludin A, Samsul D, Mira K, Syafinaz AN et al (2019) Prevalence and characteristics of e-cigarette users among Malaysian current and ex-smokers. Malays Fam Physician 4(2):10

Iqbal N, Khan ZA, Anwar SMH, Irfan O, Irfan B, Mushtaq A et al (2018) Electronic cigarettes use and perception amongst medical students: a cross sectional survey from Sindh, Pakistan. BMC Res Notes 11(1):188

Jane Ling MY, Abdul Halim AFN, Ahmad D, Ahmad N, Safian N, Mohammed Nawi A (2023) Prevalence and associated factors of e-cigarette use among adolescents in southeast asia: a systematic review. Int J Environ Res Public Health 20:3883. https://doi.org/10.3390/ijerph20053883

Joossens L, Feliu A, Fernandez E (2020) The tobacco control scale 2019 in Europe. Brussels: Association of European Cancer Leagues, Catalan Institute of Oncology; Available from: http://www.tobaccocontrolscale.org/TCS2019.Pdf
Kennedy RD, Awopegba A, De Leon E, Cohen JE (2017) Global approaches to regulating electronic cigarettes. Tob Control 26:440–445. https://doi.org/10.1136/tobaccocontrol-2016-053179
Kinnunen JM, Rimpel AH, Lindfors PL et al (2021) Electronic cigarette use among 14- to 17-year-olds in Europe. Eur J Public Health 31:402–408. https://doi.org/10.1093/eurpub/ckaa145
Kinouani S, Leflot C, Vanderkam P et al (2020) Motivations for using electronic cigarettes in young adults: a systematic review. Subst Abus 41:315–322
Kwon E, Seo D, Lin H, Chen Z (2018) Predictors of youth e-cigarette use susceptibility in a US nationally representative sample. Addict Behav 2018(82):79–85
Lee ES, Paek YJ (2020) Prevalence and correlates of the dual use of conventional and electronic cigarettes among Korean adolescents: 2016–2018 Korean youth risk behavior survey. J Korean Soc Res Nicotine Tob 11:64–74. https://doi.org/10.25055/JKSRNT.2020.11.2.64
Leventhal AM, Goldenson NI, Cho J et al (2019) Flavored e-cigarette use and progression of vaping in adolescents. Pediatrics 144:e20190789. https://doi.org/10.1542/peds.2019-0789
McKeganey N, Russell C, Katsampouris E, Haseen F (2019) Sources of youth access to JUUL vaping products in the United States. Addict Behav Rep 10:100232. https://doi.org/10.1016/j.abrep.2019.100232
Moore G, Brown R, Page N, Hallingberg B, Marynard O, McKell J, Gray L, Blackwell A, Lowthian E, Munafò M et al (2020) Young people's use of e-cigarettes in Wales, England and Scotland before and after introduction of EU Tobacco Products Directive regulations: a mixed-method natural experimental evaluation. Int J Drug Policy 85:102795
Park S, Lee H, Min S (2017) Factors associated with electronic cigarette use among current cigarette-smoking adolescents in the Republic of Korea. Addict Behav 69:22–26
Pepper JK, Ribisl KM, Brewer NT (2016) Adolescents' interest in trying flavoured e-cigarettes. Tob Control 25:ii62–ii66. https://doi.org/10.1136/tobaccocontrol-2016-053174
Pepper JK, Farrelly MC, Watson KA (2018) Adolescents' understanding and use of nicotine in e-cigarettes. Addict Behav 82:109–113. https://doi.org/10.1016/j.addbeh.2018.02.015
Pepper JK, Coats EM, Nonnemaker JM, Loomis BR (2019) How do adolescents get their e-cigarettes and other electronic vaping devices? Am J Heal Promot 33:420–429. https://doi.org/10.1177/0890117118790366
Pineiro B, Correa JB, Simmons VN, Harrell PT, Menzie NS, Unrod M et al (2016) Gender differences in use and expectancies of e-cigarettes: online survey results. Addict Behav 2016(52):91–97
Pisinger C, Døssing M (2014) A systematic review of health effects of electronic cigarettes. Prev Med 69:248–260
Puteh SEW, Manap RA, Hassan TM, Ahmad IS, Idris IB, Sham FM et al (2018) The use of e-cigarettes among university students in Malaysia. Tob Induc Dis 16:57
Qanash S, Alemam S, Mahdi E, Softah J, Touman AA, Alsulami A (2019) Electronic cigarette among health science students in Saudi Arabia. Ann Thorac Med 14(1):56
Soneji S, Barrington-Trimis JL, Wills TA, Leventhal AM, Unger JB, Gibson LA et al (2017) Association between initial use of e-cigarettes and subsequent cigarette smoking among adolescents and young adults: a systematic review and meta-analysis. JAMA Pediatr 171(8):788–797
Sun J, Xi B, Ma C, Zhao M, Bovet P (2022) Prevalence of E-cigarette use and its associated factors among youths aged 12 to 16 years in 68 countries and territories: global youth tobacco survey, 20122019. Am J Public Health 112(4):650–661. Available from: https://www.ncbi.nlm.nih.gov/pubmed/35319939
Sutfin EL, McCoy TP, Morrell HE, Hoeppner BB, Wolfson M (2013) Electronic cigarette use by college students. Drug Alcohol Depend 31(3):214–221
Tehrani H, Rajabi A, Ghelichi-Ghojogh M, Nejatian M, Jafari A (2022) The prevalence of electronic cigarettes globally: a systematic review and meta-analysis. Arch Public Health 80:240

Trucco EM, Cristello JV, Sutherland MT (2021) Do parents still matter? The impact of parents and peers on adolescent electronic cigarette use. J Adolesc Heal 68:780–786. https://doi.org/10.1016/j.jadohealth.2020.12.002

Vardavas C (2022) European tobacco products directive (TPD): current impact and future steps. Tob Control 31:198–201

Wagener TL, Floyd EL, Stepanov I, Driskill LM, Frank SG, Meier E et al (2017) Have combustible cigarettes met their match? The nicotine delivery profiles and harmful constituent exposures of second-generation and third-generation electronic cigarette users. Tob Control 26(e1):e23–e28

Wang JW, Cao SS, Hu RY (2018) Smoking by family members and friends and electronic-cigarette use in adolescence: A systematic review and meta-analysis. Tob Induc Dis. 16:05. https://doi.org/10.18332/tid/84864. PMID: 31516405; PMCID: PMC6659504

Wang TW, Neff LJ, Park-Lee E et al (2020) E-cigarette use among middle and high school students — United States, 2020. MMWR Morb Mortal Wkly Rep 69:1310–1312. https://doi.org/10.15585/mmwr.mm6937e1

Watkins NK, Ohannessian CMC (2020) Timing of vape use among adolescents: differences by family structure. Drug Alcohol Depend 209:107953. https://doi.org/10.1016/j.drugalcdep.2020.107953

Xiao L, Parascandola M, Wang C, Jiang Y (2019) Perception and current use of e-cigarettes among youth in China. Nicotine Tob Res 21(10):1401–7.

Yoong SL, Hall A, Leonard A et al (2021) Prevalenceof electronic nicotine delivery systems and electronic non-nicotine delivery systems in children and adolescents: a systematic review and meta-analysis. Lancet Public Health 6(9):e661–e673. https://doi.org/10.1016/S2468-2667(21)00106-7

Zhu S-H, Zhuang Y-L, Wong S, Cummins SE, Tedeschi GJ (2017) E-cigarette use and associated changes in population smoking cessation: evidence from US current population surveys. BMJ 358:j3262

Zhu J, Li J, Xu G, Yu J, Wang Q, He Y (2020) School-type differences in e-cigarette use and its correlates among Chinese adolescents. Tob Induc Dis 8:17

Smoking Cessation of E-cigarettes: A Current Situation Use and Overview of Cost-Effectiveness Analysis

17

Pawich Paktipat, Titawadee Pradubkham, and Wiraphol Phimarn

Abstract

Electronic cigarettes (e-cigs), also known as electronic nicotine delivery systems, have gained popularity as an alternative to conventional tobacco smoking. Despite being initially promoted as safer alternatives, mounting evidence suggests that there are potential health risks associated with their use. This chapter discusses the diverse range of health concerns linked to e-cig use, including respiratory diseases, cardiovascular morbidity, and increased cancer risk. Additionally, it explores the current landscape of e-cig use, highlighting trends among different age groups and regions. While some studies have suggested potential benefits for smoking cessation, uncertainties persist regarding their long-term safety and efficacy compared to licensed pharmaceutical interventions. Economic evaluations indicate the potential cost-effectiveness of e-cigs as a smoking cessation aid; however, regulatory frameworks and safety standards remain areas of concern. Further research is warranted to better understand the comparative effectiveness and safety of e-cigs, with an emphasis on long-term outcomes and the need for robust regulatory oversight to ensure their safe use as smoking cessation tools.

Keywords

E-cigarettes · Aerosolize e-liquid · Tobacco · Economic analysis · Smoking cessation

P. Paktipat · W. Phimarn (✉)
Social Pharmacy Research Unit, Faculty of Pharmacy, Mahasarakham University, Kantarawichai, Mahasarakham Province, Thailand
e-mail: wiraphol.p@msu.ac.th

T. Pradubkham
Health Services and Pharmacy Practice Research and Innovation, Faculty of Pharmacy, Mahasarakham University, Kantarawichai, Mahasarakham Province, Thailand

Y.-H. Leong, A. S. Mohd Samin (eds.), *E-Cigarettes*,
https://doi.org/10.1007/978-981-96-5087-3_17

17.1 Introduction

Electronic cigarettes (e-cigs), also known as electronic nicotine delivery systems (ENDS), constitute a diverse range of battery-powered devices engineered to simulate smoking by generating an inhalable aerosol (US Department of Health and Human Services (HHS) 2021). This aerosol, commonly called "vapor," is produced by heating a liquid solution containing nicotine, flavorings, and various chemicals. The primary objective of this technological innovation is to provide users with a smoking experience akin to conventional cigarettes while potentially mitigating exposure to harmful substances associated with combustible tobacco products. E-cigs operate on a sophisticated operational mechanism that distinguishes them from conventional tobacco products. The core process involves activating a battery-powered heating element within the device, which converts the liquid solution into vapor (US Department of Health and Human Services (HHS) 2021). Termed "vaping," this process eliminates the combustion of tobacco, enabling users to inhale vapor instead of smoke. Theoretically, this differentiation minimizes exposure to toxic byproducts commonly associated with conventional cigarettes.

17.2 E-cigs Induced Diseases

E-cigs, battery-powered devices that heat and aerosolize e-liquids, constitute the most prevalent form of ENDS. Introduced in the United States (US) market in 2007, e-cigs were initially considered less harmful alternatives to combustible tobacco smoking (National Center for Chronic Disease Prevention and Health Promotion (US) 2016). However, e-cigs typically contain nicotine and other toxic substances, including solvent carriers, flavorings, tobacco-specific nitrosamines, volatile organic compounds, phenolic compounds, tobacco alkaloids, aldehydes, free radicals, reactive oxygen species, furans, and metals, posing risks to both users and non-users exposed to second-hand aerosols (Banks et al. 2022; Rebuli et al. 2023). Mounting evidence links ENDS to lung injuries in the US, with Centers for Disease Control and Prevention (CDC) reporting 2807 hospitalized e-cig or vaping-associated lung injury (EVALI) cases and 60 confirmed deaths in one year, observed from August 2019 to February 2020. EVALI cases have been associated with tetrahydrocannabinol-containing e-cigs, however, 14% of cases are linked to nicotine-containing e-liquids without these constituents. This data reflects findings reported during the EVALI outbreak in the US, highlighting the diverse range of products implicated in these lung injuries (Krishnasamy et al. 2020). A recent systematic review by Banks et al. (2022) highlighted the increased risk of addiction, poisoning, inhalation toxicity (including seizures), trauma, and burns associated with ENDS use. Similarly, a review by Banks et al. underscored the risks of poisoning, immediate inhalation toxicity (including seizures), EVALI, injuries, and burns owing to malfunctioning devices (Banks et al. 2023).

Although e-cigs may offer benefits to smokers attempting to quit smoking, studies suggest that they are not significantly less harmful than tobacco cigarettes

(Bullen et al. 2013; Farsalinos et al. 2014; Harrell et al. 2014). Conversely, research has shown that e-cigs are as toxic as tobacco cigarettes and can cause adverse cardiovascular effects, blood vessel injury, platelet aggregation, microvascular thrombosis, inflammation, and lung damage (Marques et al. 2021; Reinikovaite et al. 2018). Patel et al. conducted a cross-sectional study comparing the effects of e-cig and conventional smoking on stroke prevalence. They found that conventional smokers had a higher prevalence of stroke than e-cig smokers and dual smokers (6.75% versus 1.09% versus 3.72%, $p < 0.0001$), with e-cig smokers experiencing stroke at an earlier age (Patel et al. 2022). Additionally, a single-blind crossover study by Nocella et al. examined the impact of e-cigs and conventional cigarettes on platelet function in healthy adult smokers and non-smokers. Both cigarette types had short-term effects on platelet activation, with e-cigs showing a less pronounced effect in non-smokers (Nocella et al. 2018). Hingham et al. investigated the effects of e-cig exposure on the inflammatory response of human neutrophils and revealed a pro-inflammatory response similar to that observed with platelet activation (Higham et al. 2016). Several human studies have reported increased oxidative stress, endothelial damage, impaired endothelial function, and vascular changes associated with e-cig use (Antoniewicz et al. 2016; Carnevale et al. 2016; Franzen et al. 2018; Vlachopoulos et al. 2016). Antoniewicz et al. demonstrated a significant increase in endothelial progenitor cells following exposure to e-cigs, compared to conventional cigarettes, suggesting potential vascular alterations (Antoniewicz et al. 2016). Similarly, Carnevale et al. performed a single-blind crossover study in healthy subjects and showed that both e-cigs and tobacco cigarettes were associated with increased oxidative stress and endothelial dysfunction (Carnevale et al. 2016). Moreover, Vlachopoulos et al. and Franzen et al. observed adverse effects on blood pressure, aortic stiffness, and arterial function following e-cig use (Franzen et al. 2018; Vlachopoulos et al. 2016).

The studies mentioned above suggest an association between e-cig use and adverse health outcomes including cardiovascular morbidity, respiratory diseases, and increased cancer risk (Abelia et al. 2023; Glynos et al. 2018; Wu et al. 2014). Xie et al.'s prospective cohort study of US adults revealed an elevated risk of respiratory conditions among former and current e-cig users, emphasizing the need for further research on the long-term health effects of e-cigs (Xie et al. 2020).

17.3 Current Use of E-cigs Situation

The rise in e-cigs represents a significant shift from conventional tobacco smoking to a potentially less harmful alternative (Truth Initiative 2021). E-cigs are appealing because they are perceived to offer a pathway to reduced harm, providing users with an alternative nicotine delivery system that minimizes exposure to the harmful constituents of combustible tobacco (US Department of Health and Human Services (HHS) 2021). The increasing prevalence of e-cig use among US adults, especially those aged 18–24, is concerning, as highlighted by Erhabor's study, which showed a significant trend (Erhabor et al. 2023). Ling et al.'s (2023) study in Southeast Asia also underscored e-cig use among adolescents (3.3–11.8%), emphasizing the need for comprehensive interventions (Ling et al. 2023). Azeem et al. (Azeem et al. 2022) shed light on the challenges in South Asia, advocating for revised tobacco control measures and regulations. This upward trend is indicative of a changing landscape, with an estimated 8.1 million US adults identified as current e-cig users in 2018 (Truth Initiative 2021). Importantly, this transition extends beyond adults, as evidenced by data from the 2020 National Youth Tobacco Survey, which indicates a significant increase in e-cig use among high school students (Centers for Disease Control and Prevention (CDC) 2021).

However, the transition to e-cig use is complex and controversial. Research has consistently emphasized that while these devices may expose users to fewer toxic chemicals than conventional cigarettes, they are not risk-free (Johns Hopkins Medicine 2023). While some believe that e-cigs may help reduce nicotine cravings in those attempting to quit smoking, they are not FDA-approved quitting aids, and there is no conclusive scientific evidence of their effectiveness in long-term smoking cessation (US FDA 2023). A study by Johns Hopkins University in 2021 revealed a myriad of chemical ingredients in vapor products, including some that are yet to be identified. Among the identified substances are potentially harmful elements, such as caffeine, previously undiscovered chemicals, pesticides, and flavorings associated with possible toxic effects (Johns Hopkins Medicine 2023). These findings underscore the need for a nuanced understanding of the risks associated with e-cigs and emphasize the importance of ongoing research and vigilance.

17.4 Social, Economic, and Public Health Considerations

The widespread adoption of e-cigs has raised various social, economic, and public health concerns beyond individual health implications. As users transition to this alternative from conventional smoking, questions emerge regarding its impact on daily expenditure and its potential social repercussions. A critical analysis of these factors is essential for a thorough understanding of the societal effects of e-cigs (Alotaybi et al. 2022). E-cigs, which are not licensed as medicines in certain countries, have shown effectiveness in aiding smoking cessation (Chen et al. 2023). This section outlines the basic concept of economic evaluation and current status of e-cig economic evaluation.

Economic evaluation methods can be categorized into four distinct approaches: (1) Cost-minimization analysis (CMA) compares treatment options assuming equivalent therapeutic outcomes. However, CMA has recently been debated as to whether it can be considered a method for economic evaluation because it does not compare the cost to the effectiveness of health interventions. Dakin and Wordsworth (2013) have found biases in the uncertainty estimates of CMA, which may lead to inefficient decisions. (2) Cost-benefit analysis (CBA) assigns monetary values to costs and outcomes, quantifying outcomes using economic techniques such as the human capital approach and willingness-to-pay (WTP) method. (3) Cost-effectiveness analysis (CEA) compares treatment costs with clinical effectiveness, focusing on clinically effective treatments within the same unit. (4) Cost-utility analysis (CUA) quantifies costs and clinical outcomes, primarily utilizing measures such as quality-adjusted life years (QALY) and disability-adjusted life years (Husereau et al. 2022; Phimarn et al. 2023). However, evidence suggests that e-cigs have the potential for smoking cessation. Studies have examined their impact on smoking rates and health outcomes as well as their cost-effectiveness. This section focuses on CEA and CUA, as they use the incremental cost-effectiveness ratio (ICER) framework and are employed by healthcare decision-makers in various countries.

17.5 Cost-Effectiveness Evidence of E-cigs

Smoking causes untimely deaths worldwide (Jha and Peto 2014; Peto 2000). Tobacco-related deaths are anticipated to reach 10 million per year when adolescent smokers reach middle and old age. Ten percent of young women and 50% of young males globally start smoking, and only a small number quit. This may be attributable to population expansion and the fact that certain large demographic groups are succeeding generations who smoked cigarettes disproportionately. One hundred million tobacco-related deaths occurred during the twentieth century, predominantly in wealthy nations. If current smoking trends persist, it is estimated that approximately one billion people could die from tobacco-related causes over the course of the twenty-first century, with the majority of these deaths occurring in low- and middle-income nations (Aniwada et al. 2018).

Approximately 50% of these fatalities will occur before the age of 70 years (Jha and Peto 2014). In late 2015, the UK's Medicines and Healthcare Products Regulatory Agency (MHRA) approved British American Tobacco's "e-Voke" e-cig as a smoking cessation therapy (Medicines and Healthcare Products Regulatory Agency 2015). Battery-operated e-cigs, ENDS, and vapes turn liquid nicotine, flavorings, and additives into an aerosol or "vapor," which is inhaled by users (Medicines and Healthcare products Regulatory Agency 2018).

The use of e-cigs in smoking cessation is debated globally (Newton et al. 2018). One-third of Canadian smokers have attempted vaping to quit smoking (Pound and Coyle 2022). Li et al. (2020) reported that 6% of UK adults used e-cigs in 2017. Of the 2.85 million e-cig users in the UK, the majority smoke as well, representing 5.7% of the population (McNeill et al. 2018). The US has a higher average e-cig usage rate (2%) compared to that of EU countries (Coleman et al.2017; McNeill et al. 2018). Despite the initial mistrust of the international tobacco control community, many scientists and professional bodies believe that e-cigs are safer than tobacco (Hartmann-Boyce et al. 2021). English authorities aim to "ensure that safer alternatives to smoking, including e-cigs, are as widely accessible as possible" (Newton et al. 2018). The US FDA has promoted harm-reduction tobacco laws (Fairchild et al. 2018).

Organization for Economic Co-operation and Development (OECD) countries have moderately permissive e-cig policies. In some cases, regulatory frameworks in countries such as Canada and New Zealand are aligning with or being influenced by the EU's stringent standards for e-cigs. However, it is important to note that these are not direct replacements of national laws but rather an alignment with international best practices. In contrast, Australian, Mexican, and Japanese OECD countries have banned consumer e-cigs altogether (Institute for Global Tobacco Control 2017). China and Russia impose a few restrictions. Due to the potential health risks associated with e-cig use, it is essential to implement appropriate regulatory measures to protect public health (Singh et al. 2020). E-cigs are harmful; hence, regulation is required. They should be distributed only to adults and require safe fluid and equipment refills (Newton et al. 2018). The current guidance for the Stop-Smoking Service from the UK National Institute for Health and Care Excellence states, "Smokers should not be dissuaded from transitioning to e-cigs; doing so would encourage them to continue smoking" (Li et al. 2020). Despite being harmful, e-cigs can positively affect smoking cessation (Hartmann-Boyce et al. 2021; Wang et al. 2021).

Currently, e-cigs are being introduced to help with smoking cessation, and in some countries, economic evaluations have been conducted to assess the value and impact of using e-cigs for this purpose. This section will present an economic evaluation of e-cigs based on previously published evidence.

The previous study by Pound and Coyle (2022) examined the impact of e-cigs on smoking cessation and initiation in Canada, as well as their health effects and costs. Gender-specific Markov models were used to estimate lifetime discounted life years, quality-adjusted life years (QALYs), and smoking-related healthcare expenses for 15–19-year-old males and females. The study looked at three scenarios: (1) e-cig availability, (2) full unavailability, and (3) healthcare professional prescription of e-cigs for smoking cessation with existing approaches. A public-funded healthcare system was analyzed.

The e-cig completely unavailable scenario and the ENDS available through prescription, for males, gained 15 and 17 Quality Adjusted Life Years (QALYs) but saved the cost of Canadian dollars (CAD) $3 and $40,095 respectively, compared to the e-cig widely available scenario. The findings showed a similar trend for females. Thus, situations in which ENDS are unavailable or available through prescription only are dominant over the status quo. These results mean that being without e-cigs or e-cigs available only via prescription can improve population health and reduce healthcare costs. However, the slight differences in the outcomes suggest that while e-cigs may provide some economic and health benefits, their overall impact could be improved. These results can help inform healthcare policy by providing a nuanced understanding of the economic and health implications of integrating e-cigs into smoking cessation programs.

In the England-based study by Li et al. (2020), the National Health Service (NHS) and Personal Social Services (PSS) performed both 12-month and lifetime cost-effectiveness analyses. EQ-5D-5L was utilized to measure health-related quality of life, which contributed to the calculation of QALYs. These QALYs, along with the cost data from a 12-month randomized controlled trial, were used to assess the cost-effectiveness of various therapies, supplementary smoking cessation aids, and healthcare services. The lifetime analysis was model-based, incorporating clinical trial data and secondary sources. The incremental cost-effectiveness ratio (ICER) was calculated to evaluate the cost-effectiveness of the interventions. E-cig starter kits were compared with nicotine replacement treatment (NRT) for up to 3 months, and both therapies were accompanied by standard behavioral assistance. ICER is a key measure used in health economics to assess the cost-effectiveness of a healthcare intervention. It is calculated by dividing the difference in costs between two interventions by the difference in their effectiveness, typically measured in QALYs.

The ICER was £1,100 per QALYs for 12 months after smoking cessation (with an 87% probability of being below £20,000 per QALYs). The Markov model estimates that the lifetime ICER of e-cigs is £65 per QALYs, with an 85% chance of it being below the willingness-to-pay (WTP) threshold of £20,000 per QALYs. Using e-cigs in conjunction with behavioral assistance in English stop-smoking programs is expected to be more cost-effective than using nicotine replacement medications (NRT). Thomas et al. (2021) examined smoking cessation drugs and e-cigs through a systematic review, network meta-analysis, and CEA. The base case analysis found that e-cig is more cost-effective than other therapies for any WTP per QALYs value over £56. No intervention has shown a greater than 40% chance of being the most cost-effective at a WTP per QALYs threshold above £5000; hence, these findings are uncertain. After excluding depression and self-harm safety outcomes, the varenicline standard and plus NRT standard are the most cost-effective medicines. However, optimal intervention remains questionable. For any WTP value greater than £15,665, the varenicline standard is the most economically efficient UK-licensed intervention. The next most cost-effective choice is NRT. Based on value-of-information assessments, a clinical trial comparing e-cigs with active comparators such as varenicline, bupropion, or NRT may be financially prudent.

Based on the existing research, it can be inferred that e-cigs have potential as a means of smoking cessation. However, the scarcity of available evidence gives rise to uncertainties regarding the safety profiles and comparative efficacy of licensed pharmaceutical interventions. Currently, a regulatory framework is in place for e-cigs under the 2016 Tobacco and Related Products Regulations. However, it is important to note that there are no e-cig devices currently licensed as smoking cessation medicines in Great Britain (Great Britain 2016). Medicinal e-cigs must adhere to the standards set for consumer e-cigs, in addition to meeting any supplementary criteria necessary to ensure efficacy, safety, and quality, as outlined by medicine regulations (Medicines and Healthcare products Regulatory Agency 2017). Future research should focus on investigating the efficacy and safety of e-cigs as a smoking cessation tool, with emphasis on long-term outcomes. These studies should employ active therapies as comparators for comprehensive evaluation.

Internationally, the regulatory landscape varies. In the US, for example, the Food and Drug Administration (FDA) has approved specific e-cig devices, such as the Vuse Solo, for use as smoking cessation tools under its modified risk tobacco product (MRTP) authorization (Mir et al. 2022). However, such approvals are limited and subject to strict regulations. In Canada, e-cigs are not licensed as smoking cessation aids but are regulated as consumer products, with a focus on ensuring safety and quality.

The varying international approaches highlight the complexity and challenges in establishing a unified regulatory framework for e-cigs as smoking cessation tools. Future research should focus on investigating the efficacy and safety of e-cigs as a smoking cessation tool, with an emphasis on long-term outcomes. These studies should employ active therapies as comparators for a comprehensive evaluation, which could pave the way for clearer regulatory pathways and potentially lead to the licensing of e-cigs for this use in the UK and other countries.

References

Abelia XA, Lesmana R, Goenawan H, Abdulah R, Barliana MI (2023) Comparison impact of cigarettes and e-cigs as lung cancer risk inductor: a narrative review. Eur Rev Med Pharmacol Sci 27(13):6301–6318. https://doi.org/10.26355/eurrev_202307_32990

Alotaybi M, Alzahrani SS, Algethmi AM, Alamri NS, Natto YS, Hashim ST, Altammar A, Alzubaidi AS, Alzahrani IB, Alghamdi AA (2022) E-cigarettes and vaping: a smoking cessation method or another smoking innovation? Cureus 14(12):e32435. https://doi.org/10.7759/cureus.32435

Aniwada EC, Uleanya ND, Ossai EN, Nwobi EA, Anibueze M (2018) Tobacco use: prevalence, pattern, and predictors, among those aged 15–49 years in Nigeria, a secondary data analysis. Tob Induc Dis 16:07. https://doi.org/10.18332/tid/82926

Antoniewicz L, Bosson JA, Kuhl J, Abdel-Halim SM, Kiessling A, Mobarrez F, Lundbäck M (2016) Electronic cigarettes increase endothelial progenitor cells in the blood of healthy volunteers. Atherosclerosis 255:179–185. https://doi.org/10.1016/j.atherosclerosis.2016.09.064

Azeem N, Sarfraz Z, Sarfraz A, Hange N, Sarfraz M, Cherrez-Ojeda I (2022) Vaping and smokeless tobacco control in South Asia: a policy review. Ann Med Surg (2012):81, 104285. https://doi.org/10.1016/j.amsu.2022.104285

Banks E, Yazidjoglou A, Brown S, Nguyen M, Martin M, Beckwith K, Daluwatta A, Campbell S, Joshy G (2022) Electronic cigarettes and health outcomes: systematic review of global evidence. Report for the Australian Department of Health chrome-extension://efaidnbmnnnibpcajpcglclefindmkaj/ https://www.nhmrc.gov.au/sites/default/files/documents/attachments/ecigarettes/Electronic_cigarettes_and_health_outcomes_systematic_review_of_evidence.pdf

Banks E, Yazidjoglou A, Brown S, Nguyen M, Martin M, Beckwith K, Daluwatta A, Campbell S, Joshy G (2023) Electronic cigarettes and health outcomes: umbrella and systematic review of the global evidence. Med J Aust 218(6):267–275. https://doi.org/10.5694/mja2.51890

Bullen C, Howe C, Laugesen M, McRobbie H, Parag V, Williman J, Walker N (2013) Electronic cigarettes for smoking cessation: a randomised controlled trial. Lancet 382(9905):1629–1637. https://doi.org/10.1016/S0140-6736(13)61842-5

Carnevale R, Sciarretta S, Violi F, Nocella C, Loffredo L, Perri L, Peruzzi M, Marullo AGM, De Falco E, Chimenti I, Valenti V, Biondi-Zoccai G, Frati G (2016) Acute impact of tobacco vs electronic cigarette smoking on oxidative stress and vascular function. Chest 150(3):606–612. https://doi.org/10.1016/j.chest.2016.04.012

Centers for Disease Control and Prevention (CDC) (2021) National Health Interview Survey, 2021. https://www.cdc.gov/nchs/nhis/index.htm

Chen R, Pierce JP, Leas EC, Benmarhnia T, Strong DR, White MM, Stone M, Trinidad DR, McMenamin SB, Messer K (2023) Effectiveness of e-cigarettes as aids for smoking cessation: evidence from the PATH Study cohort, 2017–2019. Tob Control 32(e2):e145–e152. https://doi.org/10.1136/tobaccocontrol-2021-056901

Coleman BN, Rostron B, Johnson SE, Ambrose BK, Pearson J, Stanton CA, Wang B, Delnevo C, Bansal-Travers M, Kimmel HL, Goniewicz ML, Niaura R, Abrams D, Conway KP, Borek N, Compton WM, Hyland A (2017) Electronic cigarette use among US adults in the Population Assessment of Tobacco and Health (PATH) Study, 2013–2014. Tob Control 26(e2):e117–e126. https://doi.org/10.1136/tobaccocontrol-2016-053462

Dakin H, Wordsworth S (2013) Cost-minimisation analysis versus cost-effectiveness analysis, revisited. Health Econ 22(1):22–34. https://doi.org/10.1002/hec.1812

Erhabor J, Boakye E, Obisesan O, Osei AD, Tasdighi E, Mirbolouk H, DeFilippis AP, Stokes AC, Hirsch GA, Benjamin EJ, Rodriguez CJ, El Shahawy O, Robertson RM, Bhatnagar A, Blaha MJ (2023) E-cigarette use among US adults in the 2021 behavioral risk factor surveillance system survey. JAMA Netw Open 6(11):e2340859. https://doi.org/10.1001/jamanetworkopen.2023.40859

Fairchild AL, Lee JS, Bayer R, Curran J (2018) E-cigarettes and the harm-reduction continuum. N Engl J Med 378(3):216–219. https://doi.org/10.1056/NEJMp1711991

Farsalinos KE, Spyrou A, Tsimopoulou K, Stefopoulos C, Romagna G, Voudris V (2014) Nicotine absorption from electronic cigarette use: comparison between first and new-generation devices. Sci Rep 4(1):4133. https://doi.org/10.1038/srep04133

Franzen KF, Willig J, Cayo Talavera S, Meusel M, Sayk F, Reppel M, Dalhoff K, Mortensen K, Droemann D (2018) E-cigarettes and cigarettes worsen peripheral and central hemodynamics as well as arterial stiffness: a randomized, double-blinded pilot study. Vasc Med (London, England) 23(5):419–425. https://doi.org/10.1177/1358863X18779694

Glynos C, Bibli S-I, Katsaounou P, Pavlidou A, Magkou C, Karavana V, Topouzis S, Kalomenidis I, Zakynthinos S, Papapetropoulos A (2018) Comparison of the effects of e-cigarette vapor with cigarette smoke on lung function and inflammation in mice. Am J Phys Lung Cell Mol Phys 315(5):L662–L672. https://doi.org/10.1152/ajplung.00389.2017

Great Britain (2016) The tobacco and related products regulations 2016. https://www.legislation.gov.uk/uksi/2016/507/contents

Harrell PT, Simmons VN, Correa JB, Padhya TA, Brandon TH (2014) Electronic nicotine delivery systems ("E-cigarettes"): review of safety and smoking cessation efficacy. Arch Otolaryngol Head Neck Surg 151(3):381–393. https://doi.org/10.1177/0194599814536847

Hartmann-Boyce J, McRobbie H, Butler AR, Lindson N, Bullen C, Begh R, Theodoulou A, Notley C, Rigotti NA, Turner T, Fanshawe TR, Hajek P (2021) Electronic cigarettes for smoking cessation. Cochrane Database Syst Rev 2021(8). https://doi.org/10.1002/14651858.CD010216.pub5

Higham A, Rattray NJW, Dewhurst JA, Trivedi DK, Fowler SJ, Goodacre R, Singh D (2016) Electronic cigarette exposure triggers neutrophil inflammatory responses. Respir Res 17(1):56. https://doi.org/10.1186/s12931-016-0368-x

Husereau D, Drummond M, Augustovski F, de Bekker-Grob E, Briggs AH, Carswell C, Caulley L, Chaiyakunapruk N, Greenberg D, Loder E, Mauskopf J, Mullins CD, Petrou S, Pwu R-F, Staniszewska S, CHEERS 2022 ISPOR Good Research Practices Task Force (2022) Consolidated health economic evaluation reporting standards 2022 (CHEERS 2022) statement: updated reporting guidance for health economic evaluations. Value Health J Int Soc Pharmacoecon Outcomes Res 25(1):3–9. https://doi.org/10.1016/j.jval.2021.11.1351

Institute for Global Tobacco Control (2017) Global tobacco control: country laws regulating e-cigarettes. Johns Hopkins Bloomberg School of Public Health, Baltimore. https://www.globaltobaccocontrol.org/sites/default/files/2021-04/SRNTAwopegba E-cigarette Scan.pdf

Jha P, Peto R (2014) Global effects of smoking, of quitting, and of taxing tobacco. N Engl J Med 370(1):60–68. https://doi.org/10.1056/NEJMra1308383

Johns Hopkins Medicine (2023) 5 Truths You Need to Know About Vaping

Krishnasamy VP, Hallowell BD, Ko JY, Board A, Hartnett KP, Salvatore PP, Danielson M, Kite-Powell A, Twentyman E, Kim L, Cyrus A, Wallace M, Melstrom P, Haag B, King BA, Briss P, Jones CM, Pollack LA, Ellington S et al (2020) Update: characteristics of a nationwide outbreak of e-cigarette, or vaping, product use–associated lung injury — United States, August 2019–January 2020. MMWR Morb Mortal Wkly Rep 69(3):90–94. https://doi.org/10.15585/mmwr.mm6903e2

Li J, Hajek P, Pesola F, Wu Q, Phillips-Waller A, Przulj D, Myers Smith K, Bisal N, Sasieni P, Dawkins L, Ross L, Goniewicz ML, McRobbie H, Parrott S (2020) Cost-effectiveness of e-cigarettes compared with nicotine replacement therapy in stop smoking services in England (TEC study): a randomized controlled trial. Addiction 115(3):507–517. https://doi.org/10.1111/add.14829

Ling MYJ, Abdul Halim AFN, Ahmad D, Ahmad N, Safian N, Mohammed Nawi A (2023) Prevalence and associated factors of e-cigarette use among adolescents in Southeast Asia: a systematic review. Int J Environ Res Public Health 20(5). https://doi.org/10.3390/ijerph20053883

Marques P, Piqueras L, Sanz M-J (2021) An updated overview of e-cigarette impact on human health. Respir Res 22(1):151. https://doi.org/10.1186/s12931-021-01737-5

McNeill A, Brose LS, Calder R, Bauld L, Robson D (2018) Evidence review of e-cigarettes and heated tobacco products: a report commissioned by Public Health England. Stationery Office. https://assets.publishing.service.gov.uk/media/5a981c6740f0b67aa27253cc/Evidence_review_of_e-cigarettes_and_heated_tobacco_products_2018.pdf

Medicines and Healthcare products Regulatory Agency (2015) Medicines and healthcare products regulatory agency. https://mhraproductsprod.blob.core.windows.net/docs-20200330/56f25daab2a2968139bc37075e194d1a5f12b33f

Medicines and Healthcare products Regulatory Agency (2017) NoGuidance for licensing electronic cigarettes and other inhaled nicotine-containing products as medicines. https://www.gov.uk/guidance/licensing-procedure-for-electronic-cigarettes-as-medicines

Medicines and Healthcare products Regulatory Agency (2018) Briefing: electronic cigarettes. https://ash.org.uk/uploads/E-Cigarettes-Briefing_PDF_v1.pdf

Mir M, Rauf I, Goksoy S, Khedr A, Jama AB, Mushtaq H, Jain NK, Khan SA, Surani S, Koritala T (2022) Electronic cigarettes: are they smoking cessation aids or health hazards? Cureus 14(5):e25330. https://doi.org/10.7759/cureus.25330

National Center for Chronic Disease Prevention and Health Promotion (U.S.) (2016) Office on smoking and health. e-cigarette use among youth and young adults: a report of the surgeon general. https://www.ncbi.nlm.nih.gov/books/NBK538680/

Newton JN, Dockrell M, Marczylo T (2018) Making sense of the latest evidence on electronic cigarettes. The Lancet 391(10121):639–642. http://doi.org/10.1016/S0140-6736(18)30202-2

Nocella C, Biondi-Zoccai G, Sciarretta S, Peruzzi M, Pagano F, Loffredo L, Pignatelli P, Bullen C, Frati G, Carnevale R (2018) Impact of tobacco versus electronic cigarette smoking on platelet function. Am J Cardiol 122(9):1477–1481. https://doi.org/10.1016/j.amjcard.2018.07.029

Patel U, Patel N, Khurana M, Parulekar A, Patel A, Ortiz JF, Patel R, Urhoghide E, Mistry A, Bhriguvanshi A, Abdulqader M, Mehta N, Arumaithurai K, Shah S (2022) Effect comparison of e-cigarette and traditional smoking and association with stroke-a cross-sectional study of NHANES. Neurol Int 14(2):441–452. https://doi.org/10.3390/neurolint14020037

Peto R (2000) Smoking smoking cessation and lung cancer in the UK since 1950: combination of national statistics with two case-control studies. BMJ 321(7257):323–329. http://doi.org/10.1136/bmj.321.7257.323

Phimarn W, Saramunee K, Leelathanalerk A, Srimongkon P, Chanasopon S, Phumart P, Paktipat P, Babar Z-U-D (2023) Economic evaluation of pharmacy services: a systematic review of the literature (2016–2020). Int J Clin Pharm 45(6):1326–1348. https://doi.org/10.1007/s11096-023-01590-0

Pound CM, Coyle D (2022) A cost-utility analysis of the impact of electronic nicotine delivery systems on health care costs and outcomes in Canada. Health Promot Chronic Dis Prev Can 42(1):29–36. https://doi.org/10.24095/hpcdp.42.1.05

Rebuli ME, Rose JJ, Noël A, Croft DP, Benowitz NL, Cohen AH, Goniewicz ML, Larsen BT, Leigh N, McGraw MD, Melzer AC, Penn AL, Rahman I, Upson D, Crotty Alexander LE, Ewart G, Jaspers I, Jordt SE, Kligerman S et al (2023) The e-cigarette or vaping product use-associated lung injury epidemic: pathogenesis, management, and future directions: an official american thoracic society workshop report. Ann Am Thorac Soc 20(1):1–17. https://doi.org/10.1513/AnnalsATS.202209-796ST

Reinikovaite V, Rodriguez IE, Karoor V, Rau A, Trinh BB, Deleyiannis FW-B, Taraseviciene-Stewart L (2018) The effects of electronic cigarette vapour on the lung: direct comparison to tobacco smoke. Eur Respir J 51(4):1701661. https://doi.org/10.1183/13993003.01661-2017

Singh S, Windle SB, Filion KB, Thombs BD, O'Loughlin JL, Grad R, Eisenberg MJ (2020) E-cigarettes and youth: patterns of use, potential harms, and recommendations. Prev Med 133:106009. https://doi.org/10.1016/j.ypmed.2020.106009

Thomas KH, Dalili MN, López-López JA, Keeney E, Phillippo D, Munafò MR, Stevenson M, Caldwell DM, Welton NJ (2021) Smoking cessation medicines and e-cigarettes: a systematic review network meta-analysis and cost-effectiveness analysis. Health Technol Assess 25(59):1–224. http://doi.org/10.3310/hta25590

Truth Initiative (2021) No e-cigarettes: facts, stats, and regulations. https://truthinitiative.org/research-resources/emerging-tobacco-products/e-cigarettes-facts-stats-and-regulations

U.S. Department of Health and Human Services (HHS) (2021) E-cigarette use among youth and young adults: a report of the surgeon general

U.S.FDA (2023) Products C for tobacco products, ingredients & components – vaporizers, e-cigarettes, and other electronic nicotine delivery systems (ENDS). https://www.fda.gov/TobaccoProducts/Labeling/ProductsIngredientsComponents/ucm456610.htm

Vlachopoulos C, Ioakeimidis N, Abdelrasoul M, Terentes-Printzios D, Georgakopoulos C, Pietri P, Stefanadis C, Tousoulis D (2016) Electronic cigarette smoking increases aortic stiffness and blood pressure in young smokers. J Am Coll Cardiol 67(23):2802–2803. https://doi.org/10.1016/j.jacc.2016.03.569

Wang RJ, Bhadriraju S, Glantz SA (2021) E-cigarette use and adult cigarette smoking cessation: a meta-analysis. Am J Public Health 111(2):230–246. https://doi.org/10.2105/AJPH.2020.305999

Wu Q, Jiang D, Minor M, Chu HW (2014) Electronic cigarette liquid increases inflammation and virus infection in primary human airway epithelial cells. PLoS One 9(9):e108342. https://doi.org/10.1371/journal.pone.0108342

Xie W, Kathuria H, Galiatsatos P, Blaha MJ, Hamburg NM, Robertson RM, Bhatnagar A, Benjamin EJ, Stokes AC (2020) Association of electronic cigarette use with incident respiratory conditions among US adults from 2013 to 2018. JAMA Netw Open 3(11):e2020816. https://doi.org/10.1001/jamanetworkopen.2020.20816

18 Updates on Interventions for E-cigarette Smoking Cessation

Nurain Afiqah Mohd Aznan,
Nur Syuhada Abqari Lutfhfi Albert,
'Atifah Najib Badri Najib, Syahrir Zaini,
Norny Syafinaz Ab Rahman,
and Mohamad Haniki Nik Mohamed

Abstract

E-cigarettes (e-cigs) are commonly believed to be safer than conventional cigarettes. There are at least thirty-four tobacco and menthol-flavored e-cig products that have been authorized by the U.S. FDA. However, concerns exist regarding safety of e-cigarettes and the need to quit vaping especially among never smokers. This chapter aims to evaluate the effectiveness of different methods for quitting e-cigs. Studies show that interactive text messaging was significantly successful in quitting e-cigs at both the 30-day and 7-day point prevalence of abstinence (PPA) at the three- and seven-month follow-up. Vape taper and varenicline were found to be helpful in aiding people to quit vaping after three months of treatment. Nicotine replacement therapy (NRT) increased the likelihood of vaping cessation at three and six months in half of the studies reviewed. In combination with other measures, counseling has successfully contributed to quitting e-cig use. The side effects from using all interventions were categorized as low. All e-cig cessation interventions were effective with no major safety issues.

Keywords

E-cigarette cessation · Pharmacological · Behavioral interventions

---

N. A. M. Aznan · N. S. A. L. Albert · 'A. N. B. Najib
Kulliyyah of Pharmacy, International Islamic University Malaysia (IIUM), Kuantan, Pahang, Malaysia

S. Zaini · N. S. A. Rahman · M. H. N. Mohamed (✉)
Department of Pharmacy Practice, Kulliyyah of Pharmacy, International Islamic University Malaysia (IIUM), Kuantan, Pahang, Malaysia
e-mail: haniki@iium.edu.my

Y.-H. Leong, A. S. Mohd Samin (eds.), *E-Cigarettes*,
https://doi.org/10.1007/978-981-96-5087-3_18

18.1 Introduction

Electronic cigarettes (e-cigs), often known as vapes, are battery-powered devices with an inhalation-activated system in which a cartridge (e-liquid) is heated to produce vapors that the users inhale. They have been advertised as a safer alternative to conventional cigarettes. However, the lower-risk alternative of e-cigs in comparison to conventional cigarettes is not risk-free due to an insufficient number of high-quality research being done on the harmful chemicals inside the e-cig. Currently, there are thirty-four menthol-flavored e-cig products that have been authorized by FDA to support their claims and potential benefits, after an extensive scientific review was conducted. The marketing authorization was done through the premarket tobacco product application (PMTA) pathway (U.S. Food and Drug Administration 2024). Due to concerns about potential health risks such as intoxication and addiction, some countries have banned the use of e-cigs (Maia et al. 2020; Da Silva and Moreira 2019; Vishal Rao et al. 2020). Some studies have also found that e-cigs have the potential to de-normalize smoking as they are easy to use and accessible compared to conventional cigarettes (McKeganey 2016). Several countries have not yet banned the use of e-cigs completely but have restricted it. For example, the United States implemented a ban on selling e-cigs to individuals under the age of 21 years, which is in line with the federal minimum purchase age (U.S. Food and Drug Administration 2019). The European Union of Tobacco Products Directive (TPD) regulates the sale of nicotine-containing e-cigs by setting a maximum concentration of nicotine (20 mg/mL) in e-liquid and sets requirements of child-resistant devices for the device-related storage containers (European Commission 2014).

In the US, the FDA has restricted the sale of e-cigs and other Electronic Nicotine Delivery Systems (ENDS) to people younger than 18 years old (Sharpless 2019). It has also shifted prevention efforts from policy to enforcing education of youth about the harmfulness of e-cigs. It has also banned manufacturers from making misleading health and cessation claims on e-cigs (Hammond et al. 2019). Additional restrictions apply to the importation, packaging, labeling, advertising, promotion, sale, and distribution of e-cigs on the market. The FDA will revoke approval if the company fails to meet regulatory standards or discovers products being sold to non-smokers, including youth (Hunt 2021). In the United Kingdom (UK), despite concerns regarding youth exposure to e-cigs, the government allows nicotine-containing e-cigs to be licensed as medicinal products (Wickware 2021). Nevertheless, the products are subject to strict restrictions on use, including a nicotine limit of 20 mg/mL of e-liquid, restrictions on certain forms of marketing, and the requirement for health warnings on the products, as required by the European Tobacco Products Directive (Hammond et al. 2019). However, there are still many young people who use e-cigs either with nicotine or without nicotine. There are different actions imposed on each of them. The number of dual users and long-term smokers has also increased over time.

E-cig smoking is very common in Malaysia, especially among young people. The popularity of e-cigs started in the 2010s and has been increasing ever since.

This is due to the lack of strict regulations and the fact that e-cigs are still not banned in Malaysia. In 2016, more than 600,000 Malaysian adults were reported to be current e-cig smokers (Ab Rahman et al. 2019) and has reached to more than 1.1 million current e-cig smokers in 2019 (Institute for Public Health 2019). Surprisingly, young adults aged 18–34 years old have a higher prevalence of e-cig use (30.2%), compared to the other population age groups (Perialathan et al. 2018). The increasing trend of e-cig use may have contributed to the outbreak of vaping-associated pulmonary injury (VAPI), also called e-cig or vaping product use-associated lung injury (EVALI) outbreak, which was first detected in the US in 2019. The outbreak reached its peak at the end of the year, resulting in over 30 reported deaths. EVALI has been associated with tetrahydrocannabinol (THC), a substance found in e-cigs. THC contains harmful chemicals like vitamin E acetate and cannabidiol (CBD) (Salzman et al. 2019). Vitamin E acetate is commonly used in nutritional supplements and skincare products. It has never been associated with any negative effects, instead, it enhances a person's well-being (Blount et al. 2019). However, when inhaled, it poses a risk. This is because the vitamin has the tendency to increase the fluidity of lung surfactant, causing the surfactant layer to collapse and contributing to shortness of breath and lung inflammation (Blount et al. 2019).

Studies show that using e-cigs to quit smoking may have harmful effects on human health. Some of these effects include respiratory problems, increased risk of heart disease, and toxicity to white blood cells (Mir et al. 2022). Another major issue is e-cig addiction. After quitting conventional smoking, a person may become addicted to e-cigs, which may be due to higher nicotine dependence levels on the current e-cigs compared to conventional cigarettes (Jankowski et al. 2019). Aside from this concern, non-smokers are increasingly trying to use e-cigs out of curiosity, especially among youth aged 11–18 years old. There is also a lack of evidence-based studies on e-cig cessation interventions (Jankowski et al. 2019).

Despite the increasing use of e-cigs among non-smokers, particularly young people aged 11–18 years, there is a significant lack of research and evidence-based studies focusing on effective strategies to quit e-cig use. To address this issue, it is crucial to prioritize research funding and support studies that investigate the most effective methods for quitting e-cig use. By filling this knowledge gap, we can develop evidence-based strategies that will empower individuals to successfully overcome their dependence on these devices and improve their overall health and well-being. Moreover, understanding the underlying psychological and social factors contributing to excessive device use can pave the way for targeted interventions and support systems to address these issues effectively.

18.2 Type of E-cig Cessation Interventions

Prior to the utilization of any intervention for e-cig cessation, people use conventional methods such as "cold turkey", willpower, and replacing e-cig smoking with another activity (Holt and Latimer 2024). The "cold turkey" method, which is related to abrupt cessation without gradual reduction, is a common method for those trying to quit e-cigs (Bluestein et al. 2023). Nowadays, interventions for e-cig cessation in adolescents and adult populations include pharmacological interventions and behavioral interventions. Studies reported the use of either single intervention or combination interventions is a common practice (Caponnetto et al. 2023). Pharmacological interventions that were used include varenicline (Fucito et al. 2024) and nicotine replacement therapy (NRT) (Palmer et al. 2023) in any dosage form (gum, lozenge, transdermal patch, or nasal spray). Varenicline reduces cravings and satisfaction from nicotine products by partially mimicking nicotine's effects on receptors (Tonstad et al. 2020). NRT delivers controlled doses of nicotine, reducing withdrawal symptoms as a result of stopping the nicotine products (Cahill et al. 2013).

Other than pharmacological therapy, interactive text messages, counseling, and vape taper were also utilized as behavioral intervention for e-cig cessation. An example of the interactive text message intervention is known as "This Is Quitting" (Graham et al. 2021a). The program tailors its messages based on a user's age, enrolment date, or quit date, and the vape product they use, which can be updated via text message. Those who set a quit date get messages for one week before and eight weeks after, offering encouragement, support, skills training, coping strategies, and information about the risks of vaping, the benefits of quitting, and cutting down to quit. On the other hand, counseling and vape taper intervention are related to a pharmacist guiding the patients through a quit attempt using behavioral support and a taper of both nicotine e-liquid concentration and vaping sessions. The vape taper was created to help patients stop using vape and be nicotine-free within 12 weeks of enrolment (Sahr et al. 2020b).

18.3 E-cigs Smoking Abstinence

One study provided data comparing the interactive text message intervention against a control group receiving only assessment-related messages using an intention-to-treat (ITT) analysis (Graham et al. 2021a). At seven months, e-cig abstinence was shown to be higher ($p < 0.001$) in the interactive text message group than in the control group, with a risk ratio (RR) of 1.29 (95% CI: 1.11–1.50) and odds ratio (OR) of 1.39 (95% CI: 1.15–1.68) of abstinence in the past 30 days. Similarly, for the past seven days at seven months follow-up, the interactive text message also surpassed the control in terms of quit rate. This study also included data on cigarette usage following seven months of follow-up. The study showed a 10.3% decrease (95% CI: 7.6, 12.9) in the past 30-day cigarette use compared to the baseline. Also, at baseline, a higher proportion of E-cig Users and Dual Users were Dual Abstinent in the interactive text message intervention than in the control, with a percentage point difference of 6.5% (95% CI 1.2, 11.8, $p = 0.019$) and 8.6% (95% CI 3.4, 13.9, $p = 0.0014$), respectively.

Another study does not compare interactive text messages with other interventions (Graham et al. 2020). However, it provided that after three months, almost 25% of participants had been able to quit vaping in the past seven days, while about 15% of participants had been able to quit vaping, reported as not having any puffs in the past 30 days. Thus, it can be noted that most participants were able to quit vaping completely with only a 10% difference between 7-day ppa and 30-day ppa after three months. Furthermore, a case report provided descriptive data about a participant who has been suggested with vape-taper intervention with counseling (Sahr et al. 2020a). Throughout the weeks, the participants were able to decrease the frequency and concentration of e-liquid. Participants had entirely stopped vaping after 10 weeks and remained nicotine-free even after six months. Another study found that varenicline paired with counseling was an effective method for quitting smoking (Barkat et al. 2019). Varenicline has been described as effectively quit vaping after three months of medication, as seen by his refusal to take any e-cig puffs and triggering decline.

Moreover, studies commonly reported that NRT has been used in combination with counseling. It was reported that the subjects in two studies were effectively vape-free after three months of treatment (Kelly et al. 2020; Silver et al. 2016). Both cases reported using a combination of NRT (patch, gum, and lozenge) and reporting health improvement during the course of treatment. Furthermore, one case series indicated that after six and eight months of treatment, more than half of the participants had stopped using e-cigs (Sikka et al. 2021). Two of them used a combination of NRT, whereas the rest of the subjects in that study solely used one type of NRT, either transdermal or nasal spray.

Another study compared three types of interventions at three months and six months following treatment initiation; NRT with counseling, vape-taper with counseling, and self-guided quit as a control (Sahr et al. 2021). The study discovered that the vape-taper group had favorable results compared to other interventions at both time points, with 75% quit rates. At three months, the self-guided group had the highest percentage quit rates, with more than 77% of participants vaping-free. However, after six months, the percentage had dropped to less than 45%, demonstrating that the participants were unable to remain vape-free with self-guided treatment. The NRT group, on the other hand, had the lowest quit rates at both time points, with less than 43% of participants.

18.4 Side Effects and Limitations of Interventions

Pharmacological interventions are commonly associated with several side effects that come with benefits. NRT (transdermal, buccal, and oral) caused side effects such as nausea, nicotine dependence, and dizziness, which might have been caused by nicotine itself (Motooka et al. 2018). The reporting odds ratio (ROR), an authorized pharmacovigilance index, was calculated using a disproportionality analysis. The ROR is the ratio of the odds of reporting specific side effects versus all other side effects for a given product, compared with the reporting odds for all other drugs or products present in the database. For transdermal nicotine exposure, top 5 side effects reported were nausea (149 cases, ROR: 2.28 (95% CI: 1.92–2.69)), dizziness (132 cases, ROR: 3.04 (95% CI: 2.54–3.63)), application site erythema (108 cases, ROR: 32.52 (95% CI: 26.74–39.55)), headache (98 cases, ROR: 1.84 (95% CI: 1.50–2.25)), and dyspnea (94 cases, ROR: 1.93 (95% CI: 1.57–2.38)). According to the findings of other studies, nicotine patches or transdermal may produce itching and redness at the site of application (Sahr et al. 2020a, 2021). One of the studies also discovered that the gum made one of the participants' heartburn worse (Sahr et al. 2021).

Furthermore, another study gave information on the participant's varenicline side effects (Barkat et al. 2019). The participant experienced one vivid dream early in treatment and gained weight after two months, which he attributed to vaping in place of meals. According to a meta-analysis (Shang et al. 2023), the common adverse events for Varenicline were nausea, vomiting, abnormal dreams, sleep disturbances, headache, depression, irritability, indigestion, and nasopharyngitis ($p < 0.05$). Varenicline had mild to moderate adverse events but was well tolerated. Among the most frequently reported adverse events were categorized as neuropsychiatric: anxiety and irritability (Ebbert et al. 2021).

For non-pharmacological interventions such as the study of vaping cessation text message program (Graham et al. 2021a, b), there are several limitations, including no biochemical verification, lack of inclusion of teens, potential bias due to timing and pandemic as well as assessment reactivity. The study did not include biochemical verification of abstinence because it is difficult to do in digital cessation studies, and they chose to use 30-day abstinence as a more rigorous primary measure instead. The study did not include teenagers (13–17 years old), even though they use e-cigs a lot. While the "This Is Quitting" program is popular with teens, more research is needed to see how effective it is for them. Recruiting during the "quitting season" and the disruptions from the COVID-19 pandemic might have influenced the results. The high motivation to quit around New Year's and the pandemic's unique conditions could have affected quit rates for both groups in the study. Finally, using incentivized text message assessments might have influenced participants' responses and behavior during the study.

Another non-pharmacological intervention, which is known as pharmacist-assisted vape taper and behavioral support for the cessation of electronic nicotine delivery system use also has several limitations, including challenges in quantifying vaping, risk of continuous exposure to e-liquid and individual variability in responses (Sahr et al. 2020a). Vaping has different usage patterns than cigarette smoking, making it hard to measure nicotine dependence accurately. Factors like the strong urge to vape in the morning and restrictions on vaping locations differ from smoking habits. Moreover, the tapering method keeps users exposed to e-liquid, which has its risks. Both the clinician and the patient need to weigh these risks against the benefits. Last but not least, the success of the tapering schedule and behavioral support varies for each person. Factors like personal habits, social situations, and addiction levels affect how well someone can quit.

18.5 Effectiveness of E-cigs Smoking Intervention

This section examines the effectiveness of interventions for individuals trying to quit e-cigs smoking in the general population, comparing findings from past studies to outcomes observed at three and six months. In order to evaluate the effectiveness of the interventions used, all studies were based exclusively on self-reported data provided by the participants. However, it should be noted that self-reported data may have limitations, such as recall bias or social desirability bias, which could affect the accuracy of the reported smoking cessation outcomes.

The Fagerstrom nicotine dependence test (FTND) was the most utilized self-reported measurement. All studies that used the FTND test demonstrated the impact of interventions by lowering the FTND score throughout the course of the treatment weeks (Graham et al. 2021a, b; Kelly et al. 2020; Sahr et al. 2020a, 2021; Silver et al. 2016). Furthermore, the FTND test at the baseline has aided in identifying the dose to be given to people who have received pharmacological therapies (Kelly et al. 2020; Sahr et al. 2020a, 2021; Silver et al. 2016). In addition, the results of the FTND test have provided valuable insights into the optimal treatment approach for people undergoing pharmacological therapies. These findings have helped to personalize and tailor the dose of medication based on individual needs and responses.

The types of interventions that have been shown to be effective in helping with e-cig cessation have primarily been pharmacist counseling. This behavioral intervention has been paired with other interventions in which this method has been proven to be effective. As nicotine concentration varies in e-liquid, consulting a pharmacist is the ideal solution for offering information on dosing the medication, NRT, or e-liquid concentration for vape-taper intervention. Counseling from a pharmacist can also assist to motivate people, as quitting e-cigs might be difficult for them owing to barriers such as social pressures, stress, and the ease with which e-cigs are available in the market.

Interactive text message serves as another behavioral intervention that has been demonstrated to be significantly effective in e-cig cessation. Studies utilized text messaging as a method to communicate, motivate, and analyze participants by portraying it as a non-judgmental and helpful buddy (Graham et al. 2021a, b, 2022). They designed it to provide participants with encouragement and support, skill training, self-efficacy-building exercises, coping methods, and information about the risks of vaping and the benefits of stopping for nine weeks after they selected their quit date. The RR and OR for e-cig smoking abstinence at seven months for 30-day ppa and 7-day ppa were both favored to an interactive text message.

Similarly, the vape taper intervention was used efficiently in the study by providing the participants with a quit plan. The pharmacist aided the participant in tapering the frequency of e-cig usage and e-liquid concentration by implementing a week-by-week strategy over a three-month period (Sahr et al. 2020a, 2021). The vape taper intervention with behavioral support also appears to be superior to NRT with behavioral assistance in terms of increased quitting rates after three and six months (Sahr et al. 2021). However, because this study was conducted solely by students on a college campus, it did not adequately reflect a vast population and various flaws were identified.

Besides that, pharmacological therapies were also successful in helping people quit e-cigs smoking (Barkat et al. 2019; Sahr et al. 2020a, 2021; Sikka et al. 2021; Silver et al. 2016). It is recommended that varenicline be taken at 0.5 mg once daily for three days, then 0.5 mg twice daily for four days, and then 1 mg twice daily to be taken as a maintenance dose. Varenicline confirmed its effectiveness during the three months follow-up by successfully preventing participants from smoking e-cigs totally, resulting in improved health. The subjects also reported having vivid dreams and gaining weight as negative effects. However, the pharmacist successfully managed the adverse effects, which did not pose any substantial safety issues.

NRT also efficiently aids in the cessation of e-cig use. Combination NRT appears to be more commonly utilized in studies than single NRT of transdermal patch (Sahr et al. 2020a, 2021; Sikka et al. 2021; Silver et al. 2016). Depending on the participants' nicotine dependence, higher dependence was first advised with a 21 mg patch, while lesser dependence was recommended with a 14 mg patch, which was subsequently tapered down every month until the participants stopped using e-cigs. Some studies utilized 4 mg of nicotine lozenge and gum with no tapering, whereas others used 2 mg of nicotine lozenge and gum with no tapering. Participants in one research were also given nicotine nasal spray as therapy. However, two studies found that participants experienced itching and redness at the site of nicotine patch application, leading to noncompliance and switching to gum to manage e-cig cravings. Only one study found that nicotine gum can cause heartburn in participants (Sahr et al. 2021). All studies that employed NRT as an intervention successfully got their participants to quit vaping after three and six months.

The best intervention is one that could assist e-cig smokers in avoiding other tobacco products as well. Nevertheless, only one study assessed the use of different tobacco products during a seven-month follow-up period, and it used an interactive text message as an intervention (Graham et al. 2021a, b, 2022). This strategy was reported to possess the ability to aid in abstinence from other tobacco products as well as minimize the likelihood of relapsing with other tobacco products. As the combination of pharmaceutical and behavioral interventions has always been considered the gold standard for smoking cessation (Bahagian Kawalan Penyakit 2016), this should also apply to e-cig cessation.

18.6 Recommendations for Future Research

There is a critical need for research to explore the efficacy of novel, technology-based strategies for promoting vaping cessation among young adults (Berg et al. 2021). This investigation should encompass the potential of expanding existing text messaging programs into more interactive formats, including smartphone applications, chatbots, and social media platforms frequented by young adults, such as X, Instagram, Snapchat, and TikTok. While a cursory examination of app marketplaces like iTunes and Google Play reveals the existence of a limited number of cessation-oriented applications, their effectiveness remains unknown.

Furthermore, it is imperative to evaluate the potential of leveraging various smartphone functionalities, such as engaging videos, interactive games, and relatable memes, to enhance user engagement among young adult vapers. Gamification techniques, for instance, hold promise in delivering key cessation information and fostering skill development to quit, potentially leading to increased user adherence – a cornerstone of program success. Consequently, research efforts should prioritize investigating whether specific technology-driven approaches and features can effectively capture the attention of young adult vapers and stimulate positive behavior change.

This emphasis on exploring diverse delivery methods aligns with established Clinical Practice Guidelines on Treatment of Tobacco Use Disorder (Ministry of Health Malaysia 2015), which indicate that the provision of cessation interventions through multiple channels can contribute to improved quit rates. However, within the context of young adult vaping cessation, the latest Cochrane review on mobile health interventions highlights the superior effectiveness of human-delivered counseling, even when delivered via text messaging, compared to automated text messages alone. Given the multifaceted nature of vaping behavior and its potential to co-occur with other substance use patterns, such as smoking, tobacco use in general, and marijuana use, personalized counseling might be essential to address the multifaceted challenges associated with cessation.

18.7 Conclusion

Despite the limitations of this review, the included studies are well-suited to assist practitioners and clinicians in identifying the most appropriate interventions for e-cig smokers who wish to quit. However, the evidence available for each intervention category remains limited, making it difficult to draw definitive conclusions about the effectiveness of any single approach. Nevertheless, this review updates and synthesizes the latest evidence in the field, indicating that all evaluated methods for achieving vaping abstinence are both successful and effective, with no significant safety risks identified. Among the strategies employed are interactive text messaging, vape tapering, varenicline, and nicotine replacement therapy (NRT). Additionally, clinical counseling has been identified as the most effective strategy when used in combination with other interventions. By integrating these methods, healthcare providers can offer a comprehensive and tailored approach to support individuals in their efforts to quit e-cig smoking.

References

Ab Rahman J, Mohd Yusoff MF, Nik Mohamed MH, Mahadir Naidu B, Lim KH, Tee GH, Mohamad MS, Kartiwi M, Draman S, Ab Rahman NS, Aris T (2019) The prevalence of e-cigarette use among adults in Malaysia. Asia Pac J Public Health 31(7_suppl):9S–21S. https://doi.org/10.1177/1010539519834735

Bahagian Kawalan Penyakit, Kementerian Kesihatan Malaysia (2016) Clinical practice guidelines: treatment of tobacco use disorder. Tobacco Control Unit & FCTC Secretariat, Non-Communicable Disease Section Disease Control Division, Ministry of Health Malaysia

Barkat SS, Tellier SM, Eloma AS (2019) Varenicline for cessation from nicotine-containing electronic cigarettes. Am J Health Syst Pharm 76(23):1894–1896. https://doi.org/10.1093/ajhp/zxz218

Berg CJ, Krishnan N, Graham AL, Abroms LC (2021) A synthesis of the literature to inform vaping cessation interventions for young adults. Addict Behav 119. https://doi.org/10.1016/j.addbeh.2021.106898

Blount BC, Karwowski MP, Shields PG, Morel-Espinosa M, Valentin-Blasini L, Gardner M, Braselton M, Brosius CR, Caron KT, Chambers D, Corstvet J, Cowan E, De Jesús VR, Espinosa P, Fernandez C, Holder C, Kuklenyik Z, Kusovschi JD, Newman C, Reis GB, Rees J, Reese C, Silva L, Seyler T, Song M-A, Connie Sosnoff CR, Spitzer DT, Wang L, Watson C, Wewers MD, Xia B, Heitkemper DT, Ghinai I, Layden J, Briss P, King BA, Delaney LJ, Jones CM, Baldwin GT, Patel A, Meaney-Delman D, Rose D, Krishnasamy V, Barr JR, Thomas J, Pirkle JL (2019) Vitamin E acetate in bronchoalveolar-lavage fluid associated with EVALI. N Engl J Med 382(8). https://doi.org/10.1056/nejmoa1916433

Bluestein MA, Bejarano G, Tackett AP, Duano JC, Rawls SG, Vandewater EA, Ahluwalia JS, Hébert ET (2023) E-cigarette quit attempts and experiences in a convenience sample of adult users. Int J Environ Res Public Health 20(3):1–11. https://doi.org/10.3390/ijerph20032332

Cahill K, Stevens S, Perera R, Lancaster T (2013) Pharmacological interventions for smoking cessation: an overview and network meta-analysis. Cochrane Database Syst Rev 2013(5):CD009329. https://doi.org/10.1002/14651858.CD009329.pub2

Caponnetto P, Campagna D, Ahluwalia JS, Russell C, Maglia M, Riela PM, Longo CF, Busa B, Polosa R (2023) Varenicline and counseling for vaping cessation: a double-blind, randomized, parallel-group, placebo-controlled trial. BMC Med 21(1):220. https://doi.org/10.1186/s12916-023-02919-2

Ebbert J, Jimenez-Ruiz C, Dutro MP, Fisher M, Li J, Hays JT (2021) Frequently reported adverse events with smoking cessation medications: post hoc analysis of a randomized trial. Mayo Clin Proc 96(7):1801–1811. https://doi.org/10.1016/j.mayocp.2020.10.046

European Commission (2014) Memo: questions & answers: new rules for tobacco products

Fucito LM, Baldassarri SR, Baker NL, Palmer AM, O'Malley SS, Carpenter MJ, Krishnan-Sarin S, Gray KM, Toll BA (2024) Varenicline for e-cigarette cessation in adults: a preliminary placebo-controlled randomized trial. Am J Prev Med. https://doi.org/10.1016/j.amepre.2024.04.007

Graham AL, Jacobs MA, Amato MS (2020) Engagement and 3-month outcomes from a digital e-cigarette cessation program in a Cohort of 27 000 teens and young adults. Nicotine Tob Res 22(5):859–860. https://doi.org/10.1093/ntr/ntz097

Graham AL, Amato MS, Cha S, Jacobs MA, Bottcher MM, Papandonatos GD (2021a) Effectiveness of a vaping cessation text message program among young adult e-cigarette users: a randomized clinical trial. JAMA Intern Med 181(7):923–930. https://doi.org/10.1001/jamainternmed.2021.1793

Graham AL, Jacobs MA, Amato MS (2021b) Engagement and 3-month outcomes from a digital e-cigarette cessation program in a cohort of 27 000 teens and young adults. Nicotine Tob Res 22(5):859–860. https://doi.org/10.1093/ntr/ntz097

Graham AL, Cha S, Papandonatos GD, Amato MS, Jacobs MA, Abroms LC, Berg CJ (2022) E-cigarette and combusted tobacco abstinence among young adults: secondary analyses from a U.S.-based randomized controlled trial of vaping cessation. Prev Med. https://doi.org/10.1016/j.ypmed.2022.107119

Hammond D, Reid JL, Rynard VL, Fong GT, Michael Cummings K, McNeill A, Hitchman S, Thrasher JF, Goniewicz ML, Bansal-Travers M, O'Connor R, Levy D, Borland R, White CM (2019) Prevalence of vaping and smoking among adolescents in Canada, England, and the United States: repeat national cross sectional surveys. BMJ 365. https://doi.org/10.1136/bmj.l2219

Holt LJ, Latimer LJ (2024) Emerging adults' experiences with e-cigarette cessation. Subst Use Misuse 59(3):405–410. https://doi.org/10.1080/10826084.2023.2275563

Hunt A (2021) FDA permits marketing of e-cigarette products, marking first authorization of its kind by the agency

Institute for Public Health (2019) "National Health and Morbidity Survey (NHMS) 2019: NCDs – Non-Communicable Diseases: Risk Factors and Other Health Problems." Institute for Public Health, National Institutes of Health (NIH), Ministry of Health Malaysia (vol 1). Retrieved October 21, 2024 https://iku.nih.gov.my/nhms-2019

Jankowski M, Krzystanek M, Zejda JE, Majek P, Lubanski J, Lawson JA, Brozek G (2019) E-cigarettes are more addictive than traditional cigarettes-a study in highly educated young people. Int J Environ Res Public Health 16(13). https://doi.org/10.3390/ijerph16132279

Kelly M, Blower N, Kelsh S (2020) Use of nicotine replacement therapy and behavioral support to quit ENDS. Qeios. https://doi.org/10.32388/67DZLK

Maia ACA, Fabriz DC, Motta TA, Zanotelli VF, Zamprogno B (2020) Ban on electronic cigarettes in Brazil: a policy in favor of human rights. Eur J Pub Health 30(Supplement_5). https://doi.org/10.1093/EURPUB/CKAA166.1223

McKeganey N (2016) Visible vaping: e-cigarettes and the further de-normalization of smoking. Int Arch Addict Res Med 2. https://doi.org/10.23937/2474-3631/1510023

Ministry of Health Malaysia (2015) Clinical practice guidelines on treatment of tobacco use disorder. 6

Mir M, Rauf I, Goksoy S, Khedr A, Jama AB, Mushtaq H, Jain NK, Khan SA, Surani S, Koritala T (2022) Electronic cigarettes: are they smoking cessation aids or health hazards? Cureus 14(5):e25330. https://doi.org/10.7759/cureus.25330

Motooka Y, Matsui T, Slaton RM, Umetsu R, Fukuda A, Naganuma M, Hasegawa S, Sasaoka S, Hatahira H, Iguchi K, Nakamura M (2018) Adverse events of smoking cessation treatments (nicotine replacement therapy and non-nicotine prescription medication) and electronic cigarettes in the Food and Drug Administration Adverse Event Reporting System, 2004–2016. SAGE Open Med 6:2050312118777953. https://doi.org/10.1177/2050312118777953

Palmer AM, Carpenter MJ, Rojewski AM, Haire K, Baker NL, Toll BA (2023) Nicotine replacement therapy for vaping cessation among mono and dual users: a mixed methods preliminary study. Addict Behav 139:107579. https://doi.org/10.1016/j.addbeh.2022.107579

Perialathan K, Rahman AB, Lim KH, Adon Y, Ahmad A, Juatan N, Jaafar N (2018) Prevalence and associated factors of ever use of electronic cigarettes: findings from a hospitals and health clinics study based in Malaysia. Tob Induc Dis 16(November):1–8. https://doi.org/10.18332/tid/99258

Rao V, Uchilla S, Arakeri G, Ravishankar S, Kar A, Thakur S, Li RJ, Dhananjay KV, Surya T, Chaturvedi P, Gomez RS, Brennan PA (2020) The e-cigarette ban in India—a step in the right direction? J Oral Pathol Med 49(7):617–620. https://doi.org/10.1111/JOP.13012

Sahr M, Blower N, Kelsh S (2020a) Use of nicotine replacement therapy and behavioral support to quit ENDS. Qeios:1–4. https://doi.org/10.32388/67dzlk

Sahr M, Kelsh SE, Blower N (2020b) Pharmacist assisted vape taper and behavioral support for cessation of electronic nicotine delivery system use. Clin Case Reports 8(1):100–103. https://doi.org/10.1002/ccr3.2599

Sahr M, Kelsh S, Blower N, Sohn M (2021) Pilot study of Electronic Nicotine Delivery Systems (ENDS) cessation methods. Pharmacy 9(1):21. https://doi.org/10.3390/PHARMACY9010021

Salzman GA, Alqawasma M, Asad H (2019) Vaping Associated Lung Injury (EVALI): an explosive United States epidemic. Mo Med 116(6):492–496

Shang X, Fenfen E, Guo K, Yang C, Zhou L, Wu Y, Yang K, Li X (2023) Effectiveness and safety of varenicline for smoking cessation: an overview and meta-analysis. J Addict Med 17(5):536–543. https://doi.org/10.1097/ADM.0000000000001171

Sharpless N (2019) How FDA is regulating e-cigarettes | FDA. US Food and Drug Administration. Retrieved October 21, 2024. https://www.fda.gov/news-events/fda-voices/how-fda-regulating-e-cigarettes

Sikka G, Oluyinka M, Schreiber R, Galiatsatos P (2021) Electronic cigarette cessation in youth and young adults: a case series. Tob Use Insights 14:1179173X211026676. https://doi.org/10.1177/1179173X211026676

Silva ALOD, Moreira JC (2019) The ban of eletronic cigarettes in Brazil: success or failure?. A proibição dos cigarros clctrônicos no Brasil: sucesso ou fracasso? Ciênc Saúde Colet 24(8):3013–3024. https://doi.org/10.1590/1413-81232018248.24282017

Silver B, Ripley-Moffitt C, Greyber J, Goldstein AO (2016) Successful use of nicotine replacement therapy to quit e-cigarettes: lack of treatment protocol highlights need for guidelines. Clin Case Reports 4(4):409–411. https://doi.org/10.1002/ccr3.477

Tonstad S, Arons C, Rollema H, Berlin I, Hajek P, Fagerström K, Els C, McRae T, Russ C (2020) Varenicline: mode of action, efficacy, safety and accumulated experience salient for clinical populations. Curr Med Res Opin 36(5):713–730. https://doi.org/10.1080/03007995.2020.1729708

U.S. Food and Drug Administration (2019) Tobacco 21 | FDA. Web Page. Retrieved October 21, 2024 https://www.fda.gov/tobacco-products/retail-sales-tobacco-products/tobacco-21

U.S. Food and Drug Administration (2024) E-Cigarettes, Vapes, and other Electronic Nicotine Delivery Systems (ENDS) | FDA. Web Page. Retrieved October 21, 2024. https://www.fda.gov/tobacco-products/productsingredients-components/e-cigarettes-vapes-and-other-electronic-nicotine-delivery-systems-ends

Wickware C (2021) UK medicines regulator to approve e-cigarettes for medical use. Pharm J 307. https://doi.org/10.1211/pj.2021.1.112822

19 E-cigarettes in Malaysia: Public Health Challenges and Economic Implications

Wency Kher Thinng Bui and Norashidah Mohamed Nor

Abstract

The surge in e-cigarette (e-cig) usage in Malaysia, particularly among the youth, poses a significant threat to public health as it falsely presents e-cigs as safer alternatives. These devices carry substantial health risks and contribute to market failures and economic burdens, including escalated healthcare costs and reduced productivity. To effectively address these concerns, it is imperative to enforce stricter regulations, enhance age verification, promote health warnings, and impose higher taxes on all tobacco products. Public awareness campaigns play a vital role in educating the public about the associated risks and economic implications. These measures are aimed at reducing Malaysia's smoking prevalence to 15% by 2025 and below 5% by 2045, ultimately leading to improved public health outcomes.

Keywords

Market failure · Externalities · E-cigarettes · Economic implications · Health impacts

19.1 Introduction

In response to the public health threat posed by smoking, the Malaysian government has implemented several tobacco control policies. Malaysia signed the Framework Convention on Tobacco Control (FCTC) in September 2003 and ratified the treaty

W. K. T. Bui (✉) · N. Mohamed Nor
School of Business and Economics, Universiti Putra Malaysia, Seri Kembangan, Malaysia
e-mail: wency@upm.edu.my

Y.-H. Leong, A. S. Mohd Samin (eds.), *E-Cigarettes*,
https://doi.org/10.1007/978-981-96-5087-3_19

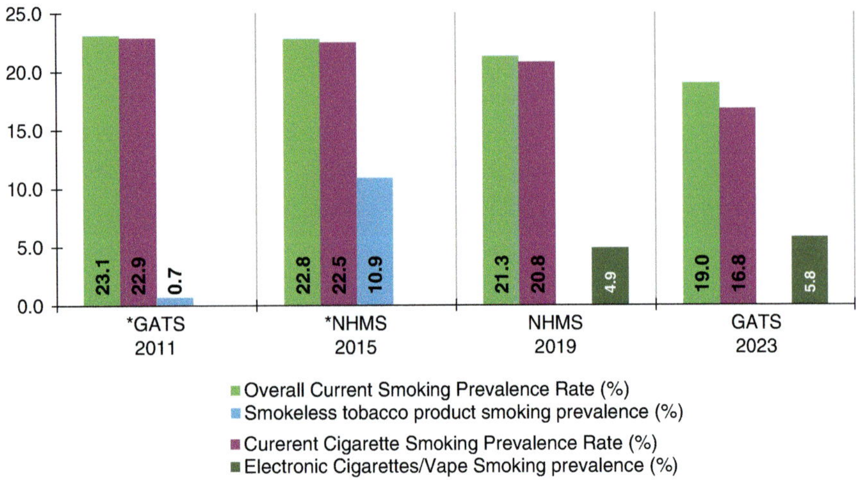

Fig. 19.1 The smoking prevalence rates aged 15 years old and above in Malaysia. Data Sources: Report of GATS Malaysia 2011 (IPH 2012a); Report of NHMS 2015 (IPH 2015); Report of GATS 2023 (Centers for Disease Control and Prevention 2023). Note: *In the NHMS 2015 and GATS 2011 reports, e-cigs were initially categorized as smokeless tobacco with bidis and other smokeless tobacco products

in 2005. The government has also adjusted the Control of Tobacco Products Regulation (CTPR) in line with the FCTC guidelines. Subsequently, the World Health Organization (WHO) introduced the affordable and achievable six practical measures for tobacco control policies in 2008, known as MPOWER. The regulation has incorporated the MPOWER[1] elements to adhere to these tobacco control policies in combating the tobacco epidemic in the country. All these tobacco control policies are working complementarily to reduce the smoking epidemic (Mackay et al. 2013).

The Malaysian government has set targets to reduce smoking rates to 15% by 2025 and less than 5% by 2045. The impact of tobacco control policies on decreasing smoking rates is evident in national health surveys in Malaysia, as illustrated in Fig 19.1. According to the surveys, the overall current smoking prevalence was about 23.1% in the GATS 2011 survey, with a gradual reduction in subsequent national surveys. The most recent NHMS survey indicates that only 19% of Malaysians smoked in 2023. Even though the data in Fig. 19.1 indicates a consistent decrease in the current smoking rate, it only reflects a 4.1 percentage-point reduction over these eight years. The slight decrease in smoking prevalence over these

[1] MPOWER is a policy package designed to help the implementation of effective interventions to reduce smoking globally at the country level. The six measures of the MPOWER tobacco control policies package include: (M) Monitoring tobacco use and prevention policies; (P) Protecting people from tobacco smoke; (O) Offering help to quit smoking; (W) Warning about the dangers of smoking; (E) Enforcing bans on tobacco advertising, promotion, and sponsorship; and (R) Raising taxes on tobacco.

eight years has raised concern in the GATS 2023 report (IPH 2024). With only two years left to reduce the percentage from 19% to 15%, the feasibility of reaching this goal is still in question.

Meanwhile, in addition to the concern about reaching the target, there is a growing worry about the increasing popularity of e-cigs in Malaysia. The prevalence of smokeless tobacco product smoking was 0.7% in GATS 2011 but surged to 10.9% in NHMS 2015. The report of the NHMS 2015 survey highlighted the significant increase in the use of e-cigs, resulting in the rise of smokeless tobacco product usage. As for specifically on the term e-cigs, the NHMS 2019 survey marked its first recording because previous surveys, due to the minimal use of smokeless products, the surveys generally included e-cigs under the smokeless tobacco product categories with other smokeless products such as bidis. The NHMS 2019 survey recorded e-cig use at 4.9%, slightly increasing to 5.0% in GATS 2023.

The data indicates an increase in the use of e-cigs, with only a slight decrease in overall smoking prevalence. When comparing data from GATS 2011–2023, the overall smoking prevalence decreased by just 4.1 percentage points, while the smoking prevalence of e-cigs rose to 5.8% in 2023. This suggests a shift in the type of smoking product from conventional cigarettes to e-cigs. This raises doubts about the industry's claims that e-cigs can help people quit smoking. This increasing trend, coupled with the absence of regulation on e-cigs in Malaysia, prompted a study by Shroff and Sreeramareddy (2024) urging the Malaysian government to enhance regulation of e-cig marketing, including stricter age verification, more explicit health warnings, and limitations on flavor varieties to reduce their appeal to youth and non-smokers. This is mainly due to the attractive nature of e-cig products and the wide variety of flavors of e-liquids, which has not only attracted non-smokers to initiate smoking but has also tempted youth to take up smoking.

Therefore, the growing use of e-cigs among youth in Malaysia is a major concern for the Malaysian government. Figure 19.2 shows data from various national surveys indicating a troubling increase in e-cig use among youth. According to the TECMA 2016 report, the prevalence of e-cig use among youth aged 13-19 in Malaysia was 11.3%. Compared to TECMA 2016, NHMS reports focused on the smaller age groups, yet the trend of e-cig use continues to rise. The NHMS 2017 revealed a slight increase in e-cig use to 9.8% in 2017, and the most recent NHMS 2022 survey reported a notable increase, with 14.9% of youth currently using e-cigs. This substantial increase in e-cig use has resulted in an overall rise in smoking prevalence to 18.5% in 2022.

The popularity of e-cigs among youth has significantly increased, with a preference for e-cigs over conventional cigarettes. While adult smokers still favor conventional cigarettes, youth are increasingly using e-cigs. As shown in Fig. 19.2, the prevalence of cigarette smoking ranged from around 11.5% to 11.7%, as reported by NHMS 2012 and TECMA 2016, later peaking at 13.8% in 2017 but dropping significantly to 6.2% in 2022 based on the subsequent NHMS surveys. These findings indicate a significant proportion of youth in Malaysia engaging in e-cig use over the past decade, highlighting the importance of public health strategies to

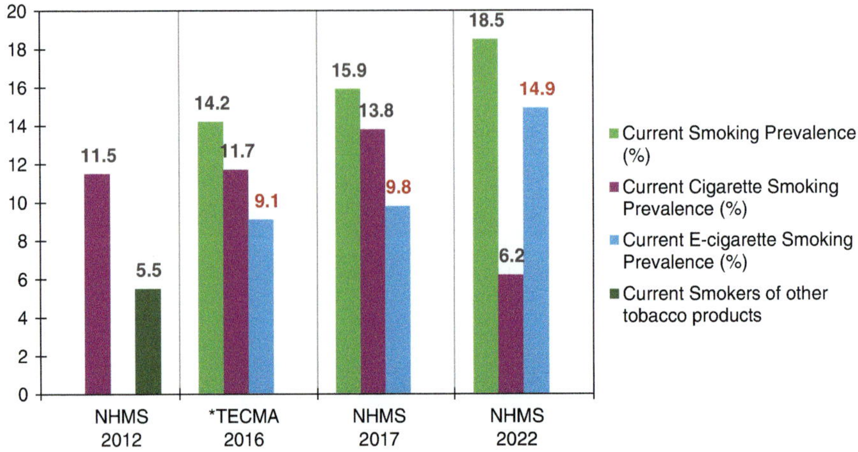

Fig. 19.2 Current E-cig Smoking Prevalence among youth in Malaysia. Data Sources: Report of NHMS Malaysia 2012 (IPH 2012b); Report of TECMA 2016 (IPH 2016); Report of NHMS 2017 (IPH 2017); and Report of NHMS 2022 (IPH 2022). Note: *The TECMA survey included youth aged 13–19, while the NHMSs included only those aged 13–17

address tobacco and e-cig use among youth. In fact, both smoking conventional cigarettes and using e-cigs present significant health risks to both smokers and non-smokers.

19.2 The Market Failure of E-cig Use

In economic theory, individuals are assumed to be rational and capable of making decisions in their best interests. This should lead to the most efficient allocation of society's resources and market fully efficient. However, this assumption does not hold true regarding the tobacco market, including e-cigs as well. In reality, there are numerous market failures due to poor decision-making, leading to market inefficiencies. For example, individuals often have incorrect perceptions of the health risks associated with smoking, as tobacco products are highly addictive, and there are externalities involved. As a result, government intervention in the tobacco market is seen as justified (Jha et al. 2000; Pekurinen 1991; Wilkins et al. 2001).

The category of market failures within the e-cig market is associated with the inaccurate perception of health risks. This is often due to imperfect information on the health impacts. As a result, individuals may not fully internalize the associated risks, which can lead to an unlikely of quitting smoking or vaping. For example, a study has identified misconceptions among young adults regarding vaping, and the majority did not perceive vaping as significantly harmful and may perceive e-cigs as a harmless alternative (Golan et al. 2023; Pepper et al. 2014; Rohde et al. 2018). In a study conducted in Malaysia by Sreeramareddy et al. (2023), it was found that out of a sample of 302 current e-cig smokers, 60.6% were male and 86% had achieved higher education. The main reasons given for using e-cigs included the belief that they were less harmful than conventional cigarettes (56.3%), enjoyment (46.7%), and the availability of various flavors (40.4%). The misleading health perspectives related to e-cigs, such as the belief that they are safer than conventional cigarettes or that they do not pose health risks, can lead to a false sense among smokers and non-smokers (Basáñez et al. 2018).

In fact, e-cigs can have adverse health effects, including immediate impacts on lung function and potential risks associated with exposure to harmful chemicals (Lee and Suzuki 2022). Studies found positive associations between e-cig use and respiratory symptoms among smokers (Alnajem et al. 2020; Osei et al. 2019; Wills et al. 2019). Research has shown that e-cig use affects vital signs, such as heart rate and blood pressure, with immediate physiological effects (McClelland 2022). In particular, vaping, particularly with certain flavors and containing nicotine, is not a safe alternative to cigarette smoking (McClelland 2022).

The spread of inaccurate information about health risks associated with vaping is often linked to the vape industry's practice of distorting and misleading information. The industry frequently promotes vaping products as a safer alternative to conventional cigarettes and as aids for smoking cessation. However, numerous studies have demonstrated that individuals using e-cigs are less likely to quit smoking compared to non-smokers and that e-cig use does not significantly assist in smoking cessation among the general adult population (Chen et al. 2023; Kalkhoran and Glantz 2016; Wang et al. 2021). For instance, Kalkhoran and Glantz (2016) found that smokers who used e-cigs had 28% lower odds of quitting smoking compared to those who did not use e-cigs. Additionally, those who are highly enthusiastic about vaping, such as customizing e-cigs, using non-tobacco flavors, and taking more puffs per use, as well as using e-liquids with higher nicotine content and tobacco flavors, may reduce their chances of quitting smoking (Brandon et al. 2019). Moreover, e-cig smokers found it more challenging to quit smoking mainly due to the easy accessibility of e-cigs for everyone (Kalkhoran and Glantz 2016). Therefore, reducing the accessibility of e-cig products in the market could potentially lead to smokers ceasing to use them.

Shroff and Sreeramareddy (2024) examined e-cig retailer websites in Malaysia in 2022 and found that most websites lacked effective age verification processes, with only 3% requiring identification proof. Additionally, marketing claims highlighted health benefits (47%) and promoted e-cigs as smoking cessation tools (37%) without supporting evidence. The study also reported various e-cigs and flavors being marketed, with most disclosing nicotine presence but with minimal health warnings. These findings point to the need for stricter regulations to prevent youth access and ensure accurate health information dissemination. This is crucial due to the rising prevalence of e-cig use among young people in Malaysia and the potential for nicotine addiction and other adverse health outcomes.

In addition, the addictive nature of e-cig use contributes to market failure. Once people start using e-cigs, their tobacco use becomes reinforced by both psychological addiction and nicotine addiction. Psychological addiction to smoking is similar to forming a habit with other addictive products or practices, yet nicotine addiction is not just a matter of choice or taste but rather a difficult-to-break addiction once it takes hold (Jha et al. 2000). Adolescents often underestimate the risk of addiction due to the lack of information compared to adults (Grossman et al. 1998; Jha et al. 2000).

Because of its highly addictive nature, nicotine poses a significant risk in e-cigs, especially those containing nicotine. Non-smokers who use e-cigs are at risk of becoming dependent on nicotine, making it difficult for them to quit and potentially leading to addiction to conventional tobacco products (WHO 2021). Leventhal et al. (2015) study indicated that adolescents who used e-cigs were more likely to start smoking combustible tobacco compared to non-smokers. The likelihood that youth who smoke e-cigs will eventually progress to regular use of conventional cigarettes is two to three times higher than that of those who do not (Berry et al. 2019). In Malaysia, dual smokers exhibited higher nicotine dependence, and more side effects compared to exclusive e-cig smokers. Those side effects included sore or dry mouth/throat (41.4%), cough (33.4%), headache (20.5%), and dizziness (16.2%) (Sreeramareddy et al. 2023).

Lastly, the externalities are another category of market failure. In an efficient condition, smokers should bear the costs and benefits of their consumption, meaning that they should not impose any costs on other people. However, studies have shown that exposure to second-hand aerosols can lead to respiratory symptoms among non-smokers (Alnajem et al. 2020; Yoshioka et al. 2023). Furthermore, exposure to second-hand aerosols from heated tobacco products has been associated with respiratory and cardiovascular abnormalities, highlighting the risks posed by such exposure (Imura and Tabuchi 2021). Another concern arises due to the presence of nicotine in e-liquids, particularly regarding pediatric exposure and the potential for nicotine poisoning (Ershad et al. 2020; Kamboj et al. 2016). For example, approximately 30 cases of vaping-related poisoning were reported between 2015 and 2020 in Malaysia, with most incidents involving children aged 1–4 years, who are at high risk of severe health outcomes, including death, after ingesting e-liquid (Das 2022).

19.3 Healthcare Cost and Economy Burden of E-cigs

Smoking behavior has imposed a significant healthcare system burden and economic cost on the country, and the emergence of e-cigs has exacerbated this issue. People who currently use e-cigs, whether exclusively or along with other tobacco products, were more likely to report poor health compared to those who never used tobacco. Those with poor health also tended to use healthcare services more and have more visits to the emergency room and doctor. In 2018, the total annual healthcare costs attributed to all current e-cig use were $15.1 billion ($2024 per user) in the United States (Wang et al. 2023).

Spending on e-cigs can be cheaper than conventional cigarettes (Sapru et al. 2020) or more (Agaku et al. 2021). Whether the expenses of e-cigs are higher or lower than those of conventional cigarettes, price sensitivity still significantly influences decision-making. Pepper and Brewer (2014) found that some smokers may perceive the cost of e-cig use as higher than they initially expected, which could lead to them discontinuing their use of e-cigs. Therefore, imposing taxation is necessary to make e-cigs and e-liquids more expensive and discourage the demand in the market. For example, the e-liquid tax in 2018 in Indonesia increased prices slightly and a decline in e-cig demand (Kowitt et al. 2022). In addition, Agaku et al. (2021) estimated that levying a tax rate of 37.5% to 75% on e-cigs in South Africa could generate up to ZAR 1.10 billion to ZAR 2.20 billion (US$0.08–US$0.15 billion) in tax revenue. If the Malaysian government adopts a similar approach to the high consumption of e-cigs in Malaysia, the tax revenue could be utilized for health promotion campaigns and programs, ultimately reducing e-cig use in the country.

From an economic perspective, the externalities associated with high spending on e-cigs can impose significant financial constraints on households, particularly those in low-income groups. The allocation of limited financial resources to e-cig products reduces the ability of these households to afford essential goods for themselves and their family members. It is crucial to raise awareness of the adverse impact of e-cig use, not only on health but also on household financial stability. By educating the public about these economic externalities, individuals can be encouraged to redirect their spending towards more essential needs, thereby enhancing overall financial well-being and lessening the socioeconomic burden of e-cig spending.

In addition to financial constraints, the use of e-cigs has been associated with a loss of productivity. Romberg et al. (2021) found that employees feel negatively impacted by others' use of e-cigs at work, which can reduce productivity even among those who do not vape. The employees taking breaks to vape could disrupt others' work as well. The employees also perceived that vaping at work disrupts their colleagues' work and causes anxiety among individuals who are concerned about the potential harm of second-hand aerosol. Therefore, it is important for public health strategies to focus on tackling the increasing use of e-cigs among adolescents and to work towards denormalizing and prohibiting the use of e-cigs in indoor places to reduce passive exposure (Alnajem et al. 2020). Similarly, in a study by Graham et al. (2020), parents who reported their child using e-cigs or being uncertain about it experienced work-related challenges. They felt anxious or worried (43.4%), had difficulty focusing (35.1%), were less productive (29.3%), and missed work (27.3%) as a result.

19.4 Conclusion

Urgent action is imperative to address the rising prevalence of e-cig use in Malaysia, to reduce the overall smoking rate to 15% by 2025 and 5% by 2045. The increasing popularity and perceived safety of e-cigs present a significant challenge, potentially undermining the progress made in reducing smoking rates. Market failures, externalities, and misinformation regarding the relative safety of e-cigs necessitate the implementation of comprehensive regulatory measures covering both e-cigs and conventional cigarettes. These strategies should include stricter marketing regulations, robust age verification processes, and explicit health warnings to mitigate their appeal, particularly among youth and non-smokers. Additionally, levying higher taxes on all tobacco products, including e-cigs and e-liquids, could help decrease demand and generate tax revenue for the government.

Public awareness campaigns are critical for educating the population about the adverse health impacts associated with both e-cigs and conventional cigarettes, as well as their addictive potential. These initiatives should aim to correct the misconceptions about e-cigs as a harmless alternative to smoking while emphasizing the health risks involved. Moreover, addressing the economic burden of using e-cigs and conventional cigarettes as well is also important. Excessive spending on either or both products can strain household finances, particularly in low-income groups, and diminish workplace productivity. Educating the public about these financial and societal impacts is able to encourage informed decision-making, potentially redirecting expenditure towards essential needs and enhancing overall well-being.

To achieve these targets, the Malaysian government needs to implement strict regulations, effective public awareness campaigns, and economic measures targeting not only e-cigs but also conventional cigarettes. With these actions, Malaysia can protect public health, reduce the burden of smoking, and ensure a healthier future for all Malaysians. In the long term, these actions will lead to sustained public health improvements, significant economic savings on healthcare costs, and a more productive and healthier society. By addressing these issues comprehensively, Malaysia can secure lasting health benefits for future generations, reduce smoking-related diseases, and create a more resilient healthcare system.

References

Agaku I, Egbe C, Ayo-Yusuf O (2021) Potential revenue from taxing e-cigarettes and comparison of annual costs of daily e-cigarette use versus daily cigarette smoking among South African adults. Tob Induc Dis 19(January):1–14. https://doi.org/10.18332/tid/131861

Alnajem A, Redha A, Alroumi D, Alshammasi A, Ali M, Alhussaini M, Almutairi W, Esmaeil A, Ziyab AH (2020) Use of electronic cigarettes and secondhand exposure to their aerosols are associated with asthma symptoms among adolescents: a cross-sectional study. Respir Res 21(1):300. https://doi.org/10.1186/s12931-020-01569-9

Basáñez T, Majmundar A, Cruz TB, Unger JB (2018) Vaping associated with healthy food words: a content analysis of Twitter. Addict Behav Rep 8:147–153. https://doi.org/10.1016/j.abrep.2018.09.007

Berry KM, Fetterman JL, Benjamin EJ, Bhatnagar A, Barrington-Trimis JL, Leventhal AM, Stokes A (2019) Association of electronic cigarette use with subsequent initiation of tobacco cigarettes in US youths. JAMA Netw Open 2(2):e187794. https://doi.org/10.1001/jamanetworkopen.2018.7794

Brandon KO, Simmons VN, Meltzer LR, Drobes DJ, Martínez Ú, Sutton SK, Palmer AM, Bullen CR, Harrell PT, Brandon TH (2019) Vaping characteristics and expectancies are associated with smoking cessation propensity among dual users of combustible and electronic cigarettes. Addiction 114(5):896–906. https://doi.org/10.1111/add.14551

Centers for Disease Control and Prevention (2023) 2023 Global Adult Tobacco Survey (GATS) Malaysia. https://drupal.gtssacademy.org/sites/default/files/country_docs/2023_Logo_GATS%20Malaysia_ES_508%20%281%29.pdf

Chen R, Pierce JP, Leas EC, Benmarhnia T, Strong DR, White MM, Stone M, Trinidad DR, McMenamin SB, Messer K (2023) Effectiveness of e-cigarettes as aids for smoking cessation: evidence from the PATH Study cohort, 2017–2019. Tob Control 32(e2):e145–e152. https://doi.org/10.1136/tobaccocontrol-2021-056901

Das M (2022) Increase in vaping and e-cigarette use by children in Malaysia. Lancet Oncol 23(11):1360. https://doi.org/10.1016/S1470-2045(22)00596-4

Ershad M, Beld J, Mostafa A, Dela Cruz M, Vearrier D, Greenberg MI (2020) Nicotine content from cigarettes submerged in soda. J Med Toxicol 16(4):452–457. https://doi.org/10.1007/s13181-020-00766-8

Golan R, Muthigi A, Ghomeshi A, White J, Saltzman RG, Diaz P, Ramasamy R (2023) Misconceptions of vaping among young adults. Cureus. https://doi.org/10.7759/cureus.38202

Graham AL, Amato MS, Jacobs MA, Romberg AR, Diaz MC, Rahman B, Schillo BA (2020) Vaping in the workplace. J Occup Environ Med 62(12):986–992. https://doi.org/10.1097/JOM.0000000000002013

Grossman M, Chaloupka FJ, Anderson R (1998) A survey of economic models of addictive behavior. J Drug Issues 28(3):631–643. https://doi.org/10.1177/002204269802800304

Imura Y, Tabuchi T (2021) Exposure to secondhand heated-tobacco-product aerosol may cause similar incidence of asthma attack and chest pain to secondhand cigarette exposure: the JASTIS 2019 study. Int J Environ Res Public Health 18(4):1766. https://doi.org/10.3390/ijerph18041766

IPH (2012a) Global Adult Tobacco Survey (GATS) Malaysia 2011. Ministry of Health Malaysia

IPH (2012b) The National Health and Morbidity Survey 2012. Malaysia global school based student health survey 2012. Ministry of Health Malaysia

IPH (2015) National Health And Morbidity Survey 2015 – Report On Smoking Status among Malaysian Adults. Ministry of Health Malaysia.

IPH (2016) Tobacco & E-Cigarette Survey Among Malaysian Adolescents (TECMA) 2016 (Aris T, Abd Ghani AA, Mohd Yusoff MF, Robert TGR, Tee HGH, Mohd Hasri NH, Manan NS, Mohd Radzi NF, Kamaruddin NA eds). Ministry of Health Malaysia

IPH (2017) National Health and Morbidity Survey (NHMS) 2017: adolescent health survey 2017. Ministry of Health Malaysia, Malaysia

IPH (2022) National Health and Morbidity Survey 2022: adolescent health survey. Ministry of Health Malaysia

IPH (2024) Key findings: national health and morbidity surveys 2023. Non-Communicable Diseases and Healthcare Demand. https://iku.gov.my/nhms-2023

Jha P, Musgrove P, Chaloupka FJ, Yurekli A (2000) The economic rationale for intervention in the tobacco market. In: Tobacco control in developing countries. Oxford University Press, New York, pp 153–174. https://doi.org/10.1093/oso/9780192632500.003.0007

Kalkhoran S, Glantz SA (2016) E-cigarettes and smoking cessation in real-world and clinical settings: a systematic review and meta-analysis. Lancet Respir Med 4(2):116–128. https://doi.org/10.1016/S2213-2600(15)00521-4

Kamboj A, Spiller HA, Casavant MJ, Chounthirath T, Smith GA (2016) Pediatric exposure to E-cigarettes, nicotine, and tobacco products in the United States. Pediatrics 137(6). https://doi.org/10.1542/peds.2016-0041

Kowitt SD, Anshari D, Orlan EN, Kim K, Ranney LM, Goldstein AO, Byron MJ (2022) Impact of an e-cigarette tax on cigarette and e-cigarette use in a middle-income country: a study from Indonesia using a pre–post design. BMJ Open 12(5):e055483. https://doi.org/10.1136/bmjopen-2021-055483

Lee C, Suzuki JB (2022) The risk of oropharyngeal cancer from E-cigarette use: an urgent public health concern. J Public Health Int 5(4):1–4. https://doi.org/10.14302/issn.2641-4538.jphi-22-4335

Leventhal AM, Strong DR, Kirkpatrick MG, Unger JB, Sussman S, Riggs NR, Stone MD, Khoddam R, Samet JM, Audrain-McGovern J (2015) Association of electronic cigarette use with initiation of combustible tobacco product smoking in early adolescence. JAMA 314(7):700. https://doi.org/10.1001/jama.2015.8950

Mackay J, Ritthiphakdee B, Reddy KS (2013) Tobacco control in Asia. Lancet 381(9877):1581–1587. https://doi.org/10.1016/S0140-6736(13)60854-5

McClelland ML (2022) Comparison of different types of vape products on vital signs following twenty minutes of vaping and exposure to vapor. J Lung Pulm Respir Res 9(1):1–6. https://doi.org/10.15406/jlprr.2022.09.00270

Osei AD, Mirbolouk M, Orimoloye OA, Dzaye O, Uddin SMI, Dardari ZA, DeFilippis AP, Bhatnagar A, Blaha MJ (2019) The association between e-cigarette use and asthma among never combustible cigarette smokers: behavioral risk factor surveillance system (BRFSS) 2016 & 2017. BMC Pulm Med 19(1):180. https://doi.org/10.1186/s12890-019-0950-3

Pekurinen MJ (1991) Economic aspects of smoking is there a case for government intervention in Finland?

Pepper JK, Brewer NT (2014) Electronic nicotine delivery system (electronic cigarette) awareness, use, reactions and beliefs: a systematic review. Tob Control 23(5):375–384. https://doi.org/10.1136/tobaccocontrol-2013-051122

Pepper JK, Emery SL, Ribisl KM, Southwell BG, Brewer NT (2014) Effects of advertisements on smokers' interest in trying e-cigarettes: the roles of product comparison and visual cues. Tob Control 23(suppl 3):iii31–iii36. https://doi.org/10.1136/tobaccocontrol-2014-051718

Rohde JA, Noar SM, Horvitz C, Lazard AJ, Cornacchione Ross J, Sutfin EL (2018) The role of knowledge and risk beliefs in adolescent e-cigarette use: a pilot study. Int J Environ Res Public Health 15(4):830. https://doi.org/10.3390/ijerph15040830

Romberg AR, Diaz MC, Briggs J, Stephens DK, Rahman B, Graham AL, Schillo BA (2021) Vaping in the workplace: prevalence and attitudes among employed US adults. J Occup Environ Med 63(1):10–17. https://doi.org/10.1097/JOM.0000000000002061

Sapru S, Vardhan M, Li Q, Guo Y, Li X, Saxena D (2020). E-cigarettes use in the United States: reasons for use, perceptions, and effects on health. BMC Public Health 20(1):1518. http://doi.org/10.1186/s12889-020-09572-x

Shroff SM, Sreeramareddy CT (2024) Marketing claims, promotional strategies, and product information on Malaysian e-cigarette retailer websites-a content analysis. Subst Abuse Treat Prev Policy 19(1):11. https://doi.org/10.1186/s13011-024-00592-z

Sreeramareddy CT, Shroff SM, Gunjal S (2023) Nicotine dependence and associated factors among persons who use electronic e-cigarettes in Malaysia – an online survey. Subst Abuse Treat Prev Policy 18(1):51. https://doi.org/10.1186/s13011-023-00558-7

Wang RJ, Bhadriraju S, Glantz SA (2021) E-cigarette use and adult cigarette smoking cessation: a meta-analysis. Am J Public Health 111(2):230–246. https://doi.org/10.2105/AJPH.2020.305999

Wang Y, Sung H-Y, Lightwood J, Yao T, Max WB (2023) Healthcare utilisation and expenditures attributable to current e-cigarette use among US adults. Tob Control 32(6):723–728. https://doi.org/10.1136/tobaccocontrol-2021-057058

WHO (2021) WHO report on the global tobacco epidemic, 2021 Addressing new and emerging products

Wilkins N, Yurekli A, Hu T-W (2001) Tool 3. Demand analysis economic analysis of tobacco demand. https://documents.worldbank.org/en/publication/documents-reports/documentdetail/390251468322463134/economic-analysis-of-tobacco-demand#:~:text=Economic%20analysis%20of%20tobacco%20demand%20(English)&text=Tool%203%20explains%20how%20to,likely%20impact%20of%20changes...

Wills TA, Pagano I, Williams RJ, Tam EK (2019) E-cigarette use and respiratory disorder in an adult sample. Drug Alcohol Depend 194:363–370. https://doi.org/10.1016/j.drugalcdep.2018.10.004

Yoshioka T, Shinozaki T, Hori A, Okawa S, Nakashima K, Tabuchi T (2023) Association between exposure to secondhand aerosol from heated tobacco products and respiratory symptoms among current non-smokers in Japan: a cross-sectional study. BMJ Open 13(3):e065322. https://doi.org/10.1136/bmjopen-2022-065322

20 Electronic Smoking Devices (ESDs): Future Challenges for Southeast Asia

Anton Oliver M. Javier, Tan Yen Lian, and E. Ulysses Dorotheo

Abstract

Rapidly evolving electronic cigarettes (e-cigs) and heated tobacco products have circumvented regulations designed for conventional tobacco products. Capitalizing on the slow development of regulations and the weak implementation of laws in countries that have legalized these products, the tobacco industry has orchestrated the mainstreaming and acceptance of these electronic smoking devices to drive and sustain nicotine addiction, primarily by claiming these products are "less harmful" alternatives to cigarettes. Over time, despite long-term harms still to be identified, immediate and chronic health impacts of these products have emerged, highlighting gaps in public policy that have failed to protect the public, particularly youth, who are most targeted by these products. Governments can do much to protect public health by strengthening existing tobacco control legislation, including banning flavored products, banning any products that mimic tobacco, plugging policy gaps that are exploited by tobacco companies, and ensuring that policy development processes are protected from the profit-driven agenda of the globalized tobacco industry.

Keywords

Regulation · Product design · Renormalization · Smoking · Youth uptake

A. O. M. Javier (✉) · T. Y. Lian · E. U. Dorotheo
Southeast Asia Tobacco Control Alliance, Bangkok, Thailand
e-mail: anton@seatca.org

Y.-H. Leong, A. S. Mohd Samin (eds.), *E-Cigarettes*,
https://doi.org/10.1007/978-981-96-5087-3_20

20.1 Introduction

The wide variety and rapidly and continually evolving sizes, shapes, and components (e.g., heating element/method, puff capacity, and use of Bluetooth technology) of electronic smoking devices (ESDs)[1]; the wide variety of substrates that are already in the market or still to be introduced[2]; and non-standardized and customizable levels of nicotine delivery[3] pose multiple regulatory challenges, leaving many countries scrambling to develop appropriate regulatory approaches to control the emergence and growing popularity of newer tobacco and recreational nicotine products (National Academies of Sciences, Engineering, and Medicine 2018).

ESDs have also been adapted to deliver a variety of substances, including illicit drugs such as cannabis, synthetic cannabinoids, methamphetamine, MDMA (ecstasy), synthetic cathinones, cocaine, and heroin (Breitbarth et al. 2018).

As these products have evolved to become easier-to-conceal over time, regulators have found it increasingly difficult to implement strategies to detect and intercept these products and subject them to appropriate regulations (National Academies of Sciences, Engineering, and Medicine 2018).

Further complicating the risk assessment of these products, the tests designed to evaluate the levels of certain priority toxic emissions are not validated as quickly as these toxicants are being identified. As of the tenth session of the Conference of the Parties to the WHO Framework Convention on Tobacco Control (FCTC) (COP10), the only tests validated by the WHO Tobacco Laboratory Network (TobLabNet) for contents of e-cigs are gas chromatography-flame ionization tests for the detection of glycerin, propylene glycol, water and nicotine (WHO 2023c).

The evolution of ESD flavors has also complicated risk assessments conducted by scientists and health regulators. Over 15,000 e-cig flavors are commercially available, including a variety of fruit, dessert/candy/sweets, coffee/tea, alcohol, tobacco, mint/menthol, nuts, spices/pepper, and other characterizing flavor categories (Krüsemann et al. 2019). In addition to toxicity of inhaled flavor additives, concerns within allied health and tobacco control communities focus on flavors attracting youth susceptible to addiction and exacerbating health inequalities among people of color (Ma et al. 2022), who are generally not reached as effectively by smoking cessation programs (Watson et al. 2023).

Given that the tobacco industry model is built on addiction, we can expect that the industry will continue to find new ways to create nicotine addiction more quickly and sustain it more deeply, such as has been done in the past with cigarettes and more recently with synthetic nicotine and nicotine salts.

[1] Including e-cigarettes and heated tobacco products (HTP).

[2] For example, non-tobacco nicotine substrates used in HTP devices and the lack of any standard formula or proportion of ingredients for e-liquids/gels.

[3] Including use of nicotine salts.

20.2 Long-Term Impacts on Health, Society, and the Environment

ESD emissions are not harmless. Despite being introduced into the global market only a little over two decades ago, the short- and medium-term harms of ESDs are already evident and well-documented, while long-term, more chronic, harms are beginning to emerge.

At COP10, the WHO reported that, contrary to industry claims, new and emerging tobacco products, particularly Heated tobacco products, produce smoke, which is "produced whenever substances are heated beyond a temperature at which pyrolysis occurs" (WHO 2023b) and which is universally known to be harmful to human health. The long-term effects of exposure to ESD smoke, including fine particles and various toxic chemicals, have not yet been fully defined but can be anticipated as producing similar harmful effects as other forms of air pollution, including tobacco smoke.

The industry's own research hides the fact that these products emit harmful and potentially harmful chemicals (HPHC) that may even exceed levels found in conventional tobacco products. Twenty-two HPHCs are over twice as high and seven are over 10 times higher in the smoke of the HTP, IQOS, compared to cigarette smoke (Glantz 2018). E-cigs, on the other hand, have high levels of heavy metals from the devices or the e-liquids themselves (Aherrera et al. 2023). ESDs also emit chemicals not found in cigarette smoke, such that their chemical risk profile is both different from cigarette smoke and not fully known.

Chronic diseases arising from e-cig use, like chronic lung injury, have been observed (Poschenrieder et al. 2020). E-cigs are also known to contain cancer-causing tobacco-specific nitrosamines and have been linked to DNA changes that cause oral cancer (Herzog et al. 2024). This is in addition to the more acute diseases like injuries, burns (Brownson et al. 2016), nicotine addiction, chemical toxicities from the contents, emissions and the devices themselves; and e-cig and vaping product use-associated lung injury (EVALI) (US Food and Drug Administration 2020).

Although e-cigs are promoted as smoking cessation products, there is limited or insufficient evidence of their efficacy as such (Banks et al. 2023). In addition, among ex-smokers, there is limited evidence that the likelihood of smoking relapses is about twice as high for those who use e-cigs than for those who do not (Baenziger et al. 2021). HTPs have also been found ineffective in quitting or preventing smokers' relapse into conventional cigarette use (Odani et al. 2023). Both products are associated with dual and poly-use of cigarettes and ESDs, which carry more risk than only smoking conventional cigarettes (Tzu-Hsuan Chen 2023; Petersen et al. 2020).

There is strong evidence that young never-smokers and non-smokers who use e-cigs are about three times as likely as non-users to start smoking tobacco and to become regular smokers (Banks et al. 2023). Population-level studies in the United States have shown that ever-users of some emerging tobacco products are more likely to smoke within as little as one year of follow-up (Berry et al. 2019). Additionally, designs that mimic toys or confectionery make e-cigs more appealing to children, potentially erasing decades of effort to prevent youth uptake. Furthermore, e-cigs are being designed so that they are either more easily concealed by or more acceptable to children (US Centers for Disease Control and Prevention 2016).

Public health experts caution, therefore, that ESD use may renormalize smoking, erasing decades of tobacco control work aimed at deformalizing smoking and portending a future where it might be "normal" for people to use tobacco products again.

20.3 Public Awareness and Industry Misinformation

Public awareness about ESDs is growing (Jackson et al. 2024), but it remains generally limited, and there is significant misinformation promoted by the industry and its front groups on a variety of topics.

Tobacco companies and advocates for the industry's version of "tobacco harm reduction" frequently use the words "vapor" and "vaping" to inaccurately describe the emissions of ESDs, claiming that a lack of combustion does not produce smoke or tar, which they point to as the cause of health harms (Tan and Bigman 2020) and that, resulting from the absence of combustion, HPHCs in ESD "vapor" are reduced by more than 95% compared to cigarette smoke. They therefore promote ESDs as less harmful alternatives to cigarettes or harm reduction choices for people who they incorrectly assume cannot or will not quit smoking.

ESD proponents also spread inaccurate information about nicotine, which they claim is not the cause of disease (some even claim that nicotine is not addictive). While it is crucial to distinguish the harmful effects of conventional cigarettes from those of nicotine alone, tobacco companies frequently downplay the risk of nicotine itself, in the marketing of their products as less harmful alternatives, a claim they often pitch through online social media (Silver et al. 2022).

While these devices were evolving rapidly, a large amount of research studies financed or influenced by the tobacco industry emerged to yield industry-favorable conclusions about their safety (Pisinger et al. 2019). The industry and its proxies often portray attempts to regulate ESDs as an anti-science policy direction.

It is essential that researchers, educators, youths, parents, health professionals, policy makers, government officials, and the media strive constantly to counter this steady stream of misinformation by the industry. For example, the European Lung Foundation and the European Respiratory Society have publicly stated that e-cigs and HTPs are not viable options for quitting tobacco and that these products only serve to sustain addiction rather than act as an off-ramp from tobacco use (ELF and ERS 2020).

20.4 Policy Gaps and Legal Challenges

More than 40 countries have taken decisive action to ban ESDs, protect public health, and avert the health burden of ESDs (including the risk of poor health, illness, poisoning, physical injury from burns and explosions, and premature death)

(GGTC 2021). Among these are six out of ten ASEAN countries (Brunei, Cambodia, Lao PDR, Singapore, Vietnam, and Thailand), which have proactively banned emerging tobacco products under the precautionary principle (GGTC 2021; Vietnam News 2024). Outside of the ASEAN, Hong Kong, which has a consistently declining prevalence for tobacco use, has implemented, since 2021, a ban on ESDs until their impact on health and smoking prevalence are better understood. This position is still held by the Hong Kong Tobacco and Alcohol Control Office (n.d.). Some countries have very strict regulations amounting to a *de facto ban* such that use is permitted under a prescription-based model or other niche circumstances. In most countries, however, ESDs have not been regulated under the existing tobacco control law or regulation (WHO 2023a).

The e-cig industry, mirroring tobacco industry tactics, has strategically circumvented tobacco advertising bans and targeted youth through eye-level kiosk placements at point-of-sale and unregulated online platforms. Reports from professional societies and media outlets worldwide indicate that these devices often resemble toys, consumer products, and cartoon characters, appealing particularly to children and thwarting detection by regulatory authorities (American Lung Association 2024; Tatum et al. 2023; The Straits Times 2023; VietnamPlus 2024).

Much of the same strategies used to promote conventional tobacco products have been used to promote e-cigs. The company, Juul, makers of the eponymous e-cig product had emulated Big Tobacco's tactics in mainstreaming harmful products (JUUL Labs 2023). E-cig companies have strategically circumvented domestic tobacco advertising, promotion, and sponsorship bans in ASEAN countries and undermined evidence-based policy to aim their products at children. These products were placed in kiosks at point-of-sale, in establishments frequented by youth, and were promoted in online selling and social media platforms that were, at that time, not as tightly regulated (Jirathanapiwat and Reyes 2020). Product designs of current-generation e-cigs and heated tobacco products from all over the ASEAN are patently youth oriented in their design and marketing, and some of these same products have been sourced from online selling platforms (SEATCA 2023; Tan and Dorotheo 2024).

ESDs represent an important investment for the tobacco industry. The *global* tobacco industry is dominated by five transnational tobacco companies (TNTCs): China National Tobacco Corporation (CNTC), Philip Morris International (PMI), British American Tobacco (BAT), Japan Tobacco Inc. (JTI), and Imperial Tobacco Group (ITG). All have ventured into electronic nicotine delivery systems (ENDS) and HTPs. In a bid for market expansion, ESDs are marketed in many Asian countries, including ASEAN countries, which are perceived as growth markets (van der Eijk et al. 2022). From a base level of USD 22.5 billion in 2023, the e-cig market is expected to grow to USD 37.7 billion in just 10 years (Spherical Insights LLP 2024).

Industry exploitation of policy gaps complicates effective regulation and enforcement. With the rapid evolution of these products, defining them for legislative purposes is an ongoing challenge. Whenever these products remain unregulated or only partially regulated, the industry continues to exploit policy gaps in prevailing legislation to evade taxation, pictorial health warnings, and other regulations.

Despite some governments taking a very strong position banning the sale, importation, distribution, and use of these ESD products, gaps still remain in all ASEAN countries that could be remedied by amendments to prevailing legislation or enhanced enforcement initiatives.

20.4.1 The Policy-Making Process Should Be Isolated from the Tobacco Industry

In many ways, dissociating the tobacco industry from the policy development process is the lynchpin of implementing good governance for tobacco control.

In the Philippines, the Food and Drug Administration was met with legal action in 2022 for attempting to regulate ENDS based on scientific evidence independent of the tobacco industry. It had set limits on nicotine concentration, defined policies around the supply chain of emerging tobacco products, and regulated product communications (Philippine FDA 2019) in line with the FDA Act of 2009 (Congress of the Philippines 2008). This strategy courted legal action by organizations promoting the interests of the tobacco industry (De Vera and Noriega 2019) thus stalling implementation until a heavily pro-industry bill regulating these products ultimately passed into law. This new law, Republic Act No. 11900, lowered the age of access for emerging tobacco products to 18 years of age (from 21) and transferred ENDS and HTPs from the FDA's purview to that of the industry-friendly Department of Trade and Industry (Congress of the Philippines 2022).

Indonesia, on the other hand, has limited guidelines (or a code of conduct with regard to how government officials should engage with the tobacco industry) that only apply to the Ministry of Health. This limited coverage and scope pose a challenge to prevent and monitor interactions of the tobacco industry with other ministries beyond the Ministry of Health. In June 2021, the Minister of Investment expressed support for Sampoerna's investment in establishing a HTP factory in Indonesia. This includes backing the implementation of National Standardize Product (SNI) for HTP, excluding pictorial health warnings for HTP, and imposing lower excise tariffs (Minister of Investment 2021). The Coordinating Ministry for the Economy and the Ministry of Industry also supported the development of special regulations for new tobacco products, with research aided by the National Research and Innovation Agency (BRIN) (Riswan 2023). Indonesia has not signaled any commitment to the WHO FCTC despite its commitment to the Sustainable Development Goals which, technically, includes the pursuit of effective tobacco control strategies (UN 2024).

For its part, the Malaysian government struck nicotine off its list of regulated substances under the Poison Act of 1952. In effect, this overturned a proposal of banning the use of ENDS. In conjunction with a rolled back generational endgame (GEG) policy to prevent children born post-2007 from purchasing tobacco products, the move had effectively legalized the sale and use of e-cigs in Malaysia.

Tobacco industry interference has scaled up in Malaysia, putting it in the upper third of all countries rated under the Global Tobacco Industry Interference Index

2023 (GGTC 2023) in terms of how much industry interference was observed in the country. Members of Parliament themselves have exposed the tobacco industry for its lobbying activities that led to nicotine being taken off the controlled poisons list and the rollback of the GEG policy to reduce smoking among Malaysian youth (CodeBlue 2024).

20.4.2 Law Enforcement Needs Capacity-Building to Improve Detection of ESDs

ESDs are banned in Cambodia. However, the Cambodia Movement for Health has identified various e-cig products available in the local market that were designed to mimic children's toys.

In other countries, such products have escaped detection despite stringent regulations around electronic smoking devices and have caused an uptick in e-cig use among children and young adults. According to the 2022 Global Youth Tobacco Survey (GYTS) conducted in Cambodia, 1.1% of boys and 0.8% of girls currently use e-cigs (Cambodia Ministry of Health 2023).

ESDs are banned in Lao PDR (Lao PDR 2021), in line with WHO recommendations (WHO 2023d). However, the country has also seen the entry of ESDs that simulate the appearance of toys, confectionery, and childhood articles which then escape enforcement of currently existing regulations. In recent times, the Laotian government has cited increasing nicotine addiction among the youth (5% of boys and 3.7% of girls currently use e-cigs) (Lao Ministry of Health 2017) as a reason for further strengthening its laws to include e-cigs in its list of addictive substances (Lapuekou 2023).

Governments can enhance internal cooperation between departments and with online selling platforms for effective enforcement in order to protect prior public health gains, Singapore pre-emptively banned "imitation tobacco products" under its legislation to protect its population from the then-unknown impacts of these emerging smoking products. Singapore's policy on ESDs covers many areas of tobacco control, including raising awareness on the harms of tobacco products. Singapore has no official data on the prevalence of e-cig and HTP use; however, e-cig use among youth is not uncommon despite the ban (CNA 2023).

While effective, Singapore's enforcement initiatives do still catch coordinated attempts at marketing ESDs to vulnerable youth; recalcitrant offenders occasionally still sell e-cigs through the internet outside the two largest online selling platforms. These fly-by-night operators are relatively more difficult to catch because their online presence is not captured by regulation. It is also possible for ESDs to be accessed through web pages outside of Singapore where selling ESDs might not necessarily be illegal. The Ministry of Communications and Information can be more actively involved to facilitate regulatory action on makeshift online spaces where these illegal products are traded.

Thailand, like Singapore, pre-emptively banned ESDs for having met the definition of "imitation tobacco products". Despite this, toy-like e-cigs have emerged in the illicit market and are available in Thailand through fly-by-night sellers who skirt marketing bans (Chaolan 2024). E-cig use more than doubled among school children aged 13 to 15 from 3.3% in 2015 to 8.1% in 2021, according to the Global School-based Student Health Survey (Thailand MOPH, WHO, US CDC 2015; Vandelaer 2023). The Global Youth Tobacco Survey 2022 reported that 20.2% of boys and 15% of girls were using ESDs (Thailand Ministry of Public Health 2022). The ease with which youth are targeted by online advertisements complicates enforcement of the ban on electronic smoking products.

Furthermore, e-cig proponents are lobbying to lift the ban on ESDs. On the other hand, civil society and youth organizations are also calling to maintain the ban to protect public health (Prime Minister News 2024). Adopting comprehensive regulatory measures is crucial in countries where electronic smoking devices remain accessible. Brunei, like Singapore and Thailand, has done well to isolate itself from tobacco industry interference and has done well in priority action areas of tobacco control. Despite this, marketing of e-cigs is not illegal in Brunei (although sales remain illegal) (CTFK 2020). When product communications are unregulated, these products can still make their way to consumers, and e-cig use in Brunei Darussalam has reached a high of 17% for adult ever-users (Brunei Ministry of Health 2016) and 13.3% among youth (Brunei Ministry of Health, US CDC, WHO 2019).

For a long time, Indonesia has not had regulations around the manufacture, importation, distribution, retail, marketing, and use of ENDS. Currently, Indonesia has some restrictions on product communications, promotion, corporate social responsibility, and age of access (Government of Indonesia 2024). Furthermore, Indonesian tobacco tax rates remain relatively low, making these harmful products easily accessible to the youth.

While the Philippines does have regulations around the use of ESDs in line with its smoke- and aerosol-free policies (Republic of the Philippines – Office of the President 2017; Republic of the Philippines – Office of the President 2020), these policies have yet to be aligned with the WHO FCTC, and there is varying commitment to enforcement across subnational levels of government.

While the Philippines' policies on taxation of conventional tobacco products are hailed as a global best practice, its taxes for ESDs are comparably low, making these products still accessible to youth. Furthermore, because e-cigs (up to the fourth generation) and HTPs are sold with cartridges and refills, and devices are treated differently under the law, the devices themselves tend to escape tobacco taxes despite being used exclusively for e-smoking (Republic of the Philippines 2019).

Almost no regulation exists for e-cigs in Myanmar. Furthermore, there is very little data with regard to the prevalence of the use of e-cigs among the Myanmar population. It is important to conduct baseline studies to estimate the harmful impact of e-cigs on the population. However, there is space for proactive measures to protect the Myanmar population, especially from emerging tobacco products. Around half of the Myanmar population (44.1%) were noted to be using conventional tobacco products in 2020 (WHO 2024) and additional public health harms could be introduced by ESDs, especially with dual use with conventional tobacco products.

ESDs have been recently banned in Vietnam after hundreds of poisoning cases arising from electronic cigarette consumption over the past two years (Vietnam Express International 2024; Vietnam News 2024). Local civil society organizations have repeatedly advocated for a proactive ban on ESDs amidst the growing use of these products by the youth. ENDS use among youth was 3.5% in 2022 (Vietnam Ministry of Health, WHO 2022) from 2.6% in 2019 (Vietnam Ministry of Health, WHO, US CDC, Vietnam Ministry of Education, Hanoi University of Public Health 2019). While this value was still low, the Vietnamese government was urged by health advocates to act swiftly because the uptake of emerging tobacco products among the youth was far quicker than the uptake of conventional tobacco. Vietnam, however, has yet to implement its ban on ESDs.

20.5 Conclusion

By banning ESDs, Indonesia, Myanmar, and Vietnam have an opportunity to prevent further uptake of nicotine and the renormalization of smoking. Indonesia, in particular, can take immediate steps to plug important gaps in its new Government Regulation on Health, to enhance tobacco control regulations and better protect the youth. The Philippines and Malaysia, on the other hand, need to address tobacco industry interference in order to protect the policy-making process from the influence of the tobacco industry. Both countries should take immediate steps to work within their respective policy environments to be more compliant with the WHO FCTC and preserve or salvage previous milestones in tobacco control. Countries like Brunei, Cambodia, Lao PDR, Thailand, Vietnam, and Singapore, which have already banned e-cigs and HTPs, should enhance their capabilities for enforcement; they can also cooperate with distributors, online selling platforms, and international authorities to help prevent, curtail, and address violations committed online and out in the open in illicit markets. With the exception of Singapore, most ASEAN countries need to undertake targeted and cost-effective surveillance initiatives to monitor the use of tobacco products among the youth. Young people are more likely to take up ESDs faster than conventional tobacco products, and higher-resolution epidemiological studies are needed to monitor the trends of youth use in response to current policies.

References

Aherrera A, Lin JJ, Chen R, Tehrani M, Schultze A, Borole A, Tanda S, Goessler W, Rule AM (2023) Metal concentrations in e-cigarette aerosol samples: a comparison by device type and flavor. Environ Health Perspect 131(12). https://doi.org/10.1289/EHP11921

American Lung Association (2024, Jan 24) 9 of the strategies big tobacco used to target kids with e-cigarettes. State of Tobacco Control. Retrieved from https://www.lung.org/research/sotc/by-the-numbers/8-things-industry-ecigs

Baenziger ON, Ford L, Yazidjoglou A, Joshy G, Banks E (2021) E-cigarette use and combustible tobacco cigarette smoking uptake among non-smokers, including relapse in former smokers:

umbrella review, systematic review, and meta-analysis. BMJ Open 11(3):e045603. https://doi.org/10.1136/bmjopen-2020-045603

Banks E, Yazidjoglou A, Brown S, Nguyen M, Martin M, Beckwith K, Daluwatta A, Campbell S, Joshy G (2023) Electronic cigarettes and health outcomes: umbrella and systematic review of the global evidence. Med J Aust 218(6):267–275. https://doi.org/10.5694/mja2.51890

Berry KM, Fetterman JL, Benjamin EJ, Bhatnagar A, Barrington-Trimis JL, Leventhal AM, Stokes A (2019) Association of electronic cigarette use with subsequent initiation of tobacco cigarettes in US youths. JAMA Netw Open 2(2):e187794. https://doi.org/10.1001/jamanetworkopen.2018.7794

Breitbarth AK, Morgan J, Jones AL (2018) E-cigarettes—an unintended illicit drug delivery system. In: Drug and alcohol dependence, vol 192. Elsevier Ireland Ltd., pp 98–111. https://doi.org/10.1016/j.drugalcdep.2018.07.031

Brownson EG, Thompson CL, Goldsberry S, Chong HJ, Friedrich JB, Pham TN, Arbabi S, Carrougher GJ, Gibran NS (2016) Explosion injuries from e-cigarettes. N Engl J Med 375(14):1400–1402. https://doi.org/10.1056/NEJMc1608478

Brunei Ministry of Health (2016) Brunei STEPS survey 2015/2016 microdata. https://extranet.who.int/ncdsmicrodata/index.php/catalog/494/data-dictionary/F1?file_name=brn2015

Brunei Ministry of Health U.S. Centers for Disease Control and Prevention, World Health Organization (2019) Global Youth Tobacco Survey: Brunei Darussalam, 2019, ages 13-15. U.S. Centers for Disease Control and Prevention. https://drupal.gtssacademy.org/wp-content/uploads/2024/11/Brunei-Darussalam-GYTS-2019-Factsheet-Ages-13-15-Final_508tag.pdf

Brunei Ministry of Health (2021) Global youth tobacco survey. Fact sheet. Brunei Darussalam

Cambodia Ministry of Health (2023) Country report: Global Youth Tobacco Survey (GYTS) Cambodia 2022. National Centre for Health Promotion, Ministry of Health, Cambodia

Campaign for Tobacco-Free Kids (2020, Feb 26) Legislation by country/jurisdiction—Brunei. Retrieved from https://www.tobaccocontrollaws.org/legislation/brunei-darussalam

Channel News Asia (2023, Apr 17) The big read: with vaping increasingly rampant among youths despite ban, what more can authorities do? Retrieved from https://www.channelnewsasia.com/singapore/vaping-youths-cigarettes-legalise-vape-regulate-smoke-3419681

Chaolan S (2024, Apr 24) Raids target e-cigarette business. Bangkok Post. https://www.bangkokpost.com/thailand/general/2780154/raids-target-e-cigarette-businesses

CodeBlue (2024, Mar 14) Deputy minister shockingly admits tobacco, vape industry killed GEG proposal. CodeBlue. Retrieved from https://codeblue.galencentre.org/2024/03/14/deputy-minister-shockingly-admits-tobacco-vape-industry-killed-geg-proposal/

Congress of the Philippines (2008) Republic Act No. 9711, otherwise known as An Act Strengthening and Rationalizing the Regulatory Capacity of the Bureau of Food and Drugs (BFAD) by Establishing Adequate Testing Laboratories and Field Offices, Upgrading Its Equipment, Augmenting Its Human Resource Complement, Giving Authority to Retain Its Income, Renaming It the Food and Drug Administration (FDA), Amending Certain Sections of Republic Act No. 3720, as Amended, and Appropriating Funds Thereof. Official Gazette of the Republic of the Philippines. Retrieved from https://www.officialgazette.gov.ph/2009/08/18/republic-act-no-9711/

Congress of the Philippines (2019) Republic Act No. 11346. An act increasing the excise tax on tobacco products, imposing excise tax on heated tobacco products and vapor products, increasing the penalties for violations of provisions on articles subject to excise tax, and earmarking a portion of the total excise tax collection from sugar-sweetened beverages, alcohol, tobacco, heated tobacco and vapor products for universal health care, Amending for this purpose sections 144, 145, 146, 147, 152, 164, 260, 262, 263, 265, 288, and 289, repealing section 288(B) and 288(C), and creating new sections 263-a, 265-B, and 288-A of the National Internal Revenue Code of 1997, as amended by Republic Act No. 10963, and for other purposes. Congress of the Philippines. Retrieved from https://elibrary.judiciary.gov.ph/thebookshelf/showdocs/2/92803#:~:text=July%2025%2C%202019%20%5D-,AN%20ACT%20INCREASING%20THE%20EXCISE%20TAX%20ON%20TOBACCO%20

PRODUCTS%2C%20IMPOSING,SWEETENED%20BEVERAGES%2C%20ALCOHOL%2C%20TOBACCO%2C

Congress of the Philippines (2022) An act regulating the importation, manufacture, sale, packaging, distribution, use, and communication of vaporized nicotine and non-nicotine products, and novel tobacco products. Retrieved from https://legacy.senate.gov.ph/republic_acts/ra%2011900.pdf

De Vera E, Noriega R (2019, Oct 18) Injunction brings FDA, DOH fight against vaping to court. Manila Bulletin. Retrieved from https://mb.com.ph/2019/10/17/injunction-brings-fda-doh-fight-against-vaping-to-court/#google_vignette

E-cigarette (n.d.) Tobacco and alcohol control office – Hong Kong Department of Health. Retrieved from https://www.livetobaccofree.hk/en/reason-to-quit/e-cigarette.html

European Lung Foundation and European Respiratory Society (2020) E-cigarettes, heat-not-burn and smokeless tobacco products. Breathe (Sheffield) 16(1):161ELF. https://doi.org/10.1183/20734735.ELF161

Glantz SA (2018) PMI's own in vivo clinical data on biomarkers of potential harm in Americans show that IQOS is not detectably different from conventional cigarettes. Tob Control 27:s9–s12. https://doi.org/10.1136/tobaccocontrol-2018-054413

Global Center for Good Governance in Tobacco Control (2023) Global tobacco industry interference index—country ranking. Retrieved May 24, 2024, from https://globaltobaccoindex.org/country-ranking

Global Center for Good Governance in Tobacco Control (GGTC) (2021) E-cigarette ban regulation: global status as of February 2021. Retrieved May 24, 2024, from https://ggtc.world/knowledge/who-fctc-article-53/e-cigarette-ban-regulation-global-status-as-of-february-2021

Government of Indonesia (2024) Government regulation on Health No. 28 of 2024. Retrieved from https://assets.tobaccocontrollaws.org/uploads/legislation/Indonesia/Indonesia-MoH-Reg.-No.-28-of-2024.pdf

Heated tobacco products (n.d.) Tobacco and alcohol control office – Hong Kong Department of Health. Retrieved from https://www.livetobaccofree.hk/en/reason-to-quit/heated-tobacco-products.html

Herzog C, Jones A, Evans I, Raut JR, Zikan M, Cibula D, Wong A, Brenner H, Richmond RC, Widschwendter M (2024) Cigarette smoking and e-cigarette use induce shared DNA methylation changes linked to carcinogenesis. Cancer Res. Advance online publication. https://doi.org/10.1158/0008-5472.CAN-23-2957

Jackson SE, Tattan-Birch H, East K, Cox S, Shahab L, Brown J (2024) Trends in harm perceptions of e-cigarettes vs cigarettes among adults who smoke in England, 2014–2023. JAMA Network Open, 7(2), e240582. https://doi.org/10.1001/jamanetworkopen.2024.0582

Jirathanapiwat W, Reyes JL (2020) Today's teens, tomorrow's customers: baiting youths with new tobacco products to create a new generation of addicts. Southeast Asia Tobacco Control Alliance. Retrieved May 24, 2024, from https://seatca.org/dmdocuments/WNTD%202020%20report-SEATCA-Final.pdf

JUUL Labs (2023, June 9) Tobacco tactics. Tobacco Tactics. Retrieved from https://tobaccotactics.org/article/juul-labs/

Krüsemann EJZ, Boesveldt S, de Graaf K, Talhout R (2019) An e-liquid flavor wheel: a shared vocabulary based on systematically reviewing e-liquid flavor classifications in literature. Nicotine Tob Res 21(10):1310–1319

Lao PDR (2021) Tobacco control law (amended), No 240/NA. 29 December 2021. Retrieved from https://assets.tobaccocontrollaws.org/uploads/legislation/Lao%20PDR/Lao-PDR-TC-Law-Amended.pdf

Laos Ministry of Health (2017) Global youth tobacco survey 2016 (Fact sheet). Vientiane (Lao PDR).

Lapuekou C (2023, Nov 30) Laos to Ban E-Cigarettes amid growing concerns. The Laotian Times. Retrieved from https://laotiantimes.com/2023/11/30/laos-to-ban-e-cigarettes-amid-growing-concerns/

Ma S, Qiu Z, Yang Q, Bridges JFP, Chen J, Shang C (2022) Expanding the e-liquid flavor wheel: classification of emerging e-liquid flavors in online vape shops. Int J Environ Res Public Health 19(21):13953. https://doi.org/10.3390/ijerph192113953

Minister of Investment (2021, June 4) Letter to PT HM Sampoerna regarding the investment plan of PT HM Sampoerna in Indonesia

National Academies of Sciences, Engineering, and Medicine (2018) Public health consequences of e-cigarettes. Stratton KR, Kwan LY, Eaton DL (eds) National Academies Press, Washington, DC

Odani S, Tsuno K, Agaku IT, et al (2023) Heated tobacco products do not help smokers quit or prevent relapse: a longitudinal study in Japan. Tob Control. Advance online publication. https://doi.org/10.1136/tc-2022-057613

Petersen A, Myers MG, Tully L et al (2020) Polytobacco use among young adult smokers: prospective association with cigarette consumption. Tob Control 29(1):43–48

Philippine Food and Drug Administration (2019) Revised rules and regulations on electronic nicotine and non-nicotine delivery systems (ENDS/ENNDS). Retrieved from https://www.fda.gov.ph/wp-content/uploads/2021/08/Administrative-NO.-2019-0007.pdf

Pisinger C, Godtfredsen N, Bender AM (2019) A conflict of interest is strongly associated with tobacco industry–favourable results, indicating no harm of e-cigarettes. In: Preventive medicine, vol 119. Academic Press Inc., pp 124–131. https://doi.org/10.1016/j.ypmed.2018.12.011

Poschenrieder F, Rotter M, Gschwendtner A, Hamer OW (2020) E-cigarette-induced lung disease: from acute to chronic. Lancet 396(10250):564. https://doi.org/10.1016/S0140-6736(20)31755-4

Prime Minister News (2024, Jan 12) เด็ก เยาวชน ส่งสารถึงนายกรัฐมนตรี ขอให้คงกฎหมายห้ามนำเข้าห้ามขายบุหรี่ไฟฟ้า เรียกร้องบังคับใช้กฎหมายอย่างเคร่งครัด เพื่อปกป้องเด็กไทยจากบุหรี่ไฟฟ้า เป็นของขวัญวันเด็ก 2567 [Children and youth send messages to the Prime Minister: "Please maintain the law prohibiting the importation and sale of electronic cigarettes," demanding strict enforcement of the law to protect Thai children from e-cigarettes as a gift for Children's Day 2024]. Royal Thai Government, Bangkok (Thailand). Retrieved from https://www.thaigov.go.th/news/contents/details/77376

Republic of the Philippines – Office of the President (2017) Executive order No. 26, Series of 2017 Providing for the establishment of smoke-free environments in public and enclosed places. Republic of the Philippines – Office of the President. Retrieved from https://www.officialgazette.gov.ph/downloads/2017/05may/20170516-EO-26-RRD.pdf

Republic of the Philippines – Office of the President (2020) Executive Order No. 106, Series of 2020 Prohibiting the Manufacture, Distribution, Marketing and Sale of Unregistered and/or Adulterated Electronic Nicotine/Non-Nicotine Delivery Systems, Heated Tobacco Products and Other Novel Tobacco Products, Amending Executive Order No. 26 (S. 2017) and for Other Purposes. Republic of the Philippines – Office of the President. Retrieved from https://www.officialgazette.gov.ph/2020/02/26/executive-order-no-106-s-2020/

Riswan K (2023, Jan 13) Airlangga lepas ekspor produk tembakau inovatif bebas asap ke Malaysia. Antara. Retrieved from https://www.antaranews.com/berita/3347607/airlangga-lepas-ekspor-produk-tembakau-inovatif-bebas-asap-ke-malaysia

Silver NA, Kierstead EC, Briggs J et al (2022) Charming e-cigarette users with distorted science: a survey examining social media platform use, nicotine-related misinformation and attitudes towards the tobacco industry. BMJ Open 12:e057027. https://doi.org/10.1136/bmjopen-2021-057027

Southeast Asia Tobacco Control Alliance (2023) Big Tobacco targets kids. Retrieved from https://seatca.org/big-tobacco-targets-kids/

Spherical Insights LLP (2024, Apr 22) Global E-cigarette market size to worth USD 37.7 billion by 2033 | CAGR of 5.3% SPHERICAL INSIG. Yahoo! Finance. Retrieved from https://finance.yahoo.com/news/global-e-cigarette-market-size-060000797.html

Tan ASL, Bigman CA (2020) Misinformation about commercial tobacco products on social media: implications and research opportunities for reducing tobacco-related health disparities. Am J Public Health 110(S3):S281–S283. https://doi.org/10.2105/AJPH.2020.305910

Tan YL, Dorotheo U (2024) ASEAN tobacco control Atlas, Sixth Edition, August 2024. Southeast Asia Tobacco Control Alliance (SEATCA), Bangkok. Thailand. Retrieved from https://seatca.org/dmdocuments/SEATCA%20Tobacco%20Atlas%206th%20Edition.pdf

Tatum Z, Leventhal A, Wipfli HL (2023) Playtime: vaping devices designed as cartoons and toys may appeal to kids. Tob Control, tc-2022-057908. https://doi.org/10.1136/tc-2022-057908

Thailand MOPH, WHO, US CDC (2015) Global School-based Student Health Survey. Retrieved April 24, 2025, from: https://extranet.who.int/ncdsmicrodata/index.php/catalog/554/download/4128

Thailand Ministry of Public Health (2022) Global youth tobacco survey. Fact sheet. Thailand 2022. Bureau of Tobacco control, Department of Disease Control, Ministry of Public Health, Thailand. https://cdn.who.int/media/docs/default-source/ncds/ncd-surveillance/data-reporting/thailand/thailand_gyts_-2022_fs_508.pdf

The Straits Times (2023, Nov 7) 'They almost look like toys': new vape designs' appeal to the youth is worrying, say experts. The Straits Times. Retrieved from https://www.straitstimes.com/world/united-states/vapes-get-a-gen-z-makeover-with-candy-shades-and-eggy-shapes

Tzu-Hsuan Chen D (2023) Dual and poly-use of novel and conventional nicotine and tobacco product use in Europe: challenges for population health, regulatory policies, and the ways ahead. Public Health 11. https://doi.org/10.3389/fpubh.2023.1093771

United Nations (2024) Goal 3: ensure healthy lives and promote well-being for all at all ages. United Nations Sustainable Development Goals. Retrieved from https://sdgs.un.org/goals/goal3

US Centers for Disease Control and Prevention (2016) Surgeon General's Report: E-Cigarette Use Among Youth and Young Adults. Bethesda, Maryland, USA: US NationalInstitutes of Health. https://stacks.cdc.gov/view/cdc/44007

US Food and Drug Administration (2020) Lung injuries associated with use of vaping products. Retrieved May 24, 2024, from https://www.fda.gov/news-events/public-health-focus/lung-injuries-associated-use-vaping-products#:~:text=FDA%20and%20CDC%20recommend%20that,cigarette%2C%20or%20vaping%2C%20products

van der Eijk Y, Tan Ping Ping G, Ong SE et al (2022) E-cigarette markets and policy responses in Southeast Asia: a scoping review. Int J Health Policy Manag 11(9):1616–1624. https://doi.org/10.34172/ijhpm.2021.25

Vandelaer J (2023, Sept 28) E-cigarettes contain hazardous substances, addictive and harmful. Retrieved from https://www.who.int/thailand/news/feature-stories/detail/e-cigarettes-contain-hazardous-substances%2D%2Daddictive-and-harmful#:~:text=In%20Thailand%2C%20according%20to%20the,children%2013%2D15%20years%20old!

Vietnam Ministry of Health, World Health Organization (2022) Global youth tobacco survey. Fact sheet. Vietnam 2022.

Vietnam Ministry of Health, World Health Organization, CDC, Ministry of Education, Hanoi University of Public Health (2019) Vietnam global school-based student health survey 2019. Fact sheet

VietnamPlus (2024, Mar 26) Thailand warns against toy-shaped vape. VietnamPlus. Retrieved from https://en.vietnamplus.vn/thailand-warns-against-toy-shaped-vape-post283494.vnp

Vietnam Express International (2024, July 10) Hospitalizations on the rise among young Vietnamese due to e-cigarette poisoning https://e.vnexpress.net/news/news/hospitalizations-on-the-rise-among-young-vietnamese-due-to-e-cigarette-poisoning-4768720.html

VietNam News (2024, December 26) Việt Nam to ban e-cigarettes and heated tobacco products by 2025. https://vietnamnews.vn/society/1689690/viet-nam-to-ban-e-cigarettes-and-heated-tobacco-products-by-2025.html

Watson CV, Puvanesarajah S, Hawkins NA, Trivers KF (2023) Racial disparities in flavored tobacco product use, curiosity, susceptibility, and harm perception, National Youth Tobacco Survey 2019–2020. Health Equity 7(1):137–147. https://doi.org/10.1089/heq.2022.0087

World Health Organization (2023a) WHO report on the global tobacco epidemic. [internet]. WHO. Retrieved from https://iris.who.int/bitstream/handle/10665/372043/9789240077164-eng.pdf?sequence=1

World Health Organization (2023b) Progress report on technical matters related to Articles 9 and 10 of the WHO FCTC (Regulation of contents and disclosure of tobacco products, including waterpipe, smokeless tobacco and heated tobacco products) [Internet]. Convention Secretariat

to the Framework Convention on Tobacco Control. Retrieved from https://storage.googleapis.com/who-fctc-cop10-source/Main%20documents/fctc-cop10-7-en.pdfchrome
World Health Organization (2023c) Technical note on call to action on electronic cigarettes. World Health Organization. Retrieved from https://cdn.who.int/media/docs/default-source/tobacco-hq/regulating-tobacco-products/ends-call-to-action-background.pdf?sfvrsn=7dd2856e_11&download=true
World Health Organization (2023d) Comprehensive report on research and evidence on novel and emerging tobacco products, in particular heated tobacco products, in response to paragraphs 2(a)–(d) of decision FCTC/COP8(22). Convention Secretariat to the Framework Convention on Tobacco Control. Retrieved from https://storage.googleapis.com/who-fctc-cop10-source/Main%20documents/fctc-cop10-10-en.pdf
World Health Organization (2024) WHO global report on trends in prevalence of tobacco use 2000–2030. https://iris.who.int/bitstream/handle/10665/375711/9789240088283-eng.pdf